EUKARYOTIC CELL CULTURES
Basics and Applications

ADVANCES IN EXPERIMENTAL MEDICINE AND BIOLOGY

Editorial Board:
NATHAN BACK, *State University of New York at Buffalo*
NICHOLAS R. DI LUZIO, *Tulane University School of Medicine*
EPHRAIM KATCHALSKI-KATZIR, *The Weizmann Institute of Science*
DAVID KRITCHEVSKY, *Wistar Institute*
ABEL LAJTHA, *Rockland Research Institute*
RODOLFO PAOLETTI, *University of Milan*

Recent Volumes in this Series

Volume 164
THROMBOSIS AND CARDIOVASCULAR DISEASES
Edited by Antonio Strano

Volume 165
PURINE METABOLISM IN MAN—IV
Edited by Chris H. M. M. De Bruyn, H. Anne Simmonds, and Mathias M. Müller

Volume 166
BIOLOGICAL RESPONSE MODIFIERS IN HUMAN ONCOLOGY
AND IMMUNOLOGY
Edited by Thomas Klein, Steven Specter, Herman Friedman, and Andor Szentivanyi

Volume 167
PROTEASES: Potential Role in Health and Disease
Edited by Walter H. Hörl and August Heidland

Volume 168
THE HEALING AND SCARRING OF ATHEROMA
Edited by Moshe Wolman

Volume 169
OXYGEN TRANSPORT TO TISSUE—V
Edited by D. W. Lübbers, H. Acker, E. Leniger-Follert,
and T. K. Goldstick

Volume 170
CONTRACTILE MECHANISMS IN MUSCLE
Edited by Gerald H. Pollack and Haruo Sugi

Volume 171
GLUCOCORTICOID EFFECTS AND THEIR BIOLOGICAL CONSEQUENCES
Edited by Louis V. Avioli, Carlo Gennari, and Bruno Imbimbo

Volume 172
EUKARYOTIC CELL CULTURES: Basics and Applications
Edited by Ronald T. Acton and J. Daniel Lynn

A Continuation Order Plan is available for this series. A continuation order will bring delivery of each new volume immediately upon publication. Volumes are billed only upon actual shipment. For further information please contact the publisher.

EUKARYOTIC CELL CULTURES
Basics and Applications

Edited by
Ronald T. Acton
and
J. Daniel Lynn
University of Alabama in Birmingham
Department of Microbiology
Birmingham, Alabama

PLENUM PRESS • NEW YORK AND LONDON

Library of Congress Cataloging in Publication Data

International Cell Culture Congress (2nd: 1981: University of Alabama)
  Eukaryotic cell cultures.

  (Advances in experimental medicine and biology; v. 172)
  Bibliography: p.
  Includes index.
   1. Cell culture—Congresses. I. Acton, Ronald T. II. Lynn, J. Daniel. III. Title. IV. Series. [DNLM: 1. Cells, Cultures—Congresses. W1 AD559 v. 172/QH 573 I61 1981c]
  QH585.I534   1981   574.87   83-26977
  ISBN 0-306-41619-0

Proceedings of the Second International Cell Culture Congress,
held September 29–October 1, 1981, at the University of Alabama
in Birmingham, Birmingham, Alabama

©1984 Plenum Press, New York
A Division of Plenum Publishing Corporation
233 Spring Street, New York, N.Y. 10013

All rights reserved

No part of this book may be reproduced, stored in a retrieval system, or transmitted, in any form or by any means, electronic, mechanical, photocopying, microfilming, recording, or otherwise, without written permission from the Publisher

Printed in the United States of America

PREFACE

The Second International Cell Culture Congress was structured as was the First Congress to bring together scientists from academia and industry to discuss the use of cell culture in support of bioscience. It was felt that a forum whereby state-of-the-art presentations were followed by informal workshops would provide opportunity for the greatest exchange of information. Within the atmosphere of the workshop, problems common to basic as well as applied research were discussed and directions for the future were brought to light. These proceedings reflect and epitomize those discussions.

Although it is difficult to cover all scientific disciplines utilizing cells in culture, we feel key areas were addressed at the Congress and are herein presented. Considerable emphasis has been given to the methods for establishing cells in culture and characterizing the cells once established as well as the improved technology for growing established cell lines. Examples of how recombinant DNA technology is being used to manipulate genes within mammalian cells, to clone mammalian genes and to insert them in prokaryotes has been included. Major emphasis has been given to the use of lymphocytes in culture for understanding immune responsiveness and the culturing of a variety of cell types as a means to understand disease states.

While not all encompassing, these presentations certainly reflect the ferment of activity that now exists within the scientific disciplines that utilize cells in culture. Moreover, since all contributors had the opportunity to recently review their manuscripts, the contents of these proceedings are timely. The Congress and these proceedings could not have been possible without the cooperation of all participants as well as the organizing committee and staff of the Department of Microbiology. We are extremely grateful to all for working so diligently.

Ronald T. Acton, Ph.D.
J. Daniel Lynn, M.S.

CONTENTS

ESTABLISHING, CHARACTERIZING AND GROWING CELLS IN CULTURE

Establishing and Characterizing Cells in Culture   1
   Lewis L. Coriell

Cell Characterization by Use of Multiple
     Genetic Markers.   13
   Bharati Hukku, David M. Halton, Michael Mally
   and Ward D. Peterson, Jr.

Characteristics of Lympho-Myeolopietic Stem Cells
     Isolated from Canine Peripheral Blood.   33
   Dennis A. Stewart, Robert B. Bolin, Barbara A. Cheney,
     Joseph T. Hawkins, Keith W. Chapman and
     Donald R. Tompkins

Growth Characteristics of A431 Human Epidermoid
     Carcinoma Cells in Serum-Free Medium: Inhibition
     by Epidermal Growth Factor   49
   David W. Barnes

<u>In Vitro</u> Senescence, Differentiated Function, and
     Transformation in Cultured Vascular Endothelial
     Cells.   67
   Judith B. Grinspan, Stephen N. Mueller, James P.
     Noveral, Eliot M. Rosen and Elliot M. Levine

Cytogenetics Effects of Cyclamates   91
   Evelyn W. Jemison, Kenneth Brown, Beverly Rivers
     and Regina Knight

Mammalian Cell Culture: Technology and Physiology 119
   Nikos K. Harakas, Charles Lewis, Ronald D. Bartram,
     Bernard S. Wildi and Joseph Feder

A High Efficiency Stirrer for Suspension Cell Culture
    With or Without Microcarriers. 139
  Norman A. de Bruyne

Alternative Surfaces for Microcarrier Culture of
    Animal Cells 151
  Christine Gebb, Julian M. Clark, Michael D. Hirtenstein,
    Goran E. Lindgren, Bjoorn J. Lundgren, Ulla Lindskog
    and Per A. Vretblad

PRODUCTS OF CELLS IN CULTURE

Interferon Production in Micrcarrier Culture of
    Human Fibroblast Cells 169
  Victor G. Edy

In Vitro and In Vivo Effects of Endothelial
    Cell-Derived Growth Factor 179
  Corrine M. Gajdusek and Sandra A. Harris-Hooker

Lymphotoxins - A Multicomponent System of Growth
    Inhibitory and Cell-Lytic Glycoproteins. 205
  G.A. Granger, J. Klostergaard, R.S. Yamamoto,
    J. Devlin, S.L. Orr, D. McGriff and K.M. Miner

Production of Gamma (Immune) Interferon by a Permanent
    Human T-Lymphocyte Cell Line 219
  Jerome E. Groopman, Ilana Nathan and David W. Golde

Thymic Inhibition of Myelopoietic Proliferation. 227
  D.F. Gruber and G.D. Ledney

Problems in the Bioassay of Products from Cultures HEK Cells:
    Plasminogen Activator. 241
  Marian L. Lewis, Dennis R. Morrison, Bernard J.
    Mieszkuc and Diane L. Fessler

A Simple Methodology for the Routine Production and
    Partial Purification of Human Lymphoblastoid
    Interferon 269
  P.J. Neame and R.T. Acton

CONTENTS

USE OF RECOMBINANT DNA TECHNOLOGY

The Cloning Isolation and Characterization of a
    Biologically Active Human Enzyme, Urokinase,
    In E. Coli 281
  P.P. Hung

Cloning of the Rat Endogenous Helper Leukemia Virus
    DNA Sequence and Expression of the Helper Activity
    Encoded by the Cloned DNA Sequence in Normal Rat
    Kidney Cells by Microinjection 295
  Stringer S. Yang, Rama Modali and Edwin Murphy, Jr.

Analysis of CAD Gene Amplification using a Combined
    approach of Molecular Genetics and Cytogenetics. 319
  Geoffrey M. Wahl, Virginia Allen, Suzanne Delbruck,
  Walter Eckhart, Judy Meinkoth, Bruno Robert de Saint
  Vincent and Louise Vitto

LYMPHOCYTES IN CULTURE: A SYSTEM FOR UNDERSTANDING IMMUNE RESPONSIVENESS

Alloreactive T Cell Clones 347
  Frank W. Fitch

Productive Murine Leukemia Virus (MuLV) Infection of EL4
    T-Lymphoblastoid Cells: Selective Elevation of H-2
    Surface Expression and Possible Association of Thy-1
    Antigen with Viruses 365
  Susanne L. Henley, Kim S. Wise and Ronald T. Acton

Construction of Human T-Cell Hybrids with Helper Function. . . 383
  Oscar Irigoyen, Philip V. Rizzolo, Yolene Thomas, Linda
  Rogozinski and Leonard Chess

Cultured Human T Lymphocyte Lines and Clones as
    Immunogenetic Tools. 405
  Dolores J. Schendel and Rudolf Wank

USE OF CELLS IN CULTURE TO STUDY DISEASE

Long-term Persistance in Experimental Animals of Components
    of Skin-Equivalent Grafts Fabricated in the
    Laboratory 419
  Euguene Bell, Stephanie Ellsworth Sher, Barbara E. Hull
  and Robert L. Sarber

Angiotensin Receptors and the Control of Na^+ and K^+
  Transport in Cultured Aortic Smooth Muscle and
  Brain Microvessel Cells. 435
  Tommy A. Brock and Jeffrey B. Smith

Antigenic Expression of Human Melanoma Cells in Serum-Free
  Medium 455
  Thomas F. Bumol, John R. Harper, Darwin O. Chee
  and Ralph A. Reisfeld

Use of Cell Culture to Identify Human Precancer. 471
  Henry C. Lyko and James X. Hartmann

Uses of Tissue Culture and Cryopreservation in Pancreatic
  Islet Transplantation. 489
  Collin J. Weber, F. Xavier Pi-Sunyer, Earl Zimmerman,
  Gajanan Nilaver, Michael Kazim, Orion Hegre and
  Keith Reemtsma

Mechanisms of Mediator Release from Neutrophils. 527
  Gerald Weissmann, Charles Serhan, Helen M. Korchak
  and James E. Smolen

INDEX 553

ESTABLISHING AND CHARACTERIZING CELLS IN CULTURE

Lewis L. Coriell

Institute for Medical Research
Copewood Street
Camden, N.J   08103

The human body is made up of approximately 10 trillion cells, each too small to be seen with the naked eye. Each cell contains enough DNA for one to three million genes, but probably not more than 50,000 different structural genes with the remainder being accounted for by redundancy and programming. In any specialized tissue such as bone, skin, thyroid, it is obvious that a limited number of genes are active and the majority are suppressed. Each tissue must perform its special functions and also be responsive to the whole body and to the demands of other tissues; for growth, response to disease, trauma, repair; to signals carried by blood, lymph and nerves. Coordination of all this is also programmed in the genes. A very complex system. With so many interactions going on simultaneously, it becomes very difficult to follow the minute details in a single cell by examining the whole animal.

Cell culture has provided an answer to this dilemma. By growing each cell type in culture, its physiology and pharmacology can be observed alone and in response to various added stimuli or other cells, their products, hormones, etc.

Use of cell cultures has spawned in the last thirty (30) years, an explosion of knowledge about life processes within the living cell and how they are controlled. The science of applied genetics has been called the most rapidly progressing area of human knowledge in the world today and a great deal of this progress can be attributed to the application of cell culture techniques.

Exploitation of these new techniques has only just begun. Human animal, plant and invertebrate cells are being manipulated in culture to study normal life processes, the changes that occur in disease, for early diagnosis, treatment and prevention of disease and the manipulation of genes for production of growth hormone and other products. The prospects for understanding and modifying the mechanisms of genetic control of aging and cancer were never better.

I will confine my remarks to some basic principles in establishing and characterizing animal cell cultures based on experience in operating the National Human Genetic Mutant Cell Repository and the Aging Cell Repository (1).

The great volume of cell culture research in the past has been carried out with cell types easiest to grow in vitro, i.e., fibroblasts, lymphoblasts and a few tumor cells.

Many investigators are now concentrating on cell culture of specialized epithelial cells which in general are more fastidious in their growth requirements and difficult to propagate in serial cultures. The majority of fatal cancers arise from epithelial cells. Others in this workshop will be describing new procedures and nutritional requirements for growth of new cell types.

My remarks will be confined to three common failures in cell culture laboratories which should be addressed no matter what cell type is being established in cell culture. They are: reproducibility of culture conditions, maintenance of sterility, and storage in liquid nitrogen.

Culture Conditions: When maintained in vitro, cells are very sensitive to changes in the culture conditions. Minor changes may cause them to gain or loose characteristics which they expressed when first established in cell culture. On the other hand, experiments using cell cultures take days, weeks or months to complete and have to be repeated several times for confirmation of results. This means that for any series of experiments, one must plan to stabilize all conditions including temperature of incubators, culture vessels and glassware preparation, pH of culture media, use of single lots of culture medium, trypsin growth supplements, and a single lot of fetal calf serum which has been pretested for toxicity, growth stimulation and sterility. Many cell culturists have observed that a lot of serum that supports excellent growth of one cell type will be only fair, poor, or toxic for another cell type. This is true to a greater or lesser extent for all the culture conditions mentioned above.

Sterility: Cell cultures provide excellent nutrition for molds, yeasts, bacteria, mycoplasma and viruses. Cell cultures must be opened to add fresh culture medium one, two or more times a week. This provides many opportunities for microbial contamination of cell cultures (2). Measures to prevent contamination are well established (3,4). They may be summarized as follows:
1. Pretest all culture media components for microbial contamination.

2. Transfer cell cultures in Hepa (high efficiency particulate air) filtered transfer hoods or rooms.
3. Use no antibiotics in cell culture media for maintaining stock cultures. Exceptions are made for first TC passage from potentially infected tissue biopsies and for short term biochemical experiments.
4. Treat every cell culture as if it were contaminated, i.e., employ measures to prevent the spread of the contaminant to other sterile cultures maintained in the laboratory. Two polices are paramount in achieving this goal:
   a. Transfer only one cell line at a time within the transfer enclosure, and:
   b. Disinfect the work surface, equipment and hands before introducing the next cell line.

If these measures are faithfully followed, cross contamination of cell lines will be eliminated.

<u>Storage in Liquid Nitrogen</u>: The third safeguard against uncontrolled change, contamination or loss of a cell culture is to store several ampules at an early passage in liquid nitrogen. The cells will remain unchanged for years and provide a standard seed stock for repetition or expansion of experiments by the original investigator or investigators in other laboratories (5).

<u>Characterization of Cell Lines</u>: Cells in culture tend to be fibroblast-like, epithelial-like or grow in suspended culture and it is not possible to make a precise identification upon inspection alone. New methods for identification of cells, cell components or products are published frequently. The traditional tests most used to identify cell cultures have been:

For Species of origin; the karyotype, (6) serology with species specific cytotoxic antibody, (7) and species specific fluorescent antibody (8).

For interspecies and more specific identity many assays are available: In Vitro lifespan (9),
Chromosome markers and polymorphisms (10, 11),
HLA pattern (12),
Isoenzyme pattern (13),
Organ specific products, e.g., secretion of collagen, albumen, insulin, and production of membrane antigens, are too numerous to list here
Clonal growth in soft agar (14), and
growth in nude mice (15) which is the most reliable indicator of malignant transformation.

I would like next to share some of the things we have learned in establishing and characterizing thousands of cell lines over the past 30 years. The mechanism of cross contamination of cell cultures was shown (4) to be via contamination of the environment during feeding and transfer of cell cultures. Prevention of cross contamination is accomplished by a combination of aseptic and antiseptic measures (4,16,17). We have confirmed the efficiency of these measures by re-examination of a group of 964 cell cultures expanded, characterized and stored in liquid nitrogen, before we knew how to detect non-cultivable strains of M. hyorhinis. By use of a specific immunofluorescence assay, 33 of the 964 cell cultures when retested were contaminated with M. hyorhinis. There was no evidence that the infection spread to other cell cultures in our laboratory. All 33 contaminated specimens were received from laboratories known to harbor M. hyorhinis (18).

During 9 years' operation of the Human Genetic Mutant Cell Repository and the Aging Cell Repository, we have processed and characterized over 5,000 biopsies or cell cultures and the 8th edition of the catalog contains 300 pages of listings and data (1). Included are fibroblasts and lymphoblasts from apparently normal individuals as well as from individuals with disorders of amino acid, carbohydrate, lipid, metal, nucleoside, or steroid metabolism and of connective tissue, muscle, bone, and unknown disorders. Included are many examples of twelve different categories of chromosome aberrations. Cell cultures with over 200 different translocations and inversions have been useful in gene mapping.

Submissions of new biopsies to the repositories and distributions to investigators are shown in Figure 1. Recipient use of cell cultures from the repositories are shown in Table 1.

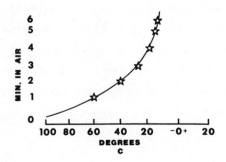

Fig. 1.

TABLE I

RECIPIENTS USE OF CELL CULTURES OBTAINED FROM
THE REPOSITORY 1979-1982

Scientific Discipline	1982	1981	1980	1979
1. Biochemistry	19.1	20.0	20.8	24.0
2. Cytogenetics	5.8	4.4	5.0	6.0
3. Immunology	8.6	9.0	4.6	3.0
4. Pharmacology	0.7	0.1	0.9	1.4
5. Aging	5.2	4.0	4.6	5.4
6. Carcinogenesis and mutagenesis	5.8	6.1	6.0	5.0
7. DNA synthesis and repair	6.4	8.0	7.0	12.0
8. Gene mapping	6.9	5.2	4.9	6.3
9. Somatic cell genetics	6.1	6.2	7.0	9.0
10. Enzyme expression	8.2	8.0	11.0	11.6
11. Clinical Diagnosis	2.4	3.0	2.7	2.5
12. Virology	2.0	1.5	1.6	2.6
13. Developmental biology	0.7	0.1	0.9	1.6
14. Physilogy	0.6	1.7	1.7	1.7
15. Cell ultrastructure	1.7	2.0	2.0	5.0
16. Interferon	2.4	5.0	5.0	5.0
17. Gene expression	6.4	6.0	7.3	9.6
18. Normal controls	7.8	6.1	6.7	5.3
19. Other uses	3.4	4.0	1.9	2.9
Totals	100.2*	100.4*	101.6*	117.9*

*The percentages of the various catagories are greater than 100% because a number of replies indicated that the cells were to be used for research in more than one scientific discipline.

Cell cultures shipped by air express arrive in good to excellent condition in 92 to 94 percent of shipments over the period of operation of the repository. At the end of 1980, 1,400 scientific publications used cell cultures from the repository in addition to 70 publications by the Repository staff in association with the submitter.

We have recently recovered a number of cell lines stored in liquid nitogen vapor or liquid phase for up to 18 years. There is no significant loss of viability. I would stress that this is achieved through frequent scheduled manual surveillance of liquid nitrogen level in addition to automatic and electronic devices to detect failure of LNR tanks.

Skin biopsies may be stored in liquid nitrogen for later establishment in cell culture without loss of viability in our experience. We chop the specimen with sharp scalpels, suspend in freeze medium with 10% glycerine and cool at the rate of 1-2°C/minute.

A caution about removing frozen ampules form the LNR is illustrated in Figure 2. When removed into room air, the temperature in the ampule rises 40°C in one minute, 60°C in two minutes. This explains why ampules explode when being lifted out of the LNR if they have a pinhole leak which permits liquid nitrogen to be asperated into the ampule while stored under the liquid phase. A rise of 50°F increases the vapor pressure of liquid nitrogen by about 5,000 PSI. To prevent explosions, check for pinhole leaks before storage in LNR, and, _always_, wear protective face shields when removing ampules from LNR. The other precaution indicated by this warming curve is to keep ampules in the vapor phase while doing recovery, inventory, or any manipulation of stored ampules. In room air, the temperature of the frozen ampule rises within 1 or 2 minutes to a level that is incompatible with long term viability of the cell culture.

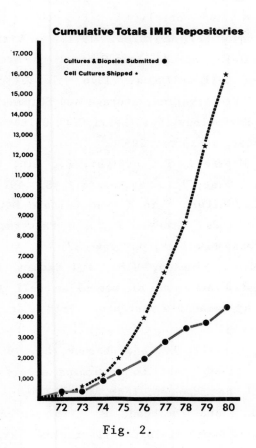

Fig. 2.

REFERENCES

1. The Human Genetic Mutant Cell Repository, Eighth Edition, October, 1980. NIH publication No. 81-2011:1-310.

2. Barile, M., Hopps, Hope, and Grabowski, M.W. 1978. Incidence and Sources of Mycoplasma Contamination: A Brief Review in Mycoplasma Infection of Cell Cultures. (G.J. McGarrity, Murphy, D.G., and Nichols, W.W., Eds.) p. 35-45, Plenum, New York, 1978.
3. McGarrity, G. and Coriell, L. Procedures to Reduce Contamination of Cell Cultures. 1971. In Vitro 6:257-265.
4. McGarrity, G.J. Spread and Control of Mycoplasmal Infection of Cell Cultures. 1976. In Vitro 643-648.
5. Coriell, L. Preservation, Storage and Shipment. 1979. In Methods in Enzymology, Vol. LVIII Cell Culture (Wm. Jakoby and Ira Pastor, eds.) pp. 29-36.
6. Hsu, T.C., Moorhead, P., Arrighi, F., Lin, C.F., and Uchida, I.A., Wang, H.C. and Federoff, S. 1973. Karyology of cells In Culture. in Tissue Culture Methods and Applications (Eds. Krouse, P., and Patterson, M.H.) Academic Press, New York, pp. 764-787.
7. Stulberg, C.S., Simpson, W.F., and Berman, L. 1961. Species-related antigens of mammalian cell strains as determined by Immunofluorescence. Proc. Soc. Exp. Biol. Med. 108:434-439.
8. Greene, A., Coriell, L., and Charney, J. 1964. A Rapid Cytotoxic antibody test to determine species of cell cultures. J. Nat. Cancer. Inst. 32:779-786.
9. Hayflick, L., and Moorhead, P. 1961. The serial cultivation of human diploid cell strains. Exp. Cell. Res. 25:585.
10. Nelson-Rees, W.A., Flendermeyer, R., and Hawthorne, P.R. 1974. Banded marker chromosomes as indicators of intra-species cellular contamination. Science 184:1093-1096.

11. Nelson-Rees, W.A., Flendermeyer, R., and Daniels, D. 1980. T-1 cells are He La and not of Normal Human Kidney Origin. Science 209:719-720.
12. Kissmeyer-Nielsen, F., and Thorsby, E. 1970. Human Transplantation Antigens. Transplant Rev. 4:1-176.
13. Beckman, L., Bergmen, S. and Lundgren, E. 1967. Isozyme Variations in Human Cells grown in vitro. Acta. Genet. Stat. Med. 17:304-310.
14. McPherson, I. 1973. Soft agar techniques, 276-280. In Tissue Culture Methods and Application (Kruse, P.F. and Patterson, M.K., eds). Academic Press, New York.
15. Shin, S., Friedman, V.H., and Risser, R. 1975. Tumorgenicity of virus-transformed cells in nude mice is correlated specifically with anchorage independent growth in vitro. Proc. Natl. As. Sci. USA 72:4435-4489.
16. McGarrity, G.J., Coriell, L. 1971. Procedures to reduce contamination of cell cultures. In Vitro 6:257-265.
17. Coriell, L. Methods of Prevention of Bacterial, Fungal and Other Contaminations. 1973. In: Contamination in Tissue Cultures (J. Fogh, ed.) Academic Press, Inc., p. 29-49.
18. McGarrity, G.J., Gamon, L., and Coriell, L. 1980. Detection of Mycoplasma hyorhinnis infection in cell repository cultures. Cytogenet. Cell Genet. 27:194-196.

CELL CHARACTERIZATION BY USE OF MULTIPLE GENETIC MARKERS

>Bharati Hukku, David M. Halton, Michael Mally
>and Ward D. Peterson, Jr.
>
>Department of Pediatrics
>Child Research Center Division
>Children's Hospital of Michigan & Wayne State University
>School of Medicine
>Detroit, Michigan

ABSTRACT

The extensive use of cell cultures for diverse research purposes is one of the truly great international growth industries. With the proliferation of cells comes a responsibility for monitoring them for inter- and intraspecies characteristics. We use multiple genetic markers for cell identification, i.e. species specific antigens, isozymic phenotypes, chromosomal complement, and HL-A haplotypes. The methodologies employed are briefly described, and various examples cited to show how these markers can be utilized for cell line monitoring.

---

[1] Supported by contract NO1-CP-9-1003, Biological Carcinogenesis Branch, Division of Cancer Cause and Prevention, National Cancer Institute

[2] Abbreviations: HL-A, histocompatibility locus; G-6-PD, glucose-6-phosphate dehydrogenase; LDH, lactate dehydrogenase; PGM_1, phosphoglucomutase 1; PGM_3, phosphoglucomutase 3; PGD, phosphogluconate dehydrogenase; Est. D, esterase D; SV-40, Simian Virus 40; Cerco. Monkey, Cercopithecus aethiops; N.A., Not applicable; Neg., Negative; CCL, Certified Cell Line (No), American Type Culture Collection

Data are summarized from 275 cultures sent to our laboratory for analysis during the past eighteen months. The data show that, overall, 35% of the cultures received were contaminated. The majority of cell cultures submitted were human cell lines. We found that 36% of these cultures were cross contaminated; 25% by cells of another species and 11% by another human cell line. This high incidence of inter- and intraspecies contamination underscores the importance of frequent monitoring of cell cultures.

The extensive use of cell cultures in biomedical research has resulted in an amazingly large inventory of cell lines and strains obtained from widely diverse species and tissues. Cell lines are constantly being initiated and, of course, well established cell lines are abundantly available for use. In individual laboratories, it is not uncommon for investigators to manipulate from several to many cell lines during the course of a day, depending on the nature of the studies being undertaken. Cell characterization and monitoring, therefore, is a necessary component of culturing activities to enable the investigator to accurately assess the current status of his cultures. It is apparent that the accuracy of the experimental results that the investigator obtains are dependent in part upon knowledge of those cell substrates. Beyond that, the frequency with which cell culture contaminations occur is a compelling argument for continuously monitoring cell lines (1,2,3,4).

We use a cell monitoring strategy as well as methodologies (5) that we feel are applicable to newly initiated cell lines as to those that have been established for some time. The strategy is to use genetically stable phenotypic markers. The methodologies involve determination of species specific cell surface antigens, isozyme phenotypes, chromosomal complement, and HL-A haplotypes.

Our procedures for using these multiple marker systems are briefly reviewed as follows:

Species specific antigens are detected by an immunofluorescent procedure (6,7). Specific antisera are prepared in rabbits or guinea pigs by inoculation of cells from culture or red blood cells of a particular species. The antisera are coupled with fluorescein isothiocyanate. The coupled antisera are mixed with living cell suspensions, incubated, washed, and mounted on slides as wet suspension preparation. The cells are examined under a UV fluorescent microscope. A positive reaction is denoted by a bright green peripheral fluorescence of the cell membrane. With antisera that have been properly titered and absorbed, one can detect a foreign cell species in a mixture of cells at a ratio of 1:1000 (8). We have available antisera specifically reactive with 20 different species of cells commonly used in cell culture studies.

For isozyme analysis, we first prepare an extract of the cells in question by freeze-thawing them six times, volume/volume in a 0.4 mM phosphate buffered saline containing 0.9% NaCl and 1 mM dithiothreitol. The extract is clarified by centrifugation at 4000 g. One microliter of extract is added to a precut slot on a sucrose-agarose electrophoresis film (9). Additional slots are filled with extracts of other cells as required for appropriate controls. The electrophoresis film is placed in an electrophoresis chamber and run for ninety minutes at 90 V. The film is then stained appropriately for the isozyme involved (10). For species identification, we use glucose-6-phosphate dehydrogenase and lactate dehydrogenase, which have readily interpretable isozyme mobilities (4,11). For those cell lines that are identified as human, we further test them for phenotypes of phosphoglucomutase 1 and 3, mitochondrial malic enzyme, esterase-D, adenosine deaminase, and 6-phospho-gluconate dehydrogenase. As Gartler first pointed out (12), and as O'Brien et al. (13) and Wright et al. (14) have further shown, isozyme analysis is a powerful tool that can be used for the precise identification of human cell lines. The frequency product derived for the seven polymorphic isozymes that are named above is 0.05. In other words, individual human cell lines can be identified with 95% certainty (14).

The cytogenetic evaluation of cell lines is the definitive identification tool, but it does have limitations imposed by the size of the sample that can be examined. Analysis begins by arresting cells in metaphase with colcemid. The cells are harvested, pelleted, placed in hypotonic solution, and fixed. The cell suspension is dropped onto cold slides and dried (5,15). The metaphases are stained in several ways. We band chromosomes with quinacrine mustard (16,17), trypsin-Giemsa (18), or Hoechst 33258 (19). In examining metaphases, we always scan at least 100 metaphases for ploidy distribution, as well as searching for metaphases of another species (20). The chromosomes of fifteen metaphases are counted precisely. Photographs are taken for record as well as for karyotype preparation. Karyotypes are prepared from all human tumor cell lines. A record of chromosome counts, percent of normal chromosomes, and distinguishing marker chromosomes are maintained on file. These chromosomal summaries, when linked with the results obtained from isozyme analysis, provide a means for precisely identifying the human cell line under study (21,22,23).

The fourth method of analyzing cell lines is by HL-A typing. It is used less frequently than the other methods of cell analysis, despite the power implied by the polymorphic antigens available. The difficulties in determining HL-A haplotypes on cells in cultures are significant (24). However, we find typing to be useful, particularly for identifying human lymphoblast cell lines (25). Antisera with 30 known haplotype specificities have been generously provided by the National Institute of Allergy and Infectious Diseases. The antisera are absorbed with lymphoblast cell line Daudi to remove B cell specificities (26). The antisera are distributed to microtiter plates and covered with oil. Cells for test are added in a standard complement dependent cytotoxicity test using trypan blue as the cell viability indicator (27). To confirm the specificity of the reactions observed, antisera are further absorbed with the test cells, and the absorbed antisera are retested with lymphoblast cell lines having known specificity of the absorption (28).

In analyzing cell cultures, it is our practice to use at least two different kinds of the above described genetic markers. The findings by one method tend to amplify and confirm results obtained by the others. To demonstrate this, we have selected four examples from the many cultures that we have examined in the past eighteen months.

The first example concerns a cell line that was sent to us in the belief that it was a hybrid between cell lines HeLa and WI-38. Our tests with immunofluorescent antisera showed that the cell line was human. In confirmation, the LDH isozyme mobility and pattern was typically human. The G-6-PD isozyme was comparable to that of a human Type A cell line. This result was unexpected, in that a hybrid between a Type A G-6-PD female cell line (HeLa) and a Type B female cell line (WI-38) would be expected to exhibit Type A G-6-PD mobility, Type B mobility and a hybrid mobility (29).

Cytogenetic examination of metaphases of the cell line by quinacrine mustard, as shown in Figure 1, quickly answered the question. In addition to characteristic HeLa marker chromosome #3 and #4, there were present one or two brightly stained acrocentric

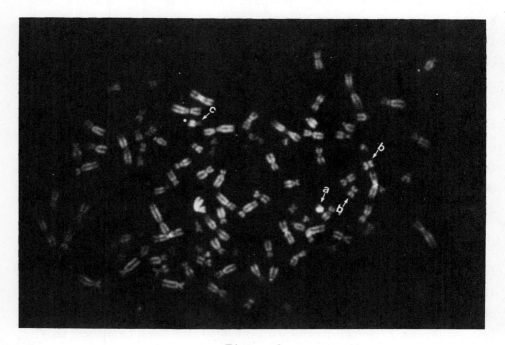

Figure 1

Hybrid between HeLa and an unknown male cell line detected by quinacrine mustard staining. a → human Y chromosome. b → HeLa marker chromosome #3. c → Hela marker chromosome #4.

chromosomes per metaphase compatible with the human Y chromosome (30). Clearly, the presence of a Y chromosome excluded WI-38 as the fusion partner in this HeLa cell hybrid.

The next example shows how multiple markers aid in the determination of interspecies contaminations. A cell line, FQ, was submitted to us as a human tumor cell line derived from a Hodgkin's lymphoma. As shown in Figure 2, the G-6-PD isozyme mobility is very similar to that of human Type A (or Rhesus and

Cercopithecus monkey) and unlike that of human Type B, horse or marmoset isozyme mobilities. The LDH isozyme mobility of the cell line, however, is very similar to that of marmoset, and is unlike that of human, monkey or horse.

As shown in Table I, carefully titered antisera against human, horse, and marmoset failed to react with the cell line, FQ. However, a weak reaction was observed with Cercopithecus monkey antiserum, and a good reaction was observed with Rhesus monkey antiserum. Absorption of Rhesus monkey antiserum with the cell line removed the reactivity of the antiserum against the cell line, and reduced the reactivity of antiserum against the homologous Rhesus monkey cell line (LLC-MK2). These results eliminated the possibility that the cell line was a mixture of human and

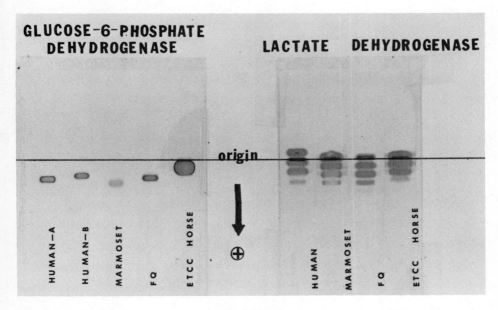

Figure 2

G-6-PD and LDH isozyme mobilities for cell line FQ compared with human (A and B G-6-PD), marmoset and horse cell lines.

Table I

REACTIONS OF CELL LINE FQ WITH SPECIES SPECIFIC ANTISERA

CELL LINES	ANTISERA					
	HUMAN	MARMOSET	HORSE	CERCO. MONKEY	RHESUS MONKEY	FQ ABSORBED RHESUS MONKEY
FQ	Neg.	Neg.	Neg.	1+	3+	Neg.
LLC-MK2 (CCL-7) (Rhesus Monkey)	Neg.	Neg.	Neg.	Neg.	4+	3+
Hela (CCL-2) (Human)	4+	Neg.	Neg.	Neg.	Neg.	Neg.
BE-MARM.* (Marmoset)	Neg.	3+	Neg.	Neg.	Neg.	Neg.
E.derm (CCL-70) (Horse)	Neg.	Neg.	4+	Neg.	Neg.	Neg.
CV-1 (CCL-70) (Cerco. Monkey)	Neg.	Neg.	Neg.	3+	Neg.	Neg.

*Obtained from Naval Biosciences Laboratory, Oakland, California

marmoset cell lines, but they strongly suggested that the cell line was of primate origin.

Banded chromosome preparations were made. A karyotype of the cell line is shown in Figure 3. Analysis of the banded chromosomes show that there are at least 26 chromosomes present in two or more copies. The evidence accumulated thus far suggested that the cell line was of primate origin with a diploid chromosome number in the fifty plus chromosome range. A favorable match then was found between the banded chromosomes in the FQ cell line karyotype and that of a karyotype prepared from owl monkey by Miller et al. (31). Further work confirmed the finding that the

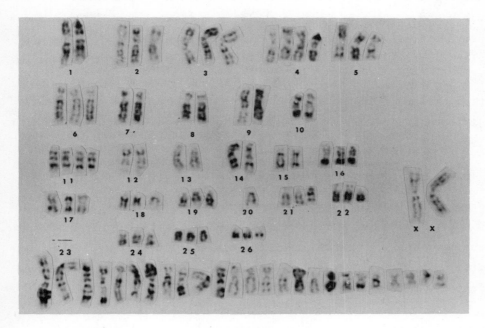

Figure 3

Karyotype of cell line FQ arranged according to that of the owl monkey, <u>Aotus trivirgatus</u>. Numbered chromosomes are normal. Unnumbered chromosomes are unassignable.

cell line was, indeed, derived from a species of owl monkey <u>Aotus trivirgatus</u> (32).

In our third example, we were called upon to examine a series of cell lines for markers to determine intraspecies relationships. We received four cell lines with a question as to whether or not they had a common origin. As shown in Table II, lines A,B,C and D were possibly related as follows: A, a parent line; B, a clone thereof; C, an SV-40 transformed derivative of A; and D, a clone of C. We first determined that all lines were human by immuno-fluorescence. All lines were further shown to have Type B G-6-PD. However, cytogenetic examination following quinacrine mustard staining, showed that lines A and B were diploid female lines while the SV-40 transformed cell lines C and D were male. As the table shows, further isozyme analysis indicated that lines A and B are identical with respect to four additional isozyme phenotypes, with a probability of identity at 90 percent. Continued isozyme analysis also showed that lines C and D were alike, with a probability of identity at 97 percent. The isozyme data clearly differentiates lines A and B from lines C and D at the esterase-D locus.

Further information was obtained by HL-A typing. The results show that lines A and B reacted with HL-A haplotypes A; 2,28: B; 5,40. Lines C and D reacted with haplotypes A; 1,3: B; 8,18. In sum, our analysis showed that the diploid lines were of the same origin, while the SV-40 transformed cell lines were both derived from a different parent line.

The last example demonstrates how the use of multiple marker system can clue one into unexpected findings. We had received a series of cell lines derived from tumor formed following the inoculation of SV-40 transformed chimpanzee cells into nude mice. The question was whether the cells from the tumor were of mouse or chimpanzee origin. On one of these cultures, we found that all cells reacted with antiserum against mouse. However, about 20 percent of the cells also reacted with mouse absorbed human antiserum. Isozyme analysis, as shown in Figure 4, revealed more G-6-PD and LDH bands than could be accounted for by either mouse, chimpanzee, or a mixture of mouse and chimpanzee cells. The immunofluorescent and isozymic analysis results suggested that a hybrid chimpanzee/mouse cell population might be present in the cell culture.

Metaphase chromosomes were banded, and, as shown in Figure 5, a population of cells was found which had identifiable mouse and

Table II

DETERMINATION OF RELATIONSHIP BETWEEN HUMAN CELL LINES
A,B,C AND D BY CYTOGENETICS, ISOZYMES AND HLA TYPING

	A	B	C	D
	Parent	Clone	SV-40 Transformed	SV-40T Clone
Cytogenetics:	Diploid	Diploid	Aneuploid	Aneuploid
	No Y Chromsome	No Y Chromsome	Y Chromosome	Y Chromosome
Isozymes:				
G_6PD	B	B	B	B
PGM_1	1-2	1-2	1-2	1-2
PGM_3	1	1	1	1
PGD	A	A	A	A
Est. D	1	1	2-1	2-1
Phenotype Frequency	0.90	0.90	0.97	0.97
HL-A[*] A	2,28	28	(1)[**],3	1,3
B	5,(40)[**]	5,40	8	8,(18)[**]

[*] HL-A reactivities not confirmed by absorbtion.

[**] ( ): weak reaction

chimpanzee chromosomes. Amongst the normal chimpanzee chromosomes found were the X and #9 chromosomes (33), which by analogy, may code and account for the chimpanzee/mouse hybrid bands observed by isozyme analysis (34). The origin of this apparently spontaneously formed chimpanzee/mouse hybrid is the subject of another investigation.

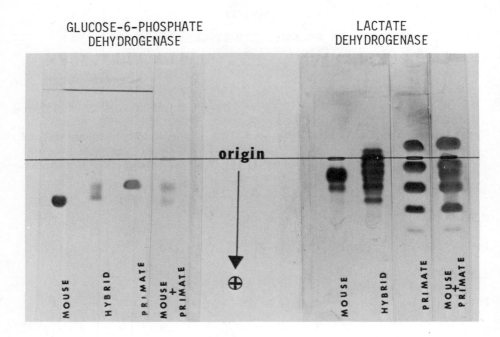

Figure 4

G-6-PD and LDH isozyme mobilities of suspected chimpanzee/mouse hybrid compared with those of mouse, primate (human Type B G-6-PD) and mouse/primate mixture.

It should be noted that, had chromosome analysis been carried out by Giemsa staining alone, it is likely that the presence of metacentric chromosomes would have been discounted as mouse metaphases formed by Robertsonian fusion between normal mouse acrocentrics. The use of the other marker systems, isozymes and immunofluorescence, directed analysis by cytogenetics to include banding techniques. As a result of this experience we now use Hoechst 33258 staining to screen potential primate/mouse hybrid cell lines. The Hoechst reagent stains mouse centromeres very brightly in contrast to the centromeres of primate chromosomes (19).

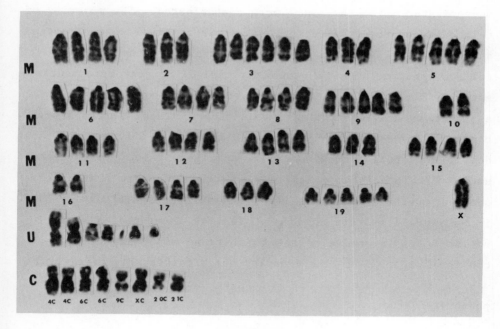

Figure 5

Karyotype of chimpanzee/mouse hybrid cell metaphase. M: normal mouse chromosome. C: normal chimpanzee chromosomes. U: unassignable chromosomes.

The examples cited above of the use of multiple marker systems for analysis and monitoring of cell cultures has demonstrated approaches to cell characterization in which reinforcement of conclusions occurs through the results obtained by the use of different methodologies.

During the past eighteen months, we have analyzed 275 cell cultures, as shown in Table III. Seventeen different species of cells were found. Of the 275 cultures examined, 65 percent were confirmed as to the species of cells, and reports were sent to investigators indicating the current status of their cell lines with respect to the characterization parameters used. However, 35 percent of the cell lines had a species or intraspecies characteristic at odds with that designated by the submitting investigator, i.e. the cells were contaminated. The majority of cell cultures received during this study interval were submitted as human cell lines. Of these, 25 percent were contaminated with cells of another species. An additional 11 percent were contaminated by human cells with a donor source incompatible with that assigned by the submitting investigator. These intraspecies contaminations were nearly evenly split (55% to 45%) between those due to HeLa cell incursions (35), and those caused by contamination with other human cell lines. The data show that interspecies contaminations continue to be a more significant problem in human cell culture management than is intraspecies contamination (4). In analyzing the total group of human cell lines, confirmation of cell line identity was significantly aided by the use of the linkage available between chromosome and isozyme phenotype markers.

The actual species of cells found in the 275 cultures examined are summarized in Table III. In all, 281 different cell lines were identified. Several cultures had more than one species of cells present.

The incidence of inter- and intraspecies contaminations in cell cultures referred to our laboratory for analysis underscores the importance of frequent monitoring of cells. It should also be noted that cells often change upon passage, even without contamination, and a record of current cell status is also of importance. We have presented methodologies and approaches to cell monitoring that are useful, and that we believe are effective in describing the current status of cell lines. We stress the need for cell line monitoring and the use of multiple genetic markers as a necessary control for experiments involving the use of cell cultures.

Table III

Summary of Number of Cell Species Monitored for Characterization and Rate of Inter- and Intraspecies Contamination Observed

Cell Species	Cultures Received No.	Cell Species Confirmed No.	%	Interspecies Contamination No.	%	Intraspecies Contamination No.	%	Actual Cell Species Found No.
Human	160	102	64	40	25	18	11	129
Mouse	27	25	92	2	8	N.A.	N.A.	76
Cat	24	23	96	1	4	N.A.	N.A.	23
Chimpanzee*	41	13	32	29	68	N.A.	N.A.	13
Rat	2	1	50	1	50	N.A.	N.A.	8
Hamster	3	3	100	0	0	N.A.	N.A.	7
C. Monkey	2	1	50	1	50	N.A.	N.A.	6
Owl Monkey	2	2	100	0	0	N.A.	N.A.	5
Chimp/Mouse Hybrid	3	1	33	2	67	N.A.	N.A.	4
Dog	2	2	100	0	0	N.A.	N.A.	2
Rabbit	2	2	100	0	0	N.A.	N.A.	2
Mink	2	1	50	1	50	N.A.	N.A.	1
Baboon	2	1	50	1	50	N.A.	N.A.	1
Deer Mouse	1	1	100	0	0	N.A.	N.A.	1
Cow	1	1	100	0	0	N.A.	N.A.	1
Goat	1	1	100	0	0	N.A.	N.A.	1
Pig	0	N.A.		N.A.		N.A.		1
TOTALS	275	180	65	78	28	18	7	281

*Most chimpanzee samples were tumor derived and the species of the tumor was in question. Therefore, these are difficult to classify as contaminants.

REFERENCES

1. K.G. Brand and J.T. Syverton, 1960, Immunology of cultivated mammalian cells. 1. Species specificity determined by hemagglutination, J. Nat. Cancer Inst. 24:1007-1019.
2. C.S. Stulberg, 1973, Extrinsic cell contamination of tissue cultures. In: Contamination in Tissue Culture, Ed. Fogh. J. Academic Press, New York and London, 1-27.
3. W.A. Nelson-Rees and R.R. Flandermeyer, 1977, Inter and intraspecies contamination of human breast tumor cell lines HBC and BrCa5 and other cell cultures, Science 195:1343-1344.
4. C.S. Stulberg, W.D. Peterson, Jr. and W.F. Simpson, 1976, Identification of cells in culture, Amer. J. Hemat. 1:237-242.
5. W.D. Peterson, Jr., W.F. Simpson and B. Hukku, 1979, Cell culture characterization: monitoring for cell identification. In: Methods in Enzymology, Cell Culture, Ed. Jakoby, W.B. and Pastan, I.H. Academic Press, New York, San Francisco and London 58:164-178.
6. C.S. Stulberg and W.F. Simpson, 1973, Serological identification of cells in culture A. Animal cells by fluorescent-labeled antibody. In: Tissue Culture: Methods and Application, Ed. Kruse, P.F. Jr. and Patterson, M.K. Jr. Academic Press, New York and London 744-749.
7. W.F. Simpson, C.S. Stulberg and W.D. Peterson, Jr., 1978, Monitoring species of cells in culture by immunofluorescence, TCA Mannual 4:771-774.
8. W.F. Simpson and C.S. Stulberg, 1963, Species identification of animal cell strains by immunofluorescence, Nature 199:616-617.
9. W.D. Peterson, Jr., C.S. Stulberg, N.K. Swanborg and A.R. Robinson, 1968, Glucose-6-phosphate dehydrogenase isoenzymes

in human cell cultures determined by sucrose-agar gel and cellulose acetate zymograms, Proc. Soc. Exp. Biol. Med. 128:772-776.

10. H. Harris and D.A. Hopkinson, 1980, Handbook of enzyme electrophoresis in human genetics. Second Edition Elsevier/North-Holland Inc. ISBN.

11. F. Montes de Oca, M.L. Macy and J.E. Shannon, 1969, Isoenzyme characterization of animal cell cultures, Proc. Soc. Exp. Biol. Med. 132:462-469.

12. S.M. Gartler, 1967, Genetic markers in cell culture, Nat. Cancer Inst. Mono. 26:167-195.

13. S.J. O'Brien, G. Kleiner, R. Olsen and J.E. Shannon, 1977, Enzyme polymorphisms as genetic signatures in human cell cultures, Science 195:1345-1358.

14. W.C. Wright, W.P. Daniels and J. Fogh, 1981, Distinction of seventy-one cultured human tumor cell lines by polymorphic enzyme analysis, J. Nat. Cancer Inst. 66:239-247.

15. P.S. Moorehead, P.C. Nowell, W.J. Mellman, D.M. Battips and D.O. Hungerford, 1960, Chromosome preparation of leukocytes cultured from human peripheral blood, Exp. Cell Res. 20:613-616.

16. T. Casperson, G. Lommaka and L. Zech, 1971, The 24 fluorescence patterns of the human metaphase chromosomes distinguishing characters and variability, Heredites 67:89-102.

17. O.J. Miller, D.A. Miller, P.W. Allerdice, V.G. Dev and M.S. Grewal, 1971, Quinacrine fluorescent karyotypes of human diploid and heteroploid cell lines, Cytogenetics 10:338-346.

18. M. Seabright, 1971, A rapid banding technique for human chromosomes, Lancet 2:971-972.

19. C.A. Kozak, J.B. Lawrence and F. Ruddle, 1977, A sequential staining technique for the chromosomal analysis of interspecific mouse/hamster and mouse/human somatic cell hybrids, Exp. Cell Research, 105:69-117.

20. T.C. Hsu and K. Benirschke, 1971-1977, An Atlas of Mammalian chromosomes, Springer Verlag, New York, Heidelberg, Berlin, Vols. 1-10.

21. P. Noguchi, R. Wallace, J. Johnson, E.M. Earley, S.J. O'Brien, S. Ferrone, M.A. Pellegrino, J. Milstein, C. Needy, W. Browne and J. Petricciani, 1979, Characterization of WIDR: a human colon carcinoma cell line, In Vitro 15:401-408.

22. L.P. Rutsky, C.L. Kaye, J.J. Siciliano, M. Chao and B.D. Kahan, 1980, Longitudinal and genetic signature analysis of cultured human colon adenocarcinoma cell line LS-180 and LS174T, Cancer Research 40:1443-1448.

23. M.J. Siciliano, P.E. Barker and R. Cailleau, 1979, Mutually exclusive genetic signatures of human breast tumor cell lines with a common chromosomal marker, Cancer Research 39:919-922.

24. C. Brantbar, M.A. Pellegrino, S. Ferrone, R.A. Reisfeld, R. Payne and L. Hayflick, 1973, Fate of HL-A antigens in aging cultured human diploid cell strains, Exp. Cell Res. 78:367-375.

25. S. Ferrone, M.A. Pellegrina and R.A. Reisfeld, 1971, A rapid method for direct HL-A typing of cultured lymphoid cells, J. Immunol. 107:613-615.

26. D.L. Mann, L. Abelson, S. Harris and D.B. Amos, 1975, Detection of antigens specific for B-lymphoid cultured cell lines with human allo antisera, J. Exper. Med. 142:84-89.

27. D.M. Amos, 1976, Cytotoxicity testing. In NIAID Manual of Tissue Typing Techniques. Ed. Ray, J.G. Jr., Hare, D.B., Pedersen, P.D. and Mullally, D.I. DHEW Publication, No. 76-545:25-28.

28. M.A. Pellegrino, S. Ferrone and A. Pellegrino, 1972, A simple microabsorption technique for HL-A typing, Proc. Soc. Exp. Biol. Med. 139:484-488.

29. M. Siniscalco, B.B. Knowles and Z. Steplewski, 1968, Hybridization of human diploid strains carrying X-linked mutants and its potential in studies of somatic cell

genetics. In: Heterospecific Genome Interaction. The Wistar Institute Symposium Monograph 9:117–136.

30. W.D. Peterson, Jr., W.F. Simpson, P.S. Ecklund and C.S. Stulberg, 1973, Diploid and heteroploid human cell lines surveyed for Y chromosome fluorescence, Nature New Biology 242:22–24.

31. C.K. Miller, D.A. Miller, O.J. Miller, R. Tantravahi and R.T. Reese, 1977, Banded chromosomes of the owl monkey, Aotus trivirgatus, Cytogenet. Cell Genet. 19:215–226.

32. N.L. Harris, D.L. Gang, S.C. Quay, S. Poppema, P.C. Zamecnik, W.S. Nelson-Rees and S.J. O-Brien, 1981, Contamination of Hodgkin's disease cell cultures, Nature 289:228–230.

33. D. Warburton, L.L. Firschein, D.A. Miller and F.E. Warburton, 1973, Karyotype of the chimpanzee, Pan troglodytes, based on measurements and banding pattern: comparison to the human karyotype, Cytogenet. Cell Genet. 12:453–461.

34. P.A. Lalley, 1977, Human biochemical genetic map, Bull. Isozyme 10:12–17.

35. W.A. Nelson-Rees, D.W. Daniels and R.R. Flandermeyer, 1981, Cross-contamination of cells in culture, Science 212:446–452.

CHARACTERISTICS OF LYMPHO-MYELOPOIETIC STEM CELLS ISOLATED FROM CANINE PERIPHERAL BLOOD

Dennis A. Stewart[1], Robert B. Bolin, Barbara A. Cheney, Joseph T. Hawkins, Keith W. Chapman and Donald R. Tompkins

Division of Blood Research
Letterman Army Institute of Research
Presidio of San Francisco, CA 94129

ABSTRACT

If hematopoietic stem cells (HSC) could be separated from peripheral blood, it might be possible to harvest these stem cells for potential clinical use. By leukapheresis techniques, we harvested mononuclear cells (MNC) from peripheral blood and then placed these cells over discontinuous stractan gradients of three densities (1.077 gm/ml, 1.071 gm/ml and 1.066 gm/ml). These separated cells were submitted to colony culture to identify colony-forming-unit activity for granulocyte-macrophage (CFU-C) and T-cell lymphocyte (CFU-L) cell lines. The lightest cells

---

[1] To whom correspondence should be addressed.

Abbreviations: Hematopoietic Stem Cells (HSC), Mononuclear Cells (MNC), Colony-Forming-Units for Lymphocytes (CFU-L), and granulocytes (CFU-C), Graft versus Host Disease (GvH).

(1.066) contained most of the CFU-C and no CFU-L activity. Heavier cells (>1.071) contained CFU-L and very little CFU-C activity. CFU-L colonies could be distinguished from CFU-C by their density and distinct morphological appearance. In addition, the amount of CFU-C could be increased in the animal by increasing the amount of blood processed (from $3.9\pm.76$ CFU-C/10^6 MNC to $6.7\pm.35$ CFU-C/10^6 MNC). This resulted in an increase of CFU-C collected from 7.6 ± 2.1 CFU-C/10^6 MNC after the first equivalent blood volume to 22.5 ± 3.4 CFU-C/10^6 MNC after the third equivalent blood volume processed. These results suggest that leukapheresis and gradient density separation may be useful procedures to obtain HSC.

INTRODUCTION

Hematopoietic stem cell (HSC) failure can be treated by transfusions of blood-forming elements (stem cells) from a donor.[1,2,3] This is usually done with stem cells derived from histocompatible donors. The harvest of bone marrow HSC is tedious since marrow access is limited. Alternative harvest techniques have not been successfully developed for human use.[3] In addition, when HSC are harvested from bone marrow, they are not isolated from lymphocytes; thus immunocompetent cells that can induce graft-versus-host disease (GvH) are infused in the recipient.[4] With current histocompatibility matching of donor cells to recipient cells, the risk of GvH can be reduced but this disease still occurs in 25 to 75% of the recipients.[3]

The use of peripheral blood derived HSC may offer an easier, more desirable alternative to the present bone marrow transplantation techniques. Modern leukapheresis techniques are being used to harvest platelets, and leukocytes selectively.[5] If leukapheresis could be adapted to harvest HSC alone, then convenient HSC products could be developed and, at the same time, risk of GvH could be reduced. Using a dog model, we have evaluated the effects of leukapheresis on HSC and lymphocyte harvests as measured by colony culture assays. We have assessed discontinuous gradient centrifugation as a technique, supplemental to the leukapheresis, to obtain a purer volume of HSC separated from T-cell lymphocytes.

MATERIALS AND METHODS

Leukapheresis

Mononuclear cells (MNC) were collected from the peripheral circulation of ten (10) healthy, normal, mongrel dogs. The dogs were anesthetized with halothane/nitrous oxide. The equivalent of three blood volumes from each dog were processed by continuous flow centrifugation, (IBM 2997 cell separator, IBM Corporation, P.O. Box 10, Princeton, NJ 08540). The mononuclear cells were collected from each equivalent blood volume, calculated from body weight and the flow rate, into separate IBM storage bags. To prevent clotting 1.0 ml of 47% trisodium citrate was added to each bag before collection. Blood was drawn from the femoral vein of the dog through the machine at a flow rate of 50-80 ml/minute and returned to the animal through the cephalic vein.

A 2% (w/v) trisodium citrate solution was administered into the collection line during the pheresis at a ratio of 10:1 or 20:1 blood/anticoagulant ratio. The whole blood ionized calcium in the animal was measured frequently during the run to monitor possible citrate toxicity. (Ionized Calcium Analyzer, Orion Biomedical, 840 Memorial Drive, Cambridge, MA 02138)

Discontinuous Density Separations

Mononuclear cell-rich suspensions from each of the three blood volume bag collections were layered over density gradients of Stractan II, an arabinogalactan (St. Regis Plywood Co., Tacoma, WA). This isopynic density gradient for platelet separation[6] was modified for the separation of the MNC. The stractan was layered into cellulose nitrate tubes (9/16 x 3 3/4, inches, Beckman model number 331101) as follows: 3 ml of 1.077 gm/ml; 1.5 ml of 1.071 gm/ml; and 1.5 ml of 1.066 gm/ml. The MNC concentration was adjusted with buffered saline glucose[6] (pH 7.4) to 10^8 cells in 7 ml, and layered onto the stractan. Tubes were spun in a SW 40Ti rotor for 25 minutes, at 20,000 rpm, ambient temperature, in an ultracentrifuge. (Model L5-75, Beckman Instruments, Palo Alto, CA 94304).

With a tubeslicer (Nuclear Supply Co., Washington, D.C., 20005) the cells collected at the gradient interface (cell fraction) were separated and placed into 12 x 75 mm tubes. Cells were counted in a hemocytometer, and total volume measured for each fraction to calculate total cell harvest. The interface cell fractions are referred to as fractions 1 through 3 (lightest to heaviest). Cell dilutions for the culture assays were made with McCoys 5A medium (Gibco Laboratories, 3175 Stanley Road, Grand Island, NY 14072). A schematic of the procedures used is outlined in Figure 1.

Colony Cultures

CFU-C and CFU-L colony forming units were assessed in soft agar cultures. The technique used for assessing CFU-C from the peripheral blood and bone marrow is a two layer agar culture

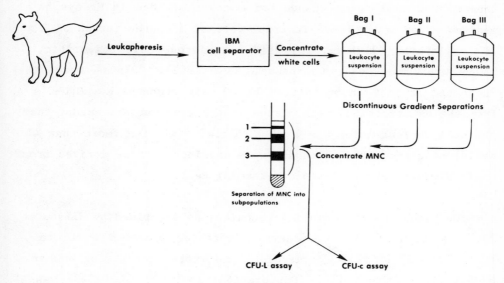

Figure 1

TECHNIQUE TO ISOLATE STEM CELLS

A schematic diagram of the apheresis, discontinuous gradient separation and subsequent assays used. Bag I, II, III represent separate MNC populations after 1,2, and 3 equivalent blood volumes processed through the cell separator.

method similar to that described by Provaznik and MacVittie.[7] However, we used as a colony-stimulating factor (CSF) (7% v/v) dog plasma collected by pheresis six hours after intravenous injection, 4 ug/kg body weight, of Salmonella typhosa. (Bacto Lipolysaccharide W, Difco Laboratories, Detroit, MI) dissolved in pyrogen-free saline.

Before the leukapheresis, bone marrow was collected from the femur of the anesthetized dog by aspiration with a Gardner needle into a syringe containing preservative-free heparin (100 u/ml in culture media). The bone marrow was filtered through six layers of nylon mesh and layered over Ficoll-Paque$^{(R)}$ according to the manufacturer's procedures. (Pharmacia Fine Chemicals, Division of Pharmacia, Inc., Piscataway, NJ 08854).

Bone marrow cells were washed twice and suspended in McCoys 5A medium. They were plated in triplicate at a concentration of 2×10^5 cells/dish (35x10mm, 2mm grid Lux plates, Cole Scientific, 23966 Craftsman Road, Calabasas, CA 91302). Plates were incubated at 37°C in humidified 5% CO_2/air atmosphere. CFU-C's from the MNC fractions were plated in the same manner, except that the cell concentration was 1×10^6/plate. Colonies (more than 40 cells) counted on the tenth day were considered to be derived from a granulocyte-macrophage colony-forming cell.

We used the CFU-L plating technique as described by Wilson et al.[8] Certain lots of concanavalin-A tested appeared toxic to growth of MNC (5×10^5 cells/dish) and whole blood. Lot number 610049 (Calbiochem-Behring Corporation, La Jolla CA 92037) was used in this study.

Electron Microscopy of Colonies

For transmission electron microscopy, the agarose containing lymphocyte colonies were processed by cutting out a wedge of agarose, lifting it gently out of the plate with a narrow spatula, and placing it into a glass vial containing 10 ml 2% glutaraldehyde. The fixed specimen was then embedded in Epon and processed for transmission electron microscopy.[7]

RESULTS

Recovery of cells after density gradient centrifugation was usually 60-90 percent of the total placed on the gradient, with the majority of the cells in fraction 3. The three fractions contained nearly pure MNCs, with only 1-2% eosinophils in the heaviest fraction (Fraction 3). However, if the dog had a high blood eosinophil count (>10%), there would be as many as 40% eosinophils in Fraction 3. All other granulocytic cells and erythrocytes formed a pellet at the bottom of the tube. Cells separated on Stractan retained their integrity and normal morphology as determined by Wright-Giemsa stain.

The CFU activity in peripheral circulation before leukapheresis compared to after leukapheresis showed a moderate increase in CFU-Cs (3.9±.76 to 6.7±1.35) and a significant decrease in CFU-L (88.9±35.8 to 11.7±3.52) (Table 1). The number of bone marrow CFU-C derived colonies (468.8±151.4) (Table 1) was similar to previous canine colony counts where endotoxin stimulated dog serum was used as the colony stimulating factor.[9] Our bone marrow counts indicated that a satisfactory CSF was present. The concentration of CFU-C in bone marrow was 117 times more than peripheral blood.

TABLE 1

CFU-C and CFU-L characteristics of peripheral blood and bone marrow

	CFU-C[1]	N	CFU-L[2]	N
Pre-pheresis	3.9±.76	11	89.0±35.8	9
Post-pheresis	6.7±1.35	10	11.7±3.52	9
Bone Marrow	468.8±151.4	7	----	--

[1]per 10^6 MNC. Values reported are the Mean ± S.E.M.
[2]per 10^5 MNC. Mean ±S.E.M.

TABLE 2

The numbers of CFU-C collected per blood volume
exchange (n=10 dogs)

	BLOOD VOLUME 1	BLOOD VOLUME 2	BLOOD VOLUME 3
Total volume[1]	80.2±3.6	91.2±4.0	92.4±5.8
Total MNC collected ($\times 10^9$)[1]	2.72±.3	3.02±.3	2.43±.2
Total CFU-C collected ($\times 10^4$)	2.3+	5.6	7.4+
Total CFU-C collected ($\times 10^3$) /kg donor wt.	1.3	3.2	4.2

[1] Mean ± S.E.M.

+ Significant difference between bag volume 1 and bag volume 3.

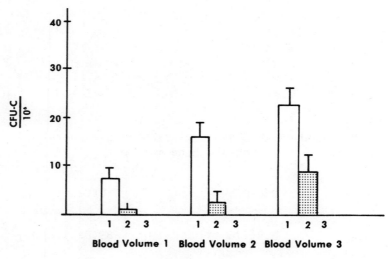

Figure 2

CFU-C in Stractan Fractions per Three Blood Volume Exchanges

CFU-C activity in the three density fractions according to the equivalent blood volume processed. Values reported are means ± SEM.

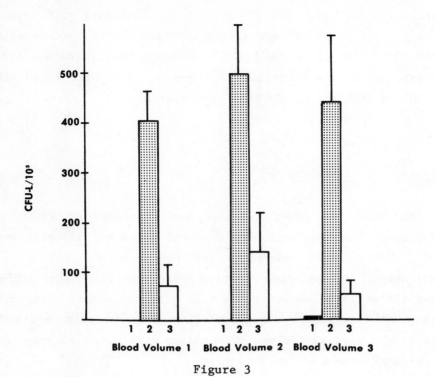

Figure 3

CFU-L in Stracton Fractions per Three Blood Volume Exchanges

CFU-L activity in the three density fractions according to the equivalent blood volume processed. Values reported are mean ± SEM.

Combined values for the total volumes, total MNC and total CFU-Cs collected for each equivalent blood volume processed from ten separate experiments on individual dogs are presented in Table 2. The total volumes and total MNCs collected per exchange do not vary significantly. However, the total CFU-Cs collected increased significantly (2.3×10^4 to 7.4×10^4). The results of CFU-C and CFU-L derived colonies for separate gradients and equivalent blood volume exchange are summarized in Figures 2 and 3. The majority of CFU-C colonies were colonies found in the lightest fraction (Fraction 1) whereas the CFU-L colonies were found in the heavier fractions (2 and 3). The CFU-C activity increased with each volume i.e., 8 CFU-C/10^6 MNC plated in blood volume 1 compared to 22 CFU-C/10^6 MNC plated in blood volume 3. This trend was not apparent for CFU-L where similar numbers of CFU-L were found with each volume (400 to 500 CFU-L/10^5 MNC plated).

Morphology and Structure

The CFU-L and CFU-C colonies were compared by light microscopy. The CFU-L colonies appeared dense and discoid; the CFU-C colonies, were amorphous and had a looser colony organization. Transmission electron microscopy (Fig 4) of a CFU-L showed the cells to be homogeneous in size and shape. They are metabolically active as demonstrated by numerous polysomes and rough endoplasmic reticulum. Filopodium were often observed on the physical boundaries of the colony.

DISCUSSION

This study confirms the observation of others that stem cells can be harvested from peripheral blood.[10-14] The harvest of peripheral blood stem cells from normal human donors has not been developed to a practical point whereby therapeutic doses could be achieved.[13,14] One reason practicality has not been achieved is that stem cells in peripheral blood constitute less than 1% of that in the bone marrow (1 to 120 ratio). Fliedner et al[11] have been able to increase yields by using dextran sulfate to mobilize stem cells in dogs. Richman et al[15] also increased harvest yield after the use of myelosuppressive drugs in oncology patients.

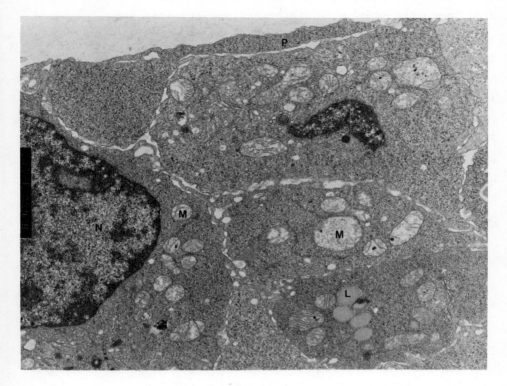

Figure 4

Transmission electron micrograph of cells in a CFU-L colony. Cells contain abundant polyribosomes, profiles of granular endoplasmic reticulum, mitochondria (M) lipid droplets (L) and a nucleus (N). Cell boundaries have filopodia (P). Magnification is 19,000x.

Neither of these procedures is practical for routine human use since dextran sulfate is a heparinoid compound and myelosuppressive drugs could be dangerous to normal volunteers. The purpose of this study was to evaluate stem cell harvest without stimulating agents to determine if improved yields could be obtained by prolonged pheresis and to evaluate the extent to which stem cell compartment shifts occur. Pheresis procedures used to obtain stem cells in humans process less than 1.5 blood volumes.[12,14] Our results show that prolonged pheresis (3 blood volumes processed) does enhance stem cell yields and these cells probably enter the circulation by compartment shift from the bone marrow pool. The concentration of CFU-C is only increased about two-fold by 1-1.5 blood volume pass whereas our results show that pheresis increases the concentration of CFU-C five-fold by the third blood volume processed. Although the enhancement of stem cells harvested by prolonged pheresis does not approximate that of bone marrow aspirations (1 to 20 ratio), pheresis may be desirable therapeutically because of easier access. Furthermore, with pheresis procedures there is the capability for repeated harvest from the peripheral pool to increase yields. This, along with the ability to remove mononuclear cells selectively from granulocytes, offers advantages in using peripheral stem cell separations. Since the mononuclear concentrates collected during the first blood volume is low in CFU-C, this product may be returned to the donor at the end of pheresis to replenish platelets and leukocytes also removed by the procedure. If $2-3 \times 10^8$ MNC/kg are required from bone marrow harvest to induce a recipient's hematopoietic recovery,[16] then about 9.4×10^4 CFU-C/kg is needed from peripheral blood based on our values of 469 CFU-C/10^6 MNC from bone marrow) (Table 1). Total CFU-C collected in this study is 8.7×10^3 CFU-C/kg, (average dog weight 18 kg), about 1/10th of a therapeutic dose for an equivalent size animal. It would then take 10 apheresis procedures to provide a therapeutic dose unless greater yields are possible. For greater yields, compartment shifts by safe drugs and/or further optimizing the yields with apheresis equipment are required. We are currently investigating these options. It appears that immunocompetent T-cell lymphocytes are not increased by prolonged pheresis as manifested by our CFU-L assay.

Fliedner et al[11], using dogs, and Dicke and van Bekkum[17], using monkeys and mice, have shown that discontinuous separation of bone marrow or peripheral blood mononuclear cells can be used to reconstitute marrow cells in aplastic animals with a reduction of GvH reactions. Our results show that this phenomenon is probably due to the physical separation of granulocytic stem cells from immunocompetent T-cell lymphocytes since cells that have densities below 1.071 gm/ml do not have CFU-L activity but have CFU-C activity. Lymphoid and myeloid precursor cells have also been separated by velocity sedimentation in mice[18] and by steep gradient zonal centrifugation in humans.[19]

These in vitro observations have practical application since the ability to harvest stem cells from peripheral blood would be more convenient than to harvest bone marrow stem cells and, therefore, increase the availability of this hematopoietic product. The ability to separate immunocompetent cells from the hematopoietic stem cells may improve the techniques of stem cell transplants by reducing GvH disease. These observations provide the basis for further investigation using peripheral blood-derived hematopoietic stem cells in the treatment of hematopoietic cell failure in vivo.

ACKNOWLEDGMENT

The authors appreciate the technical support of Dan Smith B.S. for calcium determinations and David Scott, Ph.D. for preparation and interpretation of electron micrographs. The editorial assistance of Lottie Applewhite, M.S. and the typing skills by Linda Wettstein were invaluable to the preparation of this manuscript.

REFERENCES

1. E.D. Thomas, R. Storb, R.A. Clift, A. Feder, L. Johnson, P.E. Neiman, K.G. Lerner, H. Glucksberg and C.D. Buckner, Bone-marrow transplantation (first of two parts). N. Engl. Med. 292:832-843, 1975.
2. E.D. Thomas, R. Storb, R.A. Clift, A. Feder, L. Johnson, P.E. Neiman, K.G. Lerner, H. Glucksberg and C.D. Buckner, Bone-marrow transplantation (second of two parts). N. Engl. Med. 292:895-902, 1975.
3. R.P. Gale, R.E. Champlin, S.A. Feig and J.H. Fitchen, Aplastic anemia: biology and treatment. Ann. Intern Med. 95:477-494, 1981.
4. S.C. Grebe and J.W. Streilan, Graft-versus-host reactions: a review. Adv. Immunol. 22:119-221, 1976.
5. American Association of Blood Banks, Technical Manual. Chapter 2, 1977.
6. L. Corash, H. Tan and H.R. Gralnick, Heterogeneity of human whole blood platelet subpopulations. I. Relationship between buoyant density, cell volume, and ultrastructure. Blood 49:71-77, 1977.
7. M. Porvaznik and T.J. MacVittie, Detection of gap junctions between the progeny of canine macrophage colony- forming cell in vitro. J. Cell. Biol. 82:555-564, 1979.
8. F.D. Wilson, J.A. Dyck, S.J. Knox and M. Shifrine, A "whole blood" technique for the quantitation of canine "T-lymphocyte" progenitors using a semi-solid culture system. Exp. Hematol. 8:1031-1039, 1980.
9. T.J. MacVittie and R.I. Walker, Endotoxin-induced alterations in canine granulopoiesis colony stimulating factor, colony-forming cells in culture, and growth of cells in diffusion chamber, Exp. Hematol. 6:613-618, 1978.

10. R.A. Abrams and A.M. Deisseroth, Prospects for accelerating hematopoietic recovery following myelosuppressive therapy by using autologous, cyropreserved hematopoietic stem cells collected solely from the peripheral blood. Exp. Hematol. 7:(supp. 5):1107-1115, 1979.
11. T.M. Fliedner, W. Calvo, M. Körbling, W. Nothdurft, H. Pflieger and W. Ross, Collection, storage and transfusion of blood stem cells for the treatment of hematopoietic failure. Blood Cells 5:313-328, 1979.
12. R.S. Weiner, C.M. Richman and R.A. Yankee, Semicontinuous flow centrifugation for the pheresis of immunocompetent cells and stem cells. Blood 49:391-397, 1977.
13. S.C. Sarpel, A.R. Zander, L. Horvath and R.B. Epstein, The collection preservation, and function of peripheral blood hematopoietic cells in dogs. Exp. Hematol. 7:113-120, 1979.
14. R.A. Abrams, D. Glaubiger, F.R. Appelbaum and A.B. Deisseroth, Result of attempted hematopoietic reconstitution using isologous, peripheral blood mononuclear cells: a case report. Blood 56:516-520, 1980.
15. C.R. Richman, R.S. Weiner and R.A. Yankee, Increase in circulating stem cells following chemotherapy in man. Blood 47:1031-1039, 1976.
16. E.D. Thomas and R. Storb, Technique for human marrow grafting. Blood 36:507-515, 1970.
17. K.A. Dicke and D.W. van Bekkum, Allogeneic bone marrow transplantation after elimination of immunocompetent cells by means of density gradient centrifugation. Transplant Proceedings, III:666-668, 1971.
18. R.A. Phillips and R.G. Miller, Physical separation of hematopoietic stem cells from cells causing graft-vs-host disease. I. Sedimentation properties of cells causing graft-vs-host disease. J. Immunol. 105:1168-1174, 1970.

19. J.R. Wells, G. Opelz and M.J. Cline, Characterization of functionally distinct lymphoid and myeloid cells from human blood and bone marrow. I. Separation by a buoyant density gradient technique. J. Immunol. Methods 18:63-77, 1977.

GROWTH CHARACTERISTICS OF A431 HUMAN EPIDERMOID CARCINOMA CELLS
IN SERUM-FREE MEDIUM: INHIBITION BY EPIDERMAL GROWTH FACTOR[1]

David W. Barnes

Department of Biological Sciences
University of Pittsburgh
Pittsburgh, PA 15260

ABSTRACT

A431 human epidermoid carcinoma cells in culture may be grown in the absence of serum in a one to one mixture of Ham's F12 and Dulbecco-modified Eagle's medium supplemented with 10 µg/ml bovine insulin, 10 µg/ml human transferrin, 5 µg/ml human cold-insoluble globulin and 0.5 mM ethanolamine. Growth rate of the cells in this serum-free medium is essentially identical to that of cells in medium containing 10% fetal calf serum. Addition of epidermal growth factor at concentrations which are stimulatory for the growth of other cell types in culture causes and inhibition of A431 cell growth in both serum-free and serum-containing medium. The inhibitory effect of epidermal growth factor is reversible upon replacement of epidermal growth factor-containing medium with medium containing

---

[1]This work was supported by an Anabele G. and George A. Post Memorial Grant for Cancer Research (BC-368) from the American Cancer Society.

no epidermal growth factor. Simultaneous addition to the medium of antibodyto epidermal growth factor along with epidermal growth factor prevents the inhibitory effect.

INTRODUCTION

Epidermal growth factor (EGF)[2], a powerful mitogen for a number of cell types, is a very useful tool for the study of the signals and processes controlling cell proliferation (1). The A431 human epidermoid carcinoma cell line has been of particular value in the study of early biochemical effects of EGF because these cells express an extremely large number of EGF plasma membrane receptors compared to most cell types in culture (2). The line has been used to study the internalization of EGF by cells (3-6), enhancement of protein phosphorylation by EGF treatment (7-13), and effects of EGF on cell morphology and pinocytosis (14-16), and also has been used as a source of EGF receptor for purposes of identification and purification of this membrane protein (17-19).

The work of the laboratory of Dr. Gordon Sato of the University of California at San Diego has established that serum as a supplement in culture medium can be replaced by specific combinations of (I) nutrients, (II) hormones and growth factors,

---

[2]Abbreviations: EGF, epidermal growth factor; DME, Dulbecco-modified Eagle's medium; EDTA, ethylenediaminetetraacetic acid; HEPES, 4-(2-hydroxyethyl)-1-piperazineethanesulfonic acid; FCS, fetal calf serum; CIg, cold-insoluble globulin; FGF, fibroblast growth factor; PDGF, platelet-derived growth factor; F12:DME, one to one mixture of F12 and DME supplemented with sodium bicarbonate, HEPES and antibiotics; F12:DME+FCS, F12:DME supplemented with 10% FCS; PBS, phosphate-buffered saline; F12:DME+4F, F12:DME supplemented with insulin, transferrin, CIg, and ethanolamine.

(III) binding proteins which modulate the action of nutrients, hormones and growth factors, and (IV) attachment factors (20,21). Because of the extensive use of the A431 cell line for the study of the mechanism of action of EGF, and the advantages for these kinds of studies of a serum-free medium supporting the long-term growth of the cells under hormonally and nutritionally defined conditions, a medium was developed in which the serum supplement was replaced by factors from the four classes listed above. In the course of the development of this medium, it was found that, rather than acting as a mitogen for A431 cells, EGF is highly inhibitory for the growth of these cells in serum-free or serum-containing medium (22,23). A preliminary report of this work first appeared in May 1981 (24).

MATERIALS AND METHODS

Powdered formulations of Ham's F12 and Dulbecco-modified Eagle's medium (DME) were obtained from Grand Island Biological Company. Crude trypsin, ethylenediaminetetraacetic acid (EDTA), soybean trypsin inhibitor, bovine insulin, human transferrin, ethanolamine, and 4-(2-hydroxyethyl)-1-piperazineethanesulfonic acid (HEPES) were obtained from Sigma. Fetal calf serum (FCS) was obtained from Reheis. Other sera were obtained from Grand Island Biological Company. Affinity-purified rabbit anti-mouse epidermal growth factor was obtained from LAREF. Mouse EGF, human cold-insoluble globulin (CIg), bovine fibroblast growth factor (FGF) and partially purified human platelet-derived growth factor (PDGF) were obtained from Collaborative Research, Inc.

The A431 cell line was a gift of Dr. T. Hunter, Tumor Virology Laboratory, The Salk Institute, La Jolla, CA. The culture medium used routinely as a one to one mixture of F12 and DME supplemented with 1.2 g per liter sodium bicarbonate, 15 mM HEPES and antibiotics (F12:DME). Other media supplements are as indicated in the text and figure legends. For the initiation of experiments, cells grown to confluence in F12:DME containing 10% fetal calf serum (F12:DME+FCS) were detached from flasks with a solution of 0.1% crude trypsin and 0.9 mM EDTA in phosphate-buffered saline (PBS), suspended in an equal volume of 0.1% soybean trypsin inhibitor in F12:DME, centrifuged from suspension at low speed, resuspended in the appropriate media and plated at 2×10^4 cells/35 mm diameter culture plate. Insulin, transferrin, CIg and ethanolamine were added where indicated directly to culture plates containing 1 ml of medium as small volumes of sterile, concentrated stock solutions a few minutes before adding the cell suspension in a second ml of medium. The concentrations of insulin, transferrin, CIg and ethanolamine used in the experiments described in this paper were 10 µg/ml, 10 µg/ml, 5 µg/ml and 0.5 mM, respectively. EGF was added as small volumes of sterile, concentrated stocks to give the indicated final concentrations. The final concentration of FCS in the experiments of Figure 2 through Figure 6 was 10%. For determination of cell number, cells were detached from plates with 0.1% trypsin and 0.9 mM EDTA in PBS and suspensions were counted in a Coulter particle counter.

RESULTS

The A431 cell line was established and is grown routinely in most laboratories in DME supplemented with 10% FCS (2-19). Because other media formulations or mixtures of these formulations have been shown in some situations (particularly in the absence of serum), to support better growth than that seen in DME (20,21), a survey of several media and serum supplements

was carried out on the A431 cell line to determine the best
conditions for growth in serum-containing medium and the best
basal nutrient medium to use as a base for developing a
serum-free medium. Human, horse, calf, newborn calf and fetal
calf sera were tested at various concentrations, and fetal calf
serum was found to be superior to the others for supporting A431
cell growth. Best growth using fetal calf serum as a supplement
was seen in a basal nutrient consisting of a one to one mixture
of F12 and DME. Figure 1 illustrates the growth of these cells
in F12, DME or F12:DME at various concentrations of FCS. Stock
cultures for the experiments described in this report were grown
in F12:DME supplemented with 10% FCS. As shown in this figure,
F12:DME+FCS supported growth of A431 somewhat better than the

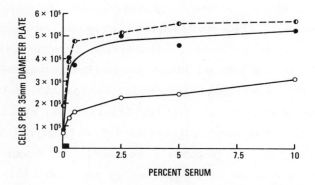

Fig. 1. Growth of A431 cells in F12, DME and F12:DME
supplemented with FCS. Cell number was determined
after incubation for 6 days in plates containing F12
(O——O), DME (O——O) or F12:DME (O---O) containing the
indicated concentrations of FCS. Initial cell
inoculum (□) was 2×10^5 cells/plate.

standard medium used previously in most laboratories (DME+10%FCS). Growth of A431 in F12 was comparatively poor at any FCS concentration. Some cell growth occurred in F12, DME or F12:DME in the complete absence of FCS, but, of the three media, F12:DME was clearly superior under conditions in which no serum was present.

A large number of hormones, binding proteins, attachment factors and supplementary nutrients previously found to be stimulatory for some cell types in serum-free media were tested over a range of concentrations on the A431 cell line, using F12:DME as the basal nutrient medium. Supplements which appeared to have no effect on A431 cell growth in serum-free medium under the conditions tested included FGF, triiodithyronine, multiplication stimulating activity, sodium selenite, aldosterone, progesterone, estradiol, testosterone, human and bovine prolactin, follicle stimulating hormone, adrenocorticotropic hormone, prostaglandin F_2 alpha, prostaglandin E_2, glucagon, thyroid stimulating hormone, thyrotropin releasing factor, luteinizing hormone releasing factor, nerve growth factor, human and bovine growth hormone and thrombin. Factors which were stimulatory for A431 cell growth in serum-free medium were insulin, transferrin, ethanolamine (25,26). CIg, serum spreading factor (27,28), glycyl-histidyl-lysine, somatostatin, parathyroid hormone, and bovine serum albumin. Factors found to be inhibitory for A431 cell growth in serum-free medium were EGF, a partially purified preparation of PDGF and cyclic AMP. Marginal inhibition of growth was also seen upon addition of human chorionic gonadotropin and porcine relaxin. In addition, polylysine was ineffective as a growth-promoting attachment factor; pretreatment of culture dishes (29) was found to be inhibitory for A431 cell growth.

The most active factors in promoting A431 cell growth in serum-free medium were insulin at 2 to 20 µg/ml, transferrin at 2 to 20 µg/ml, CIg at 2 to 10 µg/ml and ethanolamine at 0.1 to 1.0 mM. A small but consistent effect of glycyl-histidyl-lysine at 2 µg/ml was observed, and smaller, more variable effects of parathyroid hormone at 5 ng/ml and somatostatin at 100 ng/ml were sometimes observed. The stimulation of A431 cell growth by albumin at 0.5 to 2.0 mg/ml was presumed to be due to lipids, hormones or nutritional factors bound to the albumin and was not pursued further. Both CIg and serum spreading factor at similar concentrations were effective at promoting growth of A431 cells in serum-free medium in the presence of other stimulatory supplements such as insulin and transferrin. Both were also effective if the plates into which the cells were seeded were pretreated for four hours with medium containing the attachment factors and then washed with fresh medium before adding cells. Under these conditions the attachment factors adhere to the culture dish and mediate the proper attachment of cells to the substratum (20,21,27,28,30).

As shown in Figure 2, A431 cells grew with a doubling time of about 30 hours in F12:DME supplemented with 10 µg/ml transferrin, 5 µg/ml CIg and 0.5 mM ethanolamine (F12:DME+4F). This growth rate was identical to that observed for A431 cells in F12:DME+FCS. Addition of the four stimulatory factors used in F12:DME+4F to F12:DME+FCS did not improve cell growth beyond that seen in F12:DME+4F or F12:DME+FCS alone. A431 cells were also capable of growth in the absence of serum in F12:DME (without supplementation with the four factors), although the generation time was about 48 hours and a considerable time lag occurred before the onset of growth after plating. Medium changes every three days did not influence the growth rate for the first six to eight days after seeding of A431 cells in plates containing F12:DME+4F, but significantly increased the cell density in these plates at later times. Confluent cultures of A431 cells grown in F12:DME+4F could be passaged repeatedly in this serum-free medium with the use of a trypsin-EDTA

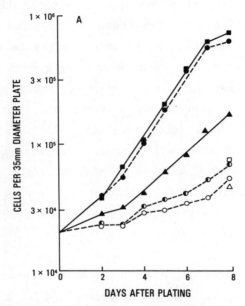

Fig. 2. Growth of A431 cells in serum-containing or serum-free medium in the presence or absence of EGF. F12:DME+FCS (□——□); F12:DME+4f (O---O); F12:DME (Δ——Δ); F12:DME+4F +10 ng/ml EGF (O---O); F12:DME+4F + 100 ng/ml EGF (O---O); F12:DME+FCS + 100 ng/ml EGF (□); F12:DME + 100 ng/ml EGF (Δ).[3]

---

[3]From Barnes, D.W. 1981. Epidermal growth factor inhibits growth of A431 human epidermoid carcinoma in serum-free cell culture. J. Cell. Biol., 93:1-4.

solution and a soybean trypsin inhibitor solution, and have been carried as long as 5 weeks in this manner.

Addition of 10 to 100 ng/ml EGF to F12:DME+4F at the time of plating the cells resulted in a strking inhibition of A431 cell growth (Figure 2). A measurable inhibition of A431 cell growth was observed after the addition to the culture medium of as little as 1 ng/ml EGF (Figure 3). It was not necessary to add EGF to culture medium at the time of plating in order to demonstrate inhibition of growth: 10 ng/ml EGF added to cells 2 or 4 days after plating resulted in a reduced cell number in those plates compared to controls when cell number was

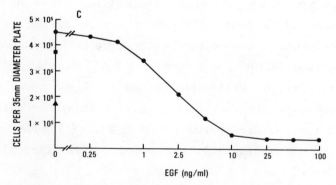

Fig. 3. Concentration dependence of EGF inhibition of A431 cell growth. Cells were counted after incubation for 6 days in plates containing F12:DME+4F with the indicated concentrations of EGF (O——O); F12:DME (Δ); F12:DME+FCS (□).[3]

determined 6 days after plating (Figure 4). As indicated in Figure 4, inhibition of growth in the presence of EGF also could

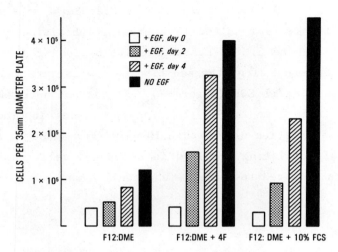

Fig. 4. Inhibition of A431 cell growth by EGF added at various times after plating. Cultures were seeded in the indicated media, either with or without 10 ng/ml EGF. To some cultures not receiving EGF initially, EGF as a small volume of a concentrated stock was added two or four days after plating to give a final concentration in the culture medium of 10 ng/ml. EGF added at the time of plating (□); EGF added 2 days after plating (□); EGF added 4 days after plating (□); no EGF added (□). Cell number in all plates was determined six days after plating.[4]

---

[4]From Barnes, D.W. 1981. Growth of A431 human epidermoid carcinoma in serum-free cell culture: inhibition by epidermal growth factor. Cold Spring Harbor Conf. on Cell Prolif., Vol. 9, in press.

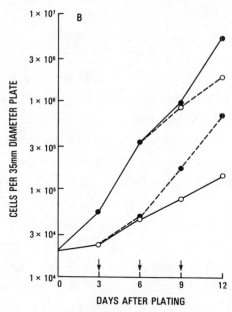

Fig. 5. Reversal of EGF inhibition of cell growth. Arrows indicate days on which medium was changed for all conditions. F12:DME+4F (O——O); F12:DME+4F 1 10 ng/ml EGF (O——O); F12:DME+4F changed to F12:DME+4F + 10 ng/ml EGF on day 6 (O---O); F12:DME+4F 1 10 ng/ml EGF changed to F12:DME+4F on day 3 (O---O).[3]

be demonstrated with cells in plates containing F12:DME+FCS. Figure 5 shows that the inhibitory effect of EGF on A431 cell growth in F12:DME+4F containing 10 ng/ml EGF and then changed to F12:DME+4F without EGF resumed growth after a lag of about three days with a generation time near that of cells not exposed to EGF. The inhibition of cell growth caused by the addition of EGF could be prevented by the simulataneous addition to the culture medium of affinity-purified rabbit anti-mouse EGF antibody (Figure 6). Addition to the medium of concentrations of the antibody higher than 10 µg/ml caused significant inhibition of A431 cell growth in F12:DME+4F.

The marked inhibition of A431 cell growth in vitro in cell culture in serum-free or serum-containing media suggested that a similar effect might be observed on these cells in vivo. To determine if the inhibition of A431 cell growth seen in culture also occurred in vivo, A431 cells were injected subcutaneously into male and female athymic mice. Male mice have elevated levels of EGF in the submaxillary compared to female mice (1), and it might be expected that this higher level of EGF is maintained in blood and interstitial fluid concentrations in male mice compared to female mice. Contrary to expectations, tumors arose with equal frequency in male and female athymic mice injected with 2×10^6 cells and arose at a greater frequency in male than in female mice injected with 2×10^5 cells.

DISCUSSION

The inhibitory effect of EGF on A431 cell growth is similar to the previously reported inhibitory effects of EGF on rat pituitary carcinoma cell lines (31,32); in both systems EGF lengthened the generation time for the cells studied but was not toxic because no decrease in cell number was observed upon EGF treatment (32). Rat pituitary carcinoma cells are also inhibited in a manner similar to that observed for EGF by FGF and thyrotropin releasing hormone (32). No inhibitory effects of FGF or thyrotropin releasing hormone on A431 cell growth were observed, although inhibitory effects of PDGF preparations and cyclic AMP were observed when these factors were included in the serum-free medium. It may be that the inhibitory effect of EGF

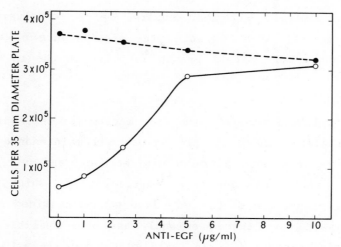

Fig. 6. Prevention of EGF inhibition of A431 cell growth by anti-EGF antibody. Cells were plated in F12:DME+4F in the presence of anti-EGF at the indicated concentrations and the presence or absence of EGF at 10 ng/ml. Cells were counted 6 days after plating. F12:DME+4F without EGF (O---O);[3] F12:DME+4F with EGF (O—O).

on A431 cells is in some way related to the large number of cell membrane receptors expressed on the surface of A431 cells. However, rat pituitary carcinoma cells which are inhibited by EGF express EGF receptors at a level near that seen for most cell types in culture, and considerably lower than that reported for A431 cells (31). The experiments described to determine the relationship of the EGF effect observed in culture to in vivo inhibition of A431 cell growth in athymic mice suggest that other factors may be involved in the control of A431 cell growth in vivo beyond those identified in culture, and additional experiments in which EGF levels are well controlled in experimental animals will be required before conclusive statements may be made regarding the significance of the observed inhibition if cell growth in vitro to the in vivo situation.

The availability of a serum-free medium for the A431 cell line which allows the long-term growth and maintenance of healthy cells in a known extracellular hormonal and nutritional environment allows the design of experiments in which the interesting properties of the A431 line can be examined in more straightforward ways than previously has been possible in serum-containing media (20,21). While it is unlikely that the data of previous reports dealing with the biochemistry of early EGF effects on A431 cells represent responses unrelated to mitogenic effects of EGF, the striking inhibition of A431 cell growth in culture by EGF suggests that future attempts to relate effects of EGF on other cell types should be approached with caution. The mechanism by which EGF inhibits the growth of the A431 cell remains to be determined.

ACKNOWLEDGEMENTS

I thank G. Sato for advice and encouragement, T. Hunter and J. Cooper for the gift of the A431 cells and P. Kelly and J. Gaudreau for helpful discussions and information regarding the EGF used in the experiments described.

REFERENCES

1. Carpenter, G. and Cohen, S. 1979. Epidermal growth factor. Ann. Rev. Biochem. 48:193-216.
2. Fabricant, R.N., DeLarco, J.E. and Todaro, G.J. 1977. Nerve growth factor receptors on human melanoma cells in culture. Proc. Nat. Acad. Sci. USA 74:565-569.
3. Haigler, H., Ash, J.F., Singer, S.J. and Cohen, S. 1978. Visualization by fluorescence of the binding and internalization of epidermal growth factor in human carcinoma cells A431. Proc. Nat. Acad. Sci. USA 75:3317-3321.
4. Cohen, S., Haigler, H.T., Carpenter, G., King, L. and McKanna, J.A. 1979. Epidermal growth factor: Visualization of the binding and internalization of EGF in cultured cells and enhancement of phosphorylation by EGF in membrane preparations in vitro. Cold Spring Harbor Conferences on Cell Proliferation 6:131-142.
5. Haigler, H.T., McKanna, J.A. and Cohen, S. 1979. Direct visualization of the binding and internalization of a ferritin conjugate of epidermal growth factor in human carcinoma cells A-431. J. Cell. Biol. 81-382-395.
6. McKanna, J.A., Haigler, H.T. and Cohen, S. 1979. Hormone receptor topology and dynamics: a morphological analysis using ferritin labelled epidermal growth factor. Proc. Nat. Acad. Sci. USA 76:5689-5693.
7. Carpenter, G., King, L. and Cohen, S. 1978. Epidermal growth factor stimulates phosphorylation in membrane preparations in vitro. Nature (Lond.) 276:409-410.
8. Cohen, S., Carpenter, G. and King, L. 1980. Epidermal growth factor-receptor-protein kinase interaction. J. Biol. Chem. 255:4834-4842.

9. King, L.E., Carpenter, G. and Cohen, S. 1980. Characterization by electrophoresis of epidermal growth factor stimulated phosphorylation using A-431 membrane. Biochem. 19:1524-1528.
10. Ushiro, H. and Cohen, S. 1980. Identification of phosphotyrosine as a product of epidermal growth factor-activated protein kinase in A431 cell membranes. J. Biol. Chem. 255:8363-8365.
11. Carpenter, G., King, L., Jr. and Cohen, S. 1979. Rapid enhancement of protein phosphorylation in A-431 cell membrane preparations by epidermal growth factor. J. Biol. Chem. 254:4884-4891.
12. Hunter, T. and Cooper, J.A. 1981. Epidermal growth factor induces rapid tyrosine phosphorylation of proteins in A431 human tumor cells. Cell 24:741-747.
13. Cooper, J.A. and Hunter, T. 1981. Similarities and differences between the effect of epidermal growth factor and Rous sarcoma virus. J. Cell Biol., in press.
14. Chinkers, M., McKanna, J.A. and Cohen, S. 1979. Rapid induction of morphological changes in human carcinoma cells A-431 by epidermal growth factor. J. Cell. Biol. 83:260-265.
15. Chinkers, M., McKanna, J.A. and Cohen, S. 1981. Rapid rounding of human epidermoid carcinoma cells A-431 induced by epidermal growth factor. J. Cell Biol. 88:422-429.
16. Haigler, H.T., McKanna, J.A. and Cohen, S. 1979. Rapid stimulation of pinocytosis in human carcinoma cells A-431 by epidermal growth factor. J. Cell. Biol. 83:82-90.
17. Wrann, M.M. and Fox, C.F. 1979. Identification of epidermal growth factor receptors in a hyperproducing human epidermoid carcinoma cell line. J. Biol. Chem. 254:8083-8086.
18. Carpenter, G. 1979. Solubilization of membrane receptor for epidermal growth factor. Life Sci. 24:1691-1698.

19. Haigler, H.T. and Carpenter, G. 1980. Production and characterization of antibody affecting epidermal growth factor: receptor interactions. Biochim. Biophys. Acta 598:314-325.
20. Barnes, D. and Sato, G. 1980. Methods for growth of cultured cells in serum-free medium. Anal. Biochem. 102:255-270.
21. Barnes, D. and Sato, G. 1980. Serum-free cell culture: a unifying approach. Cell 22:649-655.
22. Barnes, D. 1981. Epidermal growth factor inhibits growth of A431 human epidermoid carcinoma in serum-free cell culture. J. Cell Biol., 93:1-4.
23. Barnes, D. 1981. Growth of A431 human epidermoid carcinoma in serum-free cell culture: inhibition by epidermal growth factor. Cold Spring Harbor Conf. on Cell Prolif., Vol. 9, in press.
24. Barnes, D. 1981. A serum-free medium for A431 human epidermoid carcinoma cells; inhibition of growth in culture by mouse epidermal growth factor. Endocrinology 108 (suppl.):329.
25. Kano-Sueoka, T., Cohen, D.M., Yamaizumi, Z., Nishimura, S., Mori, M. and Fujiki, H. 1979. Phosphoethanolamine as a growth factor of a mammary carcinoma cell line of rat. Proc. Nat. Acad. Sci. USA 76:5741-5744.
26. Walthall, B.S., Tsao, M., Bettger, W.J., Peehl, S.M. and Ham, R.G. 1980. Differences in requirements for growth in defined media of normal human fibroblasts and keratinocytes. J. Cell Biol. 87:163a.
27. Barnes, D., Wolfe, R., Serrero, G., McClure, D. and Sato, G. 1980. Effects of a serum spreading factor on growth and morphology of cells in serum-free medium. J. Supramol. Struct. 14:47-63.

28. Barnes, D., Darmon, M. and Orly, J. 1981. Serum spreading factor: effects on RF1 rat ovary and 1003 mouse embryomal carcinoma cells in serum-free cell culture. Cold Spring Harbor Conf. on Cell Prolif., Vol. 9, in press.
29. McKeehan, W.L. and Ham, R.G. 1976. Stimulation of clonal growth of normal fibroblasts with substrates coated with basic polymers. J. Cell Biol. 71:727-734.
30. Orly, J. and Sato, G. 1979. Fibronectin mediates cytokinesis and growth of rat follicular cells in serum-free medium. Cell 17:295-305.
31. Johnson, L.K., Baxter, J.D., Vlodavsky, I. and Gospodarowicz, D. 1980. Epidermal growth factor and expression of specific genes: effects on cultured rat pituitary cells are dissociable from the mitogenic response. Proc. Nat. Acad. Sci. USA 77:394-398.
32. Schonbrunn, A., Krasnoff, M., Westenclorf, J.M. and Tashjian, A.H., Jr. 1980. Epidermal growth factor and thyrotropin-releasing hormone act similarly on a clonal pituitary cell strain. J. Cell Biol. 85:786-797.

IN VITRO SENESCENCE, DIFFERENTIATED FUNCTION, AND TRANSFORMATION
IN CULTURED VASCULAR ENDOTHELIAL CELLS*

Judith B. Grinspan, Stephen N. Mueller, James P.
Noveral, Eliot M. Rosen and Elliot M. Levine+

The Wistar Institute
Philadelphia, PA 19104

The usefulness of vascular endothelial cells as a model culture system to study disease has yet to reach the point where results obtained in vitro can provide remedies for in vivo pathologies. Although it may be premature to think that this goal soon may be achieved, investigators who are currently studying this cell culture model are probing some basic questions that are of intrinsic interest to cell culturists and cell biologists. With this as an underlying theme, our intention is to present an overview of the work being carried out in our laboratory on the proliferative and differentiated characteristics of bovine aortic endothelial cells in culture.

Before considering the isolation of endothelial cells and their in vitro characteristics it will be helpful to examine the

---

*Supported by U.S. Public Health Service grants AG-00839 and T32-CA09171
+To whom all correspondence and request for reprints should be sent

in vivo origin of the cells and the histological architecture of the blood vessels from which they are derived. The endothelium forms a cellular monolayer lining the vessel lumen and in elastic arteries rests on an acellular matrix containing collagen and other components that separates it from the next outermost layer, the media, which contains smooth muscle cells and occasional fibroblasts (1). This separation in the tissue permits detachment of the endothelial layer from the medial smooth muscle layer without major contamination by other cell types. One of the commonly used methods of separation involves treating the luminal surface of blood vessels with collagenase. Large sheets of endothelial cells are released that can be grown in cell culture (2,3,4).

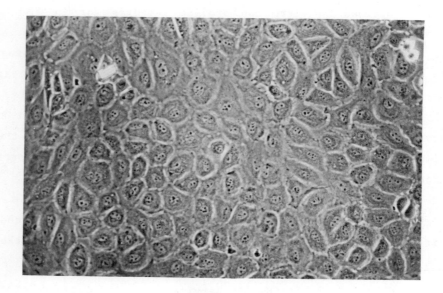

Figure 1

Representative view of a confluent, cloned bovine aortic endothelial cell culture. Clones were established from secondary mass cultures by standard dilution plating techniques. Phase contrast optics. X 142 (at 115 x 78 mm).

As a model cell system, endothelial cells have several unique advantages. Endothelial cells, which can be obtained from many species and various anatomical sites, play an important role in many pathological conditions. For example: age-related changes in the endothelium may contribute to the genesis of atherosclerosis and to the processes of tumor invasion and metastasis (5). Second in contrast to other commonly studied cultured cell types such as lung and embryo fibroblasts, endothelial cells have a defined in vivo histological origin. Third, one can obtain relatively high yields of endothelial cells for primary cultures that are nearly homogenous with respect to cell type. Fourth, in analogy to the human diploid fibroblast system that has been studied for many years, serially cultivable endothelial cells can be obtained that retain stringent growth control, normal karyotypes and exhibit finite life-spans. Finally, endothelial cells retain certain differentiated endothelial specific functions under culture conditions.

The studies presented here utilized clonal strains of endothelial cells from primary cultures obtained by collagenase digestion of fetal bovine thoracic aorta. Briefly, the lumen of the aorta was treated with 0.25% collagenase (Type I, Sigma, St. Louis, MO) and the endothelium removed by either scraping with a rubber policeman or by flushing with medium. Clones were derived from primary cultures by either inoculating cells into 2-sq cm wells at a density of one cell per well or by placing glass cloning rings around single cells that had been inoculated at low densities into tissue culture dishes (3,4). Figure 1 illustrates a classic cobblestone array of monolayered cells reminiscent of the intact endothelium in vivo. Proliferation in such monolayers was not stimulated by refeeding with fresh medium. Cultures were subcultured weekly by trypsinization (0.25% trypsin-0.09% EDTA in PBS), cell density was determined using a Coulter counter (Coulter

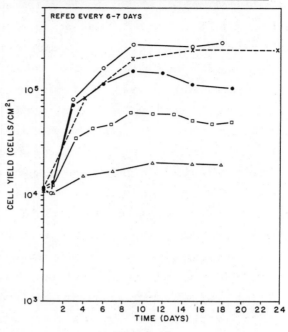

Figure 2

Growth curves for bovine fetal endothelial cell cultures. Cell density as a function of time for endothelial clone BFA-34j at four different points in the life-span (o———o, CPDL 31, 48% life-span completed; ●———●, CPDL 45, 69% life-span completed; □———□ CPDL 57, 88% life-span completed; △———△, CPDL61, 94% life-span completed), and for an uncloned endothelial culture at CPDL 4 (X-----X). Cultures were refed with fresh medium every 6 or 7 days. Each point is the average of two determinations, which agreed within 10%.

Electronic, Hialeah, FL, Model ZBI), and 1×10^4 cells/sqcm were inoculated into 75-sq cm tissue culture flasks containing 40 mls of F12 medium (6) with 20% fetal calf serum (FCS). The number of population doublings (PDs) undergone by the culture prior to each subcultivation (a measure of the relative in vitro age of the culture) was calculated from the harvest cell density by the formula PD = $\log_2$ (cell density at subcultivation ÷ cell density at inoculation) (4). The cumulative population doubling level (CPDL) was then the sum of all previous PDs.

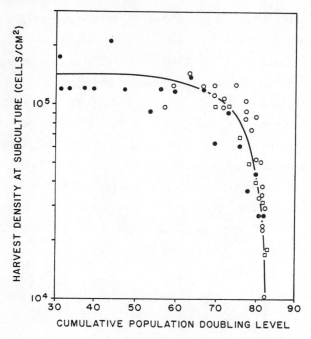

Figure 3

Cell density at subculture versus CPDL for endothelial cell clone BFA-1c. Data were obtained from (○) individual A (inoculation density, 2.5×10^4 cell/cm^2); (●) individual B (1.0×10^4 cell/cm^2); and (□) individual C (1.0×10^4 cell/cm^2). Weekly subculture cell density data could not be obtained until CPDL 30, because approximately 30 PDs occurred during the cloning procedure. Subcultivation was continued until inoculation cell density did not double after 2 weeks with weekly refeeding.

Figure 2 illustrates the growth characteristics of a clonal strain of endothelial cells in a series of growth curves determined at different points in the culture life-span. Initially, there was a period of rapid proliferation termed Phase II (7) where cell densities of greater than 10^5 cells/sq cm per week were achieved. After continued subculture, the proliferation rate slowed and harvest cell densities decreased. Eventually, the culture density would not double over a three week period even with refeeding. This decrease in culture density was accompanied by morphological changes typical of in vitro senescence. This point was designated as the end of the normal proliferative life-span or Phase III (7). Cells in such cultures were much larger than those in Phase II cultures and had an increased by cytoplasm to nucleus ratios. Figure 3 depicts the proliferative changes in an alternate manner. Here, for a different endothelial cell clone, harvest density is plotted against CPDL. Although the exact point at which senescence occurs differed from clone to clone, it was reproducible for each clone and the phenomenon has been observed more than 200 times in our laboratory for 12 clones. As another measure of changes in proliferative capacity, we

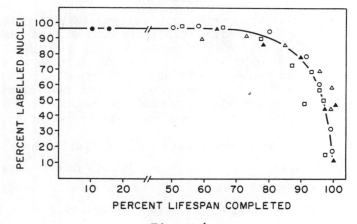

Figure 4

Proliferating cells as a function of percent life-span completed for cultures of bovine fetal endothelial cells. The percentage of proliferating cells (or rapidly cycling cells) was determined as the percentage of cells able to incorporate [^{3}H]-thymidine into their nuclei during a 24-hour labeling period. Data is presented for four endothelial cell clones (□———□), BFA-34a; o———o, BFA-34j; ▲———▲, BFA-34l; △———△, BFA-34m) and an early uncloned endothelial culture (●———●). Each point is the average of duplicate determinations which agreed within 10%.

employed tritiated thymidine labeling of cell nuclei and found changes in the growth fraction of cultures during in vitro senescence. Figure 4 depicts these alterations as the percent labeled nuclei as a function of percent life-span completed for four endothelial clones and early passage uncloned cultures. All cultures maintained a large proportion of rapidly cycling cell until about 75% life-span completed at which point there was a rapid decrease in the percentage of cells that entered S (i.e., the "Percent Labeled Nuclei"). This sudden rapid decrease paralled the sudden decrease in cell density shown in Figure 3. This was in contrast to the uncloned WI-38 cell system (8) where there was a more gradual decrease in growth fraction. Studies in the WI-38 system showed that the decrease in the number of cells entering S was caused by arrest of cells in G_0/G_1 (9).

Repeated evidence for the finite life-span of bovine endothelial cells in culture has led us to ask whether culture life-span is limited by the number of cell divisions ("genomic" time) or by a fixed amount of chronologic time ("metabolic" time). Experiments to distinguish between these possibilities were conducted by varying the length of time between subcultivations in endothelial clones. In one such experiment, cultures were

Chart I

Life-span as a function of population doublings ("genomic" time) and Calendar Time ("metabolic" time) for Endothelial Clone BFA-12d

	Initial Population Doubling	Final Population Doubling	Calendar time elapsed (weeks) from initial population doubling
Weekly Subculture	38	68	13
Triweekly	38	67	31

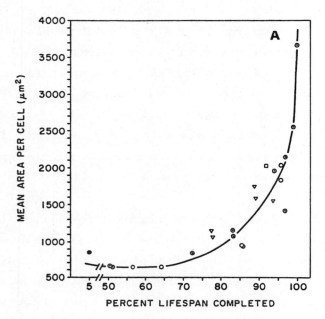

Figure 5

Mean cell-attachment area as a function of percentage life-span completed for uncloned endothelial cultures. (A) Mean cell area of silver nitrate-stained endothelial cultures was determined by ocular micrometry. Each point represents a single determination. Data are presented for endothelial clones (o———o), BFA-1c; □———□, BFA-34a; o———o, BFA-34j; ▽-----▽, BFA-34m) as well as an uncloned endothelial culture at early passage (●———●). (B) Photomicrographs of clone BFA-1c stained with silver nitrate at CPDL 41, 51% life-span completed, and (C) at CPDL 80, 100% life-span completed. x 470.

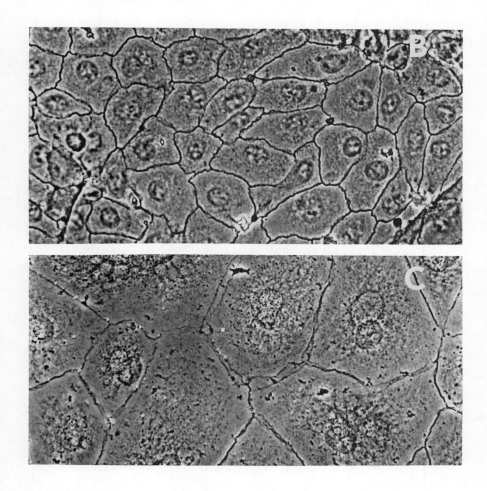

subcultivated either triweekly or weekly, thus varying the ratio of genomic/metabolic time. These experiments were begun at 50% to 60% life-span completed and carried on until the cultures senesced. The data (Chart 1) indicates that increasing the metabolic or chronologic life-span of a culture more than two-fold by increasing the time between subcultivations did not affect the total number of population doublings that culture would undergo. This suggest that culture life-span is determined by a mechanism that regulates or is linked to the number of cell generations.

In addition to changes in cellular proliferation, we have investigated other characteristics of endothelial cells which change during in vitro senescence. For example, it has been reported in other cell systems that increases in cellular attachment area were correlated with a decline in proliferative capacity (10). We have determined attachment area by a procedure based upon the ability of confluent endothelial cultures to take up silver nitrate stain at cell borders, forming black intercellular lines (11). Mean attachment areas were determined by counting the number of cells in known areas of stained cultures; frequency distributions of attachment areas were determined by weighing cut-outs of cells from photomicrographs of stained cultures (4). In Figure 5A mean attachment area is expressed as a function of percent life-span completed; two photomicrographs taken at early and late CPDLs are shown in Figures 5B and C. The data illustrates the dramatic increase in attachment area beginning at 75% life-span completed. Frequency distributions or attachment area (not shown) indicate that this change did not represent a uniform increase in cell size, but reflected the appearance of an increasing subpopulation of larger cells. Studies on human diploid fibroblasts in other laboratories have shown that this subpopulation contains a high proportion of nondividing cells (12,13) suggesting that cells lose their ability to divide while increasing in size (14).

Changes in other parameters also demonstrates an increase in cell size with senescence. We measured cell volume with a Coulter counter to insure that increases in cell attachment area were not due merely to a greater degree of flattening senescent cells. The volume distributions demonstrated shifts to larger cell volumes in senescent cultures (4). In addition, cellular protein content as a function of life-span completed was measured for five clones. Endothelial cell protein content was relatively constant at 400-500 ug/10^6 cells until 75-80% life-span completed, after which it increased to as much as 2,000 to 2,400 ug/10^6 cells (4).

Karyotype analyses of cultures using Giemsa banding techniques (16) showed that as endothelial cells senesced, they remained diploid, but an increasing fraction of cells contained translocations. In a representative clone, translocations were detected in 20% of the cells by 85% life-span completed. Chromosomal abnormalities accompanying the loss of proliferative capacity have been observed in other in vitro models of cellular senescence (16).

Because endothelial cells retain the expression of certain differentiated functions in vitro, we have monitored these functions as cultures senesced. Various investigators have suggested (13,17,18) that fetal diploid fibroblasts complete their differentiation as they proliferate, and the transition from Phase II to Phase III may represent "terminal differentiation". This is difficult to evaluate in fibroblasts, whose histologic origin is often uncertain and which have few specific markers of differentiation. In endothelial cells, we have assayed the expression of Factor VIII antigen, which can be detected in endothelial cells in vivo and in vitro but not in vascular smooth muscle cells or in fibroblasts. Employing indirect immunofluorescence with rabbit antibody to bovine Factor VIII antigen (a gift of Dr. Edward Kirby, Temple University,

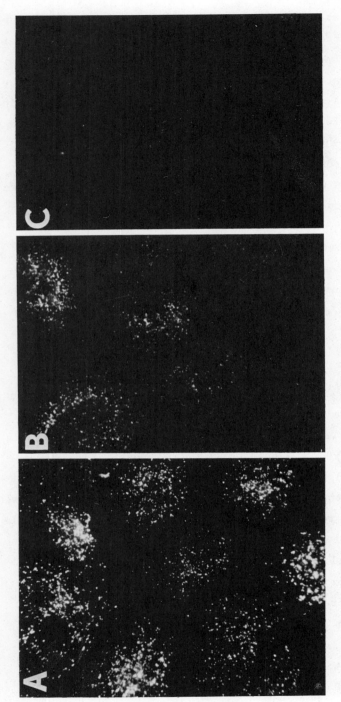

Figure 6

Detection of Factor VIII antigen by indirect immunofluorescence in cloned endothelial cells. Cloned endothelial cultures at various CPDLs were grown to confluency, extracted with acetone, and incubated with rabbit anti-bovine Factor VIII antigen, 1:1000 dilution (A and B) or normal rabbit serum, 1:500 dilution (C), followed by goat anti-rabbit IgG-FITC (1:80). Panel A, Phase II culture; Panel B, Phase III culture, Panel C, Phase II culture, x 856. Note that Phase III cells had larger cell-attachment areas then did Phase II cells (see Fig. 5B,C).

Figure 6 illustrates that both Phase II and Phase III cells expressed Factor VIII antigen, although this technique cannot be quantitated adequately enough to rule out some changes in antigen expression with age. Bovine vascular smooth muscle and lung cells, and human WI-38 lung cells were negative for Factor VIII antigen as were preimmune controls in which normal rabbit serum (1:5000 dilution) were substituted for immune serum. Thus, one differentiated characteristics, the expression of Factor VIII antigen, was retained throughout the life span of the cultured endothelial cells.

Philadelphia, PA) at 1:1000 dilution and goat anti-rabbit IgG-fluorescein isothiocyanate at 1:80 dilution, we have followed the expression of Factor VIII antigen through the culture life-span. Positive cells exhibited yellow-green, distinctly granular fluorescence distributed throughout the cytoplasm.

We also have begun to examine age-related changes in another differentiated function, angiotensin converting enzyme activity (ACE). ACE is an endothelial cell-specific activity involved in the homeostatic regulation of blood pressure, blood volume and electrolyte balance *in vivo*. Its activity was measured *in vitro* by incubating cell lysates with the synthetic substrate tritiated hippurylglycyl-glycine (20). ACE cleaved tritiated hippuric acid from this substrate, and the tritiated acid was extracted by ethylacetate and measured by liquid scintillation counting. Studies of ACE activity as a function of culture life-span have been complicated by the observation that ACE activity varied after

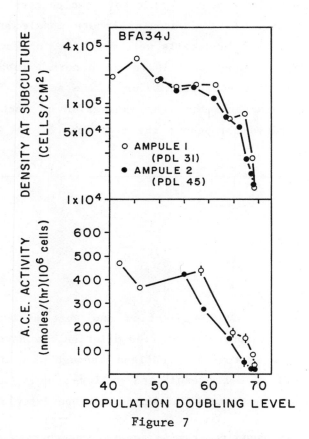

Figure 7

Cell density at subculture and ACE activity as a function of cumulative population doubling level for endothelial clone, BFA-34j. Cells were subcultured at weekly intervals using a constant inoculation density of 1×10^4 cells/cm^2. At various times throughout the life-span of the cells, cultures were trypsinized and seeded at 1×10^4 cells/cm^2 at the time of routine subculture, incubated for 3 weeks with weekly refeeding, and then assayed for ACE activity. Less than 2×10^5 cells were assayed for a one hour incubation period. Points for ACE activity represent mean ± range of S.D. (n=2-4).

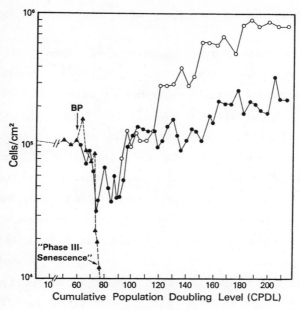

Figure 8

Life-span curves for endothelial cell clone BFA-1c cultured in the absence of BP or in the presence of BP in either F12 or MEM-CMF. Data expressed as cell density at subculture (cells/sqcm) as a function of proliferative life-span (CPDL) for control cultures propagated in F12 without BP (▲) that senesced at CPDL 76, for cultures treated with 5 μg BP/ml in F12 medium from CPDL 60 onward (flat transformants, ●), and for cultures treated with 5 μg BP/ml in F12 from CPDL 60-90 and in MEM-CMF from CPDL 90 onward (multilayered transformants, o). Cell densities and CPDLs were determined on replicate flasks which agreed within 10%.

each subcultivation depending on the cellular growth state. The enzyme activity was very low during the first few days after subcultivation and was maximal in confluent cultures; the expression of ACE appeared to require either high cell densities or a significant time for cell-cell contact in culture. Despite this complication, preliminary results (Figure 7) indicated that ACE activity, measured as nmoles/hour10^6 cells, decreased with CPDL as the proliferative capacity of the cells declined. Thus, while one specialized function, the presence of Factor VIII antigen, persisted throughout the life-span of the culture, another activity (ACE) declined with increasing population doublings.

The finite life-span or senescence of normal cells in culture can be modulated under appropriate conditions. Numerous growth factors and hormones have been found to extend the life-span of various cells in culture (39,40,41); and depending on the animal species from which the cells were derived, some cultures will transform "spontaneously" into cell lines with indefinite life-spans (21,22,23). In addition, exogenous agents, such as transforming viruses and chemical carcinogens, have been employed to transform normal cell strains into cell lines with indefinite life-spans. We have employed treatment with benzo(a)pyrene, a carcinogenic hydrocarbon, to circumvent the programmed senescence of normal endothelial cell strains. Metabolism studies have shown that bovine endothelial cells are capable of converting [^{3}H]BP to aqueous- and organic-soluble metabolites, including the 7,8-dihydrodial, a potent mutagenic and tumorigenic agent (24).

The effect of BP treatment on the life-span of a cloned strain of bovine endothelial cells is shown in Figure 8, which shows biweekly harvest cell densities as a function of CPDL. BP treatment was begun at CPDL 60 in replicate cultures. At CPDL 70 control, untreated cultures and BP-treated cultures began to show symptoms of senescence: increased cellular attachment size and decreased harvest densities. Control, untreated cultures senesced as expected at CPDL 76 (see also Figure 3). In the BP-treated cultures, a population of smaller cells began to emerge at CPDL 76 (Figure 9C), and the culture continued to proliferate. One set of these cultures has achieved over 439 population doublings and is still rapidly proliferating. These cells resemble morphologically normal untreated endothelial cells (Figures 9c and d). Confluent cultures show stringent density-dependent inhibition of growth and do not proliferate in response to fresh serum. These cells have been termed "flat transformants".

At CPDL 90, due to difficulty removing the transformed cells from the tissue culture surface by trypsinization, some cultures were switched to MEM-CMF plus 10% FCS, a culture medium containing less than half the calcium and magnesium of F12 plus 20% FCS. After an additional 30 PDLs in MEM-CMF medium, these cells began to increase in density (Figure 8) and form extensive multilayers. Cells in these cultures overlapped to such an extent that the cellular borders were indistinct in some areas (Figure 9e). By CPDL 210, harvest cell densities were almost 10^6 cells/sqcm. These cells have been called "multilayered transformants", and they also continue to proliferate beyond CPDL 400.

Multilayered and flat transformants were evaluated for some of the phenotypic alterations characteristic of transformation. As noted above, both have an apparently indefinite life-span, and this is perhaps the most important evidence of a departure from normalcy (25,26), especially in bovine cells, which have a low frequency of spontaneous transformation (3,4,27,28,29,30). Multilayered cells were anchorage independent, a cellular characteristic strongly correlated with transformation and tumorigenicity (31,32,33). They formed colonies in 0.4% soft agar or agarose with colony forming efficiencies of 2% and 14%, respectively, and also grew in liquid suspension medium. Flat transformants did not grow in agar, agarose or liquid suspension medium. Multilayered cells had a bovine karyotype which was diploid, but almost 100% of the cells contained Robertsonian translocations. Flat transformants exhibited a normal karyotype which gradually became almost 100% polyploid with a low level of translocations. Unstable and altered karyotype are typically found in transformed cells (34,35). Finally, multilayered transformant produced tumors in nude mice within three weeks of subcutaneous inoculation of 10^7 cells. In contrast, flat transformants failed to produce tumors six months after injection. Figure 9f shows a histological section of a tumor produced by

Figure 9

<u>9A</u> Phase contrast photomicrograph at CPDL 55 of normal Phase II fetal bovine aortic endothelial cells (clone BFA-1c). x 150.
<u>9B</u> Phase contrast photomicrograph at CPDL 76 of normal Phase III BFA-1c cells. x 150.
<u>9C</u> Phase contrast photomicrograph at CPDL 76 of BP-treated BFA-1c cells cultured in F12 ("early flat transformants"). Note the open areas of flask substratum in the lower center and right, and a remaining large senescent cell in the upper left. x 150.
<u>9D</u> Phase contrast photomicrograph at CPDL 155 of BF-treated BFA-1c cells cultured in F12 ("flat transformant"). Note the similarity to Figure 9A. x 150.
<u>9E</u> Phase contrast photomicrograph at CPDL 170 of BP-treated BFA-1c cells cultured in MEM-CMF ("multilayered transformants"). Note the smaller cells compared to Figures 9A & D and the multilayered area of growth running from the upper left to the lower right. x 150.
<u>9F</u> Section of tumor produced after 8 weeks by injection of 10^7 multilayered transformed cells at CPDL into a nude mouse. Hematoxylin and eosin. x 600. Histologig examination of freshly excised tumor tissue revealed that the mass was composed of irregularly shaped cells that were not organized into any particular pattern, except for the frequent occurance of cavities that may have been vascular channels. Masson trichome stain revealed some collagen interspersed within the tumor, and Gomori reticulin stain showed considerable amounts of argyrophilic material surrounding the tumor cells. The overall histologic picture was reminiscent of the "vasformative sarcomas" produced when 3T3 cells were implanted <u>in vivo</u> on solid substrates (36).

TRANFORMATION IN ENDOTHELIAL CELLS 85

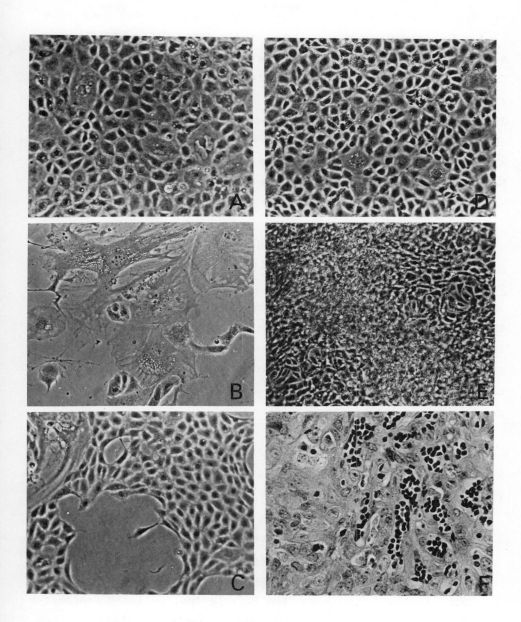

injection of multilayered cells. The histological pattern was reminiscent of the "vasoformative sarcomas" generated when 3T3 cells attached to solid substrates were injected into nude mice (36). To summarize the results of these experiments, treatment of a clonal strain of endothelial cells with BP produced two types of transformed cells that both exhibited indefinite life-spans but differed with respect to other phenotypic characteristics. Recently, these transformation experiments have been repeated, and the resultant cell lines are currently being characterized. Although transformation of endothelial cells by SV-40 virus has been reported (37,38), this is the first reported transformation of endothelial cells by a chemical carcinogen.

In conclusion, we have demonstrated that alterations in proliferative capacity and specialized function can be studied in vitro in the vascular endothelial cell, a differentiated cell type of physiologic importance. We have shown that the phenomenon of in vitro cellular senescence occurs in these cells and that at least one specialized function is altered as these cell age in culture. We have also shown that treatment with a chemical carcinogen can rescue cells from senescence, circumventing their pre-programmed senescence.

REFERENCES

1. E.P. Benditt, The origin of atherosclerosis. Sci. Amer. 236:74-85, 1977.
2. M.A. Gimbrone, Jr., Culture of vascular endothelium. In Progress in Hemostasis and Thrombosis Spaet, T.H. ed. 13:1-28, Grune & Stratton, New York, 1976.
3. S.N. Mueller, E.M. Rosen and E.M. Levine, Cellular senescence in a cloned strain of bovine fetal aortic endothelial cells. 207:889-891, 1980.

4. E.M. Rosen, S.N. Mueller, J.P. Noveral and E.M. Levine, Proliferative characteristics of clonal endothelial cell strains. J. Cell. Physiol. 107:123-127, 1981.
5. G. Thorgeirsson and A.L. Robertson, The vascular endothelium - pathobiologic significance. Am. J. Pathol. 93:803-848, 1978.
6. R.G. Ham, Clonal growth of mammalian cells in a chemically derined, synthetic medium. Proc. Natl. Acad. Sci., U.S.A. 53:288-293, 1965.
7. L. Hayflick, The cellular basis for biological aging. In Handbood of the Biology of Aging Hayflick, L. and Finch, C.E., eds. 159-186, Van Nostrand Reinhold Co., 1977.
8. V.J. Cristofalo and B. Sharf, Cellular senescence and DNA synthesis. Exp. Cell Res. 76:419-427, 1973.
9. G.L. Grove and V.J. Cristofalo, Characterization of the cell cycle of cultured human diploid cells: Effects of Aging and Hydrocortisone. J. Cell Physiol. 90:415-422, 1977.
10. S.G. Greenberg, G.L. Grove and V.J. Cristofalo, Cell size in aging monolayer cultures. In Vitro 13:297-300, 1977.
11. D.G. Fryer, G. Birnbaum and C.N. Luttrell, Human endothelium in cell cultures. J. Atheroscler. Res. 6:151-163, 1966.
12. M.P. Absher, R.G. Absher and W.D. Barnes, Geneologies of clones of diploid fibroblasts. Cinemicrophotographic observations of cell division patterns in relation to population age. Exp. Cell Res. 88:95-104, 1974.
13. G.M. Martin, C.A. Sprague, T.H. Norwood and W.R. Pendergrass, Clonal selection, attenuation, and differentiation in an in vitro model of hyperplasia. Am. J. Pathol. 74:137-154, 1974.
14. V.J. Cristofalo and D. Kritchevsky, Cell size and nucleic acid content in diploid human cell WI-38 during aging. Med. Exp. 19:313-320, 1969.
15. M. Bradford, A rapid and sensative method for the quantitation of microgram quantities of protein utilizing the principle of protein - dye binding. Anal. Biochem. 72:248-254, 1976.

16. R.C. Miller, W.W. Nichols, J. Pottash and M.M., Aronson, In Vitro aging Cytogenetic comparison of diploid human fibroblast and epithelial culture. Exp. Cell Res. 110:63-73, 1977.

17. E. Bell, L.F. Marek, D.S. Levinstone, S. Merill, S. Sher, I.T. Young and M. Eden, Loss of division potential in vitro: Aging or differentiation? Science 202:1158-1163, 1978.

18. V.J. Cristofalo, Animal cell cultures and model systems for the study of aging. Adv. Geront. Res. 4:45-78, 1972.

19. E.A. Jaffee, Endothelial cells and the biology of Factor VIII. N. Engl. J. Med. 296:377-383, 1977.

20. J.W. Ryan, A. Chung, C. Ammons and M.L. Carlton, A simple radioassay for angiotensin converting enzyme. Biochem. J. 167:501-504, 1977.

21. L. Diamond and W.M. Baird, Chemical carcinogenesis in vitro. In Growth, Nutrition and Metabolism of Cells in Culture Rothblast, G. and Cristofalo, V.J., eds. III:421-270, 1977.

22. C. Heidelberger, Chemical carcinogenesis. Ann. Rev. Biochem. 44:79-121, 1975.

23. S.H. Yuspa, P. Hawley-Nelson, B. Koehler and J.R. Stanley, A survey of transformation markers in differentiating epidermal cell lines in culture. Cancer Res. 40:4694-4703, 1980.

24. W.M. Baird, R. Chemerys, J.B. Grinspan, S.N. Mueller and E.M. Levine, Benzo(a)pyrene metabolism in bovine aortic endothelial cells and bovine lung fibroblast-like cell cultures. Cancer Res. 40:1781-1786, 1980.

25. A.E. Freeman and R.J. Huebner, Problems in interpretation of exxperimental evidence of cell transformation. J. Nat. Cancer Inst. 50:303-306, 1973.

26. R.L. Ruben and K.A. Rafferty, Limited in vitro replicative lifespan of cultured human epithelial cells: a survey of recent literature. Growth 42:357-368, 1978.

27. G.S. Duthu and J.R. Smith, In vitro proliferation and lifespan of bovine aorta endothelial cells: Effect of culture conditions and fibroblast growth factor. J. Cell. Physiol. 103:385-392, 1980.

28. S.G. Eskin, H.D. Sybers, L. Trevino, J.T. Lie and J.E. Chimoskey, Comparison of tissue-cultured bovine endothelial cells from aorta and saphenous vein. In Vitro 14:903-910, 1978.

29. A. Fenselau and R.J. Mello, Growth stimulation of cultured endothelial cells. Cancer Res. 36:3269-3273, 1976.

30. S. Schwartz, Selection and characterization of bovine aortic endothelial cells. In Vitro 14:966-980, 1978.

31. V.H. Freedman and S.I. Shin, Cellular tumorigenecity in nude mice: correlation with cell growth in semi-solid medium. Cell 3:355-389, 1974.

32. I. MacPherson and L. Montaigner, Agar suspension culture for the selective assay of cells transformed by polyoma virus. Virology 23:291-294, 1964.

33. S.I. Shin, V.H. Freeman, R. Rissa and K. Pollack, Tumorigenecity of virus transformed cells in nude mice is correlated specifically with anchorage independent growth in vitro. Proc. Nat. Acad. Sci., U.S.A. 72:4435-4439, 1975.

34. J.C. Barrett and P.O. Ts'o, Relationship between somatic mutation and neoplastic transformation. Proc. Natl. Acad. Sci., U.S.A. 75:3297-3301, 1978.

35. W.F. Benedict, J.G. Gielen and D.W. Nerbert, Polycyclic hydrocarbon-produced toxicity transformation, and chromosomal aberrations as a function of aryl hydrocarbon hydroxylase activity in cell cultures. Int. J. Cancer 9:435-451, 1972.

36. C.W. Boone, N. Takeichi, M. Paranjpe and R. Gilden, Vasoformative sarcomas arising from Balb/3T3 cells attached to solid substrates. Cancer Res. 36:1626-1633, 1976.

37. M. Gimbrone and G. Fareed, Transformation of cultured human vascular endothelium by SV40 DNA. Cell 9:685-693, 1976.

38. C.A. Reznikoff and R. DeMars, In vitro chemical mutagenesis and viral transformation of a human endothelial cell strain. Cancer Res. 41:1114-1126, 1981.
39. V.J. Cristofalo, Metabolic aspects of aging in diploid human cells. In Aging in Cell and Tissue Culture Holeckova, E. and Cistrofalo, V.J., eds. 83-119,Plenum Press, New York.
40. G.S. Duthu and J.R. Smith, In vitro proliferation and life-span of bovine aortic endothelial cells: effect of culture conditions and fibroblast growth factor. J. Cell Physiol. 103:385-392, 1980.
41. D. Gospodarowicz, A.L. Mescher and C.R. Birdwell, The control of cellular proliferation by the fibroblast and epidermal growth factors. Natl. Cancer Inst. Monogr. 48:109-130, 1978.

CYTOGENETIC EFFECTS OF CYCLAMATES[*]

Evelyn W. Jemison[1,2], Kenneth Brown[2], Beverly Rivers[2], and Regina Knight[2]

Department of Life Sciences
Virginia State University
Petersburg, Virginia 23803

ABSTRACT

PHA-stimulated human peripheral lymphocytes were used as a model system for assessing the in vitro effects of calcium cyclamate. Techniques of autoradiography, cytological staining, cell counting, liquid scintillation and karyotyping were used to study the cytogenetic damage and biochemical effects of calcium cyclamate when assayed in 24 hour intervals for 96 hours. The cells were exposed to 10^{-2} and 10^{-3} molar concentrations of calcium cyclamate in TC 199 medium with fetal calf serum and antibiotics.

---

[*] A portion of this paper was presented before the Second International Cell Culture Congress – In Support of BioScience – University of Alabama in Birmingham, September, 1981.

Supported by grants from NIH/MBS Program #2-S06-RR-08090.

[1] Professor at Virginia State University to whom requests for reprints should be sent.

[2] Students in Biology at Virginia State University. A portion of this material was taken from Graduate (M.S.) theses submitted to the Graduate School of Virginia State University by K. Brown and B. Rivers.

These studies were carried out in three (3) phases. Phase I was primarily orientation studies of the effects of cyclamates and included running preliminary test checks, the establishment of parameters of dosage, assessing growth patterns and selecting key chromosomal aberrations. Sixty four (64) of the metaphase spreads showed morphologically detectable changes and aberrations. It was also noted that the addition of cyclamate increased mitotic rate of lymphocyte cells in cultures.

Phase III arranged research designs to determine more precise characterization of chromosomal observations and morphological effects. Among other findings it was noted that of 13 types of observations only ten were found in the experimental group. The introduction of cyclamates increased the stability of the leucocyte cultures. These studies reinforced the findings on the increase of mitotic rate.

Phase III extended protocols to include autoradiography and scintillation counting. It was determined that calcium cyclamate impaired the synthesis of deoxribonunucleic acid (as depicted by decreased incorporation of tritiated thymidine), reduced grain counts in autoradiographs and increased chromosome aberrations in cyclamate treated PHA stimulated peripheral blood lymphocytes in vitro. Morphological changes and growth rates showed significant effects.

These studies indicate that calcium cyclamate has variable significant effects on leucocytes growth and chromosome morphology.

CYTOGENETIC EFFECTS OF CYCLAMATES

SUMMARY

PHA-stimulated human peripheral lymphocytes were used as a model system for assessing the _in vitro_ effects of calcium cyclamate. Techniques of autoradiography, cytological staining, cell counting, liquid scintillation and karyotyping were used to study the cytogenetic damage and biochemical effects of calcium cyclamate when assayed in 24 hour intervals for 96 hours. The cells were exposed to 10^{-2} and 10^{-3} molar concentrations of calcium cyclamate in TC 199 medium with fetal calf serum and antibiotics.

These studies were carried out in three (3) phases. Phase I was primarily orientation studies of the effects of cyclamates and included running preliminary test checks, the establishment of parameters of dosage, assessing growth patterns and selecting key chromosomal aberrations. Sixty four percent (64%) of the metaphase spreads showed morphologically detectable changes and aberrations. It was also noted that the addition of cyclamate increased mitotic rate of lymphocyte cells in cultures.

Phase II arranged research designs to determine more precise characterization of chromosomal observations and morphological effects. Among other findings it was noted that of 13 types of observations only 10 were found in the experimental group. The introduction of cyclamates increased the stability of the leucocyte cultures. These studies reinforced the findings on the increase of mitotic rate.

Phase III extended protocols to include autoradiography and scintillation counting. It was determined that calcium cyclamate impaired the synthesis of deoxyribonucleic acid (as depicted by decreased incorporation of tritiated thymidine), reduced grain counts in autoradiographs and increased chromosome aberrations in cyclamate treated PHA stimulated peripheral blood lymphocytes in vitro. Morphological changes and growth rates showed significant effects.

These studies indicate that calcium cyclamate has variable significant effects on leucocytes growth and chromosome morphology.

INTRODUCTION

For many years this laboratory has been concerned with the effects of cyclamates on chromosomes. In 1970[1] we reported the detection of a 5.3% mutation rate in the X-linked lethals of Drosophila melanogaster after feeding sodium and calcium cyclamate to these fly larvae.

The 1970 ban on cyclamates, based on their carcinogenicity, gave further impetus to studies of this type. Prior to the ban, Stone[2] in 1969 reported the characterization of cell population of exposed lymphocytes to calcium cyclamate under in vitro condition by chromosomal analysis and they reported chromatid breaks. Other workers[3] carried out multiple in vivo and in vitro studies trying to elucidate the effects of cyclamates on mammalian cells. Their work provided suggestive but uncertain relationships between cyclamates and mutagenicity and carcinogenicity. Stoltz[4] reported that purified samples of cyclamates do not cause mutagenic and carcinogenic responses in mammalian cells in vitro, but that commercial cyclamates, mixed with food substances, do cause responses.

Key Words: cyclamate, leucocytes, cytogenetic effects, chromosomal aberrations, leucocyte test system

It was in the hope of adding clarification and further documentation that these studies, extending over a seven year period, were undertaken. Our primary objective was to determine whether or not repeatable effects from calcium cyclamate exposed PHA-stimulated lymphocyte cultures would be obtained. Four major factors studied were: (1) the morphological changes during blastogenic responses to phytohemagglutinin; (2) the growth rate pattern; (3) the effect of cyclamate of chromosomes; and (4) the rate of the incorporation of tritiated thymidine into DNA as a function of time. These studies were carried out in three phases. Phase I was concerned with development of techniques, the assessment of growth patterns and extensive chromosomal analysis by karotyping. Phase II combined a model leucocyte test system and effective research design to the study of growth patterns. For this experimental phase, the selected calcium cyclamate concentrations were 10^{-2M} and 10^{-3M} and leucocyte concentrations of 10^{-3} or 10^{-4} cells per ml. Phase III introduced autoradiography and liquid scintillation counting as means of assaying morphological changes and tritiated thymidine incorporation.

MATERIALS AND METHODS

Lymphocyte Preparation

Concentrated lymphocyte suspensions was obtained using lymphocyte separation medium. (Ficoll, obtained from Pharmacia Fine Chemical, New Jersey). Buffy-coat rich plasma for this research has been received from the American Red Cross Blood Service (Norfolk, Virginia) and the local hospital laboratory. Buffy coat enriched plasma was mixed with an equal volume of sterile physiological saline and carefully layered over the Ficoll (ratio 4:3) in a centrifuge tube. The tube was centrifuged at 20°C for 40 minutes at a speed to yield 400 g at the interface. The band of cells collected at the interface was removed and washed twice in physiological saline.

Lymphocyte Culture

The concentrated suspension of lymphocytes was placed in TC 199 culture medium supplemented with 10% (v/v) fetal calf serum, PHA M 0.1 ml/5 ml medium (Difco), penicillin 160 units/1 ml medium, streptomycin 160 units/1 ml medium and mixed.[5] The initial lymphocyte suspension culture was dispersed into sixteen cultures of equal aliquots of 10^3 and 10^4 cells/ml and placed into four groups. Four cultures were designated as a control (untreated) and four cultures were designated as a positive control which had .05 mg/ml of cycloheximide added. The remaining cultures were treated with calcium cyclamate and divided into two groups. The final concentrations of cyclamates in the two groups were 10^{-2} and 10^{-3} respectively.

Lymphocyte Morphology

Starting at time 0, in 24 hour intervals, lymphocytes from each of the four cultured groups were obtained by centrifugation, fixed in acetic acid-methanol (1:3) and prepared on slides. These slides were stained with Feulgen and Giemsa stains. The slides of the treated lymphocytes were observed under the microscope at 450x for morphological differences.

Lymphocyte Proliferation

At time 0 and in 24 hour intervals, for a 96 hour period, a hemacytometer count was taken of each of the four culture groups. Trypan blue was used as a diluting fluid and also as a vital stain.[6]

Pulse-labeling

Samples of each of the four culture groups were pulse-labeled for 1 hour with tritiated thymidine (New England Nuclear, specific activity = 17.5 mmCi) in intervals of 24 hours beginning at the 24th culture hour. The final concentration of tritiated thymidine added was 1 micro ci/ml.

Autoradiography Preparation of Samples

Upon completion of pulse-labeling, one pulse-labeled sample of each of the four culture groups was used for autoradiographic analysis. In the autoradiography preparation, the labeled cells were centrifuged and washed in saline, fixed in 1:3 acetic acid-methanol, and dropped on a slide. The slides were coated with autoradiography emulsion (Kodak NTB-2), and exposed for 10 days. After the exposure of the slides for 10 days, they were developed in Kodak D-19 developer and stained with Giemsa.

Light microscope scans were made of the stained autoradiographs. A minimum of 150 cells were scored for each time point per slide of each of the culture groups. The scoring consisted of counting lymphocytes showing the label (grains lying above the nucleus) against lymphocytes unlabeled. The labeling index (labeled cells/labeled cells plus unlabeled cells x 100) was plotted against time.

Liquid Scintillation Counting

At the end of the 30 minute pulse-labeling period, lymphocytes of each culture sample were collected on a glass fiber filter. The trapped cells were washed with cold saline. The nucleic acids were precipitated with cold 10% trichloroacetic acid and 70% methanol.

The cells were dissolved in tissue solubilizer and the radioactivity determined in a toluene-based scintillation fluid. Scintillation data of the four culture groups were obtained in counts per minute in a Beckman LS 100 liquid scintillation counter. The counts per minute were plotted against time.[7]

Chromosomal Analysis Techniques

The cytological methods used were adapted from the procedures outlined by Moorhead and co-workers.[8] Metaphase cells were blocked by colchicine treatment. Four hours before the culture was to be fixed at 72 hours, colchicine was added to the medium to give a final concentration of 10^{-7} M. Just prior to fixation, the cells were centrifuged out of the medium and resuspended in Hank's saline. After an additional centrifugation, all of the supernatant was removed. The cells were resuspended in 1 ml of fresh Hank's saline. With constant agitation, 3 ml of distilled water was added in 1 ml aliquots. The cells were left in this hypotonic solution for 10 minutes in a 37°C incubator. Then the lymphocytes were centrifuged for 5 minutes and the hypotonic solution were discarded. Approximately 5 ml of freshly prepared (1:3) glacial acid-methanol fixative was carefully added without disturbing the button of cells. After 30 minutes, the button of cells was suspended in the fixative with a Pasteur pipette. The cells were washed by centrifugation if fresh fixative and finally resuspended in 0.5 ml of fixative. The fixed cells were mounted on slides by dropping single drops of cell suspension on cold slides. The slides were allowed to air dry and stained in Giemsa stain for 25 minutes. The slides were scanned for metaphase spreads of chromosomes at 1000x and photographed with a 35mm full frame camera. The chromosomes were analyzed for chromosomal aberrations according to the chromosomal aberration pattern of Heneen.[10]

CYTOGENETIC EFFECTS OF CYCLAMATES

RESULTS: Phase I

Phase I experiments were designed to test the effects of calcium cyclamates on human leucocyte cultures and secondly to identify the types of chromosomal aberrations. The characterizations were based upon work reported by Heneen.[10]

A total of 223 metaphase spreads were analyzed, out of which 131 cells had changes and aberrations. Thus, the frequency of cells showing detectable aberration in about 64%. 10 kinds of chromosomal aberrations were scored from the metaphase spreads and selected metaphase plates were karyotyped.

The chromosomes often appeared divided into segments due to presence of secondary constrictions which are not found in normal

Table 1. Effect of Calcium Cyclamates on Human Chromosomes

Kinds of Abberations	Calcium Cyclamate Concentration			% Total
	10^{-2}	10^{-3}	10^{-4}	
I. Iso-locus breaks	.00	.046	.00	.046
II. Erosions	.00	.00	.00	.00
III. Chromatid breaks	.106	.415	.52	1.04
IV. Bridges	.00	.00	.00	.00
V. Dicentric	.00	.155	.26	.38
VI. Acentric	2.89	.577	.30	3.77
VII. Ring-formation	.00	.100	.26	.38
VIII. Triradial	.00	.038	.13	.17
IX. Quadradial	.00	.030	.09	.12
X. Secondary Constrictions	.00	.215	.00	.215
Total %	3.00	1.54	1.82	6.36

Frequency is expressed out of the total number of cells with detectable changes. The total sum of frequencies is expressed per/100 cells with detectable changes. Cells showing aberrations usually had more than one type of aberration per cell.

chromosomes. Other aberrations identified were: iso-locus breaks, erosions, chromatid breaks, bridges, discentric, acentric, ring formation, triradials, and quadradials.

Stolz[4] reported in 1970 that the effects of sodium and calcium cyclamates on human cells have not been conclusive, but that chemically induced chromosomal aberrations are being used as rapid indicators of potential mutagenetic activity within the cell. Stolz, however, did not identify the kinds of aberrations induced by the cyclamates. Our work showed conclusively that the time of exposure to the cyclamate made no significant difference in the percentage of chromosomal aberrations; it also indicated that at high concentration cyclamates induce chromosomal damage in human leucocyte cultures.

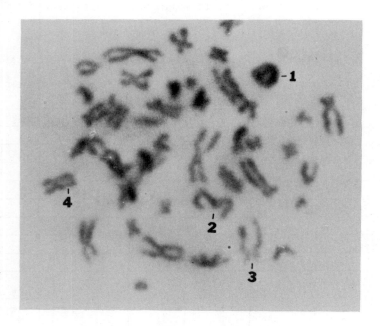

Figure 1.

Mitotic metaphase spread of abnormal chromosomes from a culture of cyclamate exposed leucocytes. Shown are: (1) ring formation, (2) symetrical exchange, (3) indistinct centromere and (4) chromatid break photograph 1000X. B.A. Rivers.

Table 2. Sums and Percent of Chromosomal Aberrations by Treatment[a]

Aberration	Sums		Percent	
	Control Group	Experimental Group	Control Group	Experimental Group
Acentric	0	3118	0.000	7.531
Rounded Chromosome Ends	59	2708	0.142	6.541
Ring Formation	3	2260	0.007	5.458
Extra Secondary Constriction	92	3890	0.222	9.396
Symmetrical Exchange	0	1427	0.000	3.416
Chromosome Balls	110	4556	0.265	11.004
Chromatid Break	0	2893	0.000	6.987
Indistinct Centromere	439	3197	1.060	7.722
Non-separation of Chromatid	437	4038	1.055	9.753
Condensed Chromosomes	9930	8940	23.985	21.594
Erosion	89	0	0.214	0.000
Dicentric	141	0	0.341	0.000
Elongated	100	0	0.241	0.000

[a] Cyclamate exposed lymphocytes and non-cyclamate exposed lymphocytes

[b] Since many cells contain more than one type of aberration, the total of the percents will not equal 100.

RESULTS: Phase II

Thirteen types of chromosome modifications were scored: highly condensed chromosomes balls and twelve aberrations. Table 2 lists the sums and the percentages of chromosome aberrations for the control group and the cyclamate exposed group. These data were obtained from microscopic examination of 12 cultures. A total of 1800 metaphase plates of peripheral fractions were examined. Of the aberrations the dicentric, erosion, and elongated chromosome appeared only in the control.

Aberrations found in the cyclamate exposed group are illustrated in Figure 1. Figures 2 and Figure 3 are photographs by Rivers[11] of metaphase plates and karyotypes to illustrate some of the types of aberrations produced in leucocyte cells from TCC 199 medium containing 10^{-2} M calcium cyclamate.

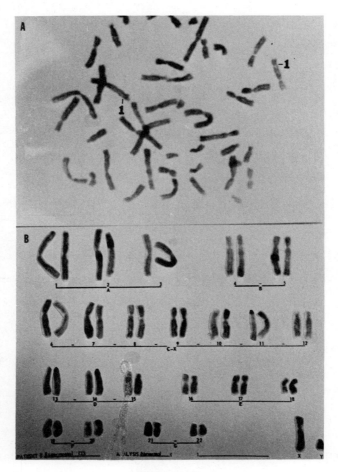

Figure 2.

Mitotic metaphase spread and karyotype of chromosomes from cyclamate exposed leucocytes. Shown are: (1) condensed chromosome, (2) indistinct centromere. Photographed at 1000X. B.A. Rivers.

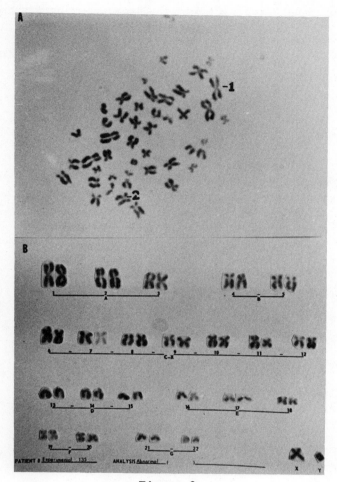

Figure 3.
Mitotic metaphase spread and karyotype of chromosomes from cyclamate exposed leucocytes. Shown are: (1) extra secondary constrictions. Photographed at 1000X. B.A. Rivers.

Table 3 lists the sums and percents of normal chromosomes found in 900 metaphase plates of the non-cyclamate exposed lymphocytes (control group) and 900 metaphase plates of the cyclamate exposed lymphocytes (experimental group).

Table 3. Sums and Percent of Normal Chromosomes by Treatment

	Sums		Percent	
	Control Groups	Experimental Group	Control Group	Experimental Group
	29999	4373	72.461	10.562

The means of each type of chromosomal aberration are listed in Table 4 by control group and cyclamate exposed group.

Table 4. Means of Chromosomal Aberrations by Treatment

Aberration	Mean	
	Control Group	Experimental Group
Acentric	0.000	3.468
Rounded Chromosome Ends	0.006	3.009
Ring Formation	0.003	2.511
Extra Secondary Constrict	0.102	4.322
Symmetrical Exchange	0.000	1.586
Chromosome Balls	0.122	5.062
Chromatid Break	0.000	3.214
Indistinct Centromere	0.488	3.552
Non-separation of Chromatid	0.486	4.487
Condensed Chromosomes	11.033	9.933
Erosion	0.099	0.000
Dicentric	1.580	0.000
Elongated Chromosomes	0.111	0.000

The non-occurrence of erosion, dicentric, and elongated chromosome resulted in no computed mean values for these aberrations. Therefore no comparison could be made with the means in the control group. For each group the means for condensed chromosomes are approximately equal.

The mean of each type of chromosomal aberration is listed in Table 4 by control group and cyclamate exposed group.

Table 5 shows that the mean of the normal chromosomes found in the cyclamate exposed group is lower than the mean of the subjects in the control groups.

Table 5. Means of Normal Chromosomes by Treatment

Control Group	Experimental Group
33.332	4.859

Table 6 records the analysis of variance summary for the Phase II study, of the three types of aberrations found in the only group dicentric aberration was the only one found to be significant. This finding agrees with the results in Table 1 (i.e., dicentric chromosomes are the larger of the two). Table 7 gives a summary of the analysis of variance of normal chromosomes.

The difference in condensed chromosomes is significant at the .05 level. The later finding suggests that some other factor influenced the degree of condensation of the chromosomes. The mean difference between each of the other types of aberration was significant.

Table 6. Analysis of Variance Summary of Aberrations

Mean Squares (df)			Residual (df)	F-Ratio (Significance of F)		
Treatment	Sex	Interaction		Treatment	Sex	Interaction
364842.438 (1)	111.649 (1)	225.039 (1)	306.716 (1796)	1189.514 (0.001)	0.034 (0.999)	0.737 (0.999)

Table 7. Analysis of Variance Summary of Normal Chromosomes

	Mean Squares (df)			Residual (df)	F-Ratio (Significance of F)		
	Treatment	Sex	Interaction		Treatment	Sex	Interaction
Acentric	5403.742 (1)	269.939 (1)	269.841 (1)	59.428 (1796)	90.929 (0.001)	4.541 (0.031)	4.541 (0.031)
Rounded Chromosome Ends	3898.000 (1)	8.958 (1)	0.045 (1)	52.508 (1796)	74.237 (0.001)	0.171 (0.999)	0.001 (0.999)
Ring Formation	2831.060 (1)	76.889 (1)	77.719 (1)	42.556 (1796)	66.526 (0.001)	1.807 (0.175)	1.826 (0.173)
Extra Secondary Constriction	8011.859 (1)	92.188 (1)	92.189 (1)	79.132 (1796)	101.247 (0.001)	1.165 (0.280)	1.166 (0.280)
Symmetrical Exchange	1131.800 (1)	46.135 (1)	46.135 (1)	27.829 (1796)	40.670 (0.001)	1.658 (0.195)	1.658 (0.195)
Chromatid Break	4647.922 (1)	135.890 (1)	135.890 (1)	55.945 (1796)	83.081 (0.001)	2.429 (0.115)	2.429 (0.115)
Indistinct Centromere	4220.563 (1)	1346.628 (1)	576.756 (1)	72.426 (1796)	58.274 (0.001)	18.274 (0.001)	7.96 (0.006)
Non-separation of Chromatid	7206.969 (1)	246.370 (1)	362.681 (1)	86.734 (1796)	87.734 (0.001)	3.022 (0.078)	4.395 (0.034)
Condensed Chromosomes	544.289 (1)	13.842 (1)	3227.022 (1)	347.893 (1796)	1.565 (0.208)	0.040 (0.999)	9.276 (0.003)
Erosion	4.405 (1)	1.027 (1)	1.578 (1)	1.578 (1706)	1.792 (0.091)	0.651 (0.651)	0.651 (0.651)
Dicentric	11.227 (1)	11.202 (1)	11.202 (1)	2.756 (1796)	4.073 (0.041)	4.064 (0.041)	4.064 (0.041)
Elongated Chromosomes	5.568 (I)	5.556 (1)	5.556 (1)	2.883 (1796)	1.932 (0.161)	1.927 (0.161)	1.927 (0.161)

RESULTS: Phase III

Buffy-coat rich human blood plasma was used as the lymphocyte source in this third series of studies. Differential counts of peripheral blood cells were as follows: polymorphs 62%, lymphocytes 26%, monocytes 6%, basophils 4%, and eosinophils 2%. Giemsa stain analysis of the Ficoll-hypaque lymphocyte preparation indicated that the monocyte lymphocyte ratio was 15 to 85 respectively. Among the concentrated lymphocyte preparation, trypan blue dye exclusion studies indicated that greater than 90% of the recovered cells were viable.

In order to ascertain the effect of calcium cyclamate on peripheral blood lymphocytes <u>in vitro</u>, the pattern of lymphocytes proliferation was established. Figure 4 is a graphic representation of the number of lymphocytes in culture with time. Superimposed on this graph are the cell numbers from parallel conducted experiments wherein 10^{-2} and 10^{-3} molar concentration of calcium cyclamate and cycloheximide .05 mg/ml were added. Initially there was a decrease in the total number of cells among all groups (0 to 18 hours). Thereafter the number of cells increased and peaked at approximately 65 to 72 hours. The peak lymphocyte titer among the cyclamate treated groups was less than the control group and showed a dose inhibitory response. Cycloheximide, a potent metabolic inhibitor, inhibits protein and deoxyribonucleic acid synthesis[11] thereby significantly reducing lymphocyte proliferation beyond that of the calcium cyclamate treated groups.

Cytochemical and morphological studies of lymphocyte samples from the untreated control, calcium cyclamate (10^{-2} and 10^{-3} M) and cycloheximide treated positive control groups were evaluated throughout the culture period (92 hours). In the assessment of the gross cellular morphological and the molecular events attributed to the calcium cyclamate and cycloheximide, the following

period (92 hours). In the assessment of the gross cellular morphological and the molecular events attributed to the calcium cyclamate and cycloheximide, the following parameters were parameters were examined: enlarged nucleus, dispersed chromatin material, reduced cytoplasm, and absence of nucleolus on blastogenesis. Consistently among the untreated control group, the greater percentage of cells displayed evidence of blastogenesis. Among the calcium cyclamate treated groups, there was evidence of a dose response. Groups treated with 10^{-3} molar concentration of calcium cyclamate displayed blastogenesis, but less than the control group. However, the 10^{-3} molar concentration of calcium cyclamate treated culture showed less proliferation than the cultures treated with 10^{-2} molar concentration of calcium cyclamate.

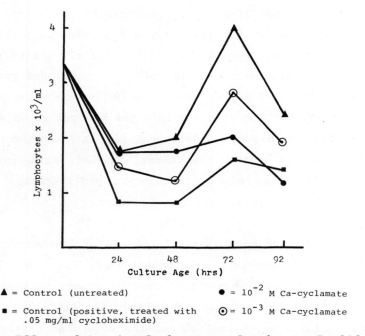

Figure 4. Effect of Calcium Cyclamate on Lymphocyte Proliferation

Additional evidence supporting the inhibitory effect of calcium cyclamate and cycloheximide is depicted in Figure 5. Figure 5 is a graphic representation of the percentage of lymphocytes in blastogenesis throughout 48 hours cultures. Calcium cyclamate, 10^{-3} molar concentration did not markedly alter the percent blastogenesis from that of the untreated control, while 10^{-2} molar concentration of calcium cyclamate and cycloheximide, the positive control markedly reduced the percent of lymphocytes undergoing blastogenesis.

To delineate further the effect of calcium cyclamate on primary cultures of PHA-stimulated blood lymphocytes in vitro, data from

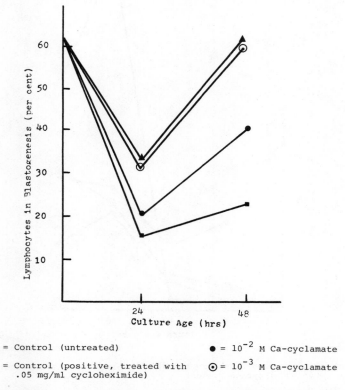

▲ = Control (untreated) ● = 10^{-2} M Ca-cyclamate
■ = Control (positive, treated with .05 mg/ml cycloheximide) ⊙ = 10^{-3} M Ca-cyclamate

Figure 5. Calcium Cyclamate Induced Inhibition of Blastogenesis Among Lymphocytes in vitro.

the uptake and incorporation of tritiated thymidine in calcium cyclamate treated lymphocyte cultures was analyzed. The results from the labeled uptake studies were then compared with radiographic preparations which paralleled treated lymphocyte cultures. Figure 6 is a graphic illustration of radioactive labeled thymidine incorporated in cultured lymphocytes over a period of 92 hours. Among the control, minimal incorporation was seen in the initial 48 hours. By 72 hours, peak labeled incorporation was seen, and decline

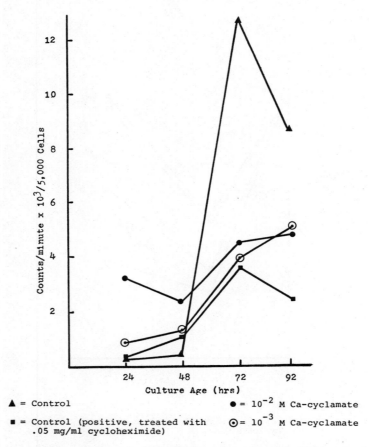

Figure 6. Effect of Calcium Cyclamate on the Incorporation of Tritiated Thymidine in Lymphocytes in vitro.

thereafter. These observations complement those of Green.[9] Calcium cyclamate and cycloheximide, respectively, inhibited the incorporation of radioactive thymidine. However, cycloheximide displayed a greater inhibitory response than calcium cyclamate. The calcium cyclamate treated cultures in this study did not display a dose inhibitory response. However, incorporation of radioactive labeled thymidine among the cyclamate treated groups coincide with the inhibitory response displayed by cycloheximide. This data suggest that calcium cyclamate may exert an inhibitory response similar to that of cycloheximide. The dual statistical classification procedure was employed to assess the incorporation of radioactive labeled thymidine into the acid precipitable fraction of human PHA-stimulated peripheral blood lymphocytes in vitro. The results from this study indicate that there was a significant

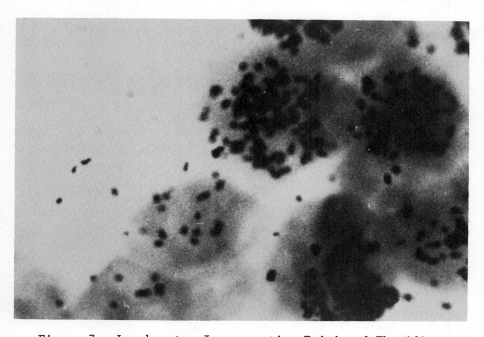

Figure 7. Lymphocytes Incorporating Tritiated Thymidine

Photomicrograph of the incorporation of tritiated thymidine into PHA-stimulus human peripheral blood lymphocytes in vitro. The autoradiograph was stained with Giemsa stain. Magnification - 1000X.

difference in the radioactive labeled thymidine incorporation into the cultured lymphocytes treated with 10^{-2}, 10^{-3} molar concentration of calcium cyclamate and .05 mg/ml cycloheximide at the respective time periods with a confidence level of 0.01. Figure 7 is a photomicrograph of the incorporation of tritiated thymidine into PHA-stimulated human lymphocytes. The exposed grains in this picture of an autoradiograph demonstrate the incorporation of radioactive labeled thymidine in PHA-stimulated lymphocytes. One notices that some of the cells are heavily labeled while others are not. The different amounts of labeled thymidine found in each of these cells (figure 7) indicate different stages of deoxyribonucleic acid synthesis. Figure 8 depicts the labeling index of PHA-stimulated purified human peripheral blood lymphocytes in vitro.

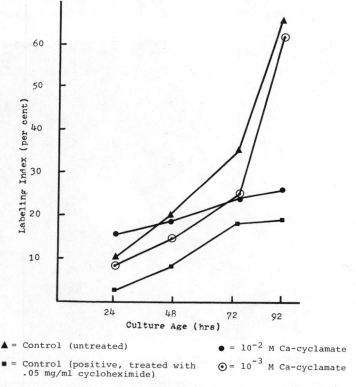

▲ = Control (untreated) ● = 10^{-2} M Ca-cyclamate
■ = Control (positive, treated with .05 mg/ml cycloheximide) ⊙ = 10^{-3} M Ca-cyclamate

Figure 8. Effect of Calcium Cyclamate on the Labeling Index of Lymphocytes in vitro.

DISCUSSION

An array of adverse cytological effects has been associated with the utilization of calcium cyclamate in vitro.[3,4,12,13,14] The primary site of action of calcium cyclamates and their bio-degraded metabolites remains undefined. Accumulated evidence on cyclamates inducing chromosomal aberrations indicated that the cyclamates have teratogenic, mutagenic, and carcinogenic activity. Stoltz, Khera, Bendall and Gunner[4] suggest that the primary sites of action of cyclamates are at the genetic level.

The experiments described in this work demonstrated that PHA-stimulated human peripheral blood lymphocytes may be employed as a test model system for assessing the effects of calcium cyclamates on eucarayotic cells. Experiments were not done to delineate the nature of the existing peripheral blood lymphocyte. However, on the basis of published reports, wherein identification procedures were used, greater than 95% of the cells were thymus lymphocytes.[14]

Data obtained from studies described herein indicated that in vitro cultured primary PHA-stimulated peripheral blood lymphocytes displayed unusual morphological characteristics upon exposure to 10^{-2} and 10^{-3} molar concentration of calcium cyclamate. In this system, calcium cyclamate was shown to inhibit the proliferation of cultured lymphocytes. Cytokinetic studies of PHA-stimulated peripheral blood lymphocytes in vitro showed proliferation. The onset of deoxyribonucleic acid synthesis was 20 to 30 hours. Deoxyribonucleic acid was assessed by the incorporation of radioactive thymidine in to acid precipitable material and autoradiographic analysis of radioactive labeled cultured lymphocytes. The detection of exposed grains in autoradiographs of labeled lymphocytes coincided with the occurrence of blastogenesis. The above morphological cell characteristics were interpreted as an indication that the lymphocytes were in the synthesis phase of the cell cycle.

Among PHA-stimulated lymphocyte cultures to which was added calcium cyclamate or cycloheximide, in each of the treated cultures morphological characteristics were abated. The increased incorporation of radioactive thymidine into acid precipitable material as exposed grains in autoradiographs of untreated cultures, post 24 to 48 hours, was reduced by addition of 10^{-2} and 10^{-3} molar concentration of calcium cyclamate. The maximum difference in the number of lymphocyte cells, labeled thymidine uptake, exposed grains of autoradiographs, and chromosome aberrations was seen at about 72 hours (Figure 8).

Jemison, Rivers, Boone and Knight,[16] reported that the most frequent occurring chromosomal aberration found in PHA-stimulated lymphocytes in vitro, exposed and unexposed to calcium cyclamate, was condensed chromosomes. Their statistical results indicated that at the .05 confidence level, there was no significant difference between the occurrence of condensed chromosomes in cyclamate exposed and unexposed lymphocytes. The results of Jemison, Rivers, Boone and Knight[16] research study on the effects of calcium cyclamates on chromosomes in PHA-stimulated lymphocytes in vitro also reported the significance of other chromosome aberrations. These observations support the work of Stone[2] and others, including Legator, Palmer and Petersen,[14] who reported that calcium cyclamate induces chromosomal breaks (isochromatid breaks and quadriradials), all of which were concentration dependent. These observations were interpreted to indicate that the primary site of action of calcium cyclamate is at the genetic level. Additional support for this hypothesis was obtained in chromosomal analysis among PHA-stimulated non-treated and cyclamate treated peripheral blood lymphocytes. A number of chromosomal aberrations were seen among the cyclamate treated PHA-stimulated lymphocyte cultures. The frequency of many of the chromosomal aberrations was directly related to the concentration of calcium cyclamate used. Nevertheless, some were not. It remains to be determined whether each chromosome has equal affinity for calcium cyclamate, which may account for the disproportionate array of aberrations seen. This could possibly be addressed in an extensive analysis of specific chromosomes among cyclamate treated lymphocytes.

Data obtained in this study indicate that calcium cyclamate impaired the synthesis of deoxyribonucleic acid (depicted by decreased incorporation of radioactive thymidine into acid precipitable material), reduced grain counts in autoradiographs and increased chromosome aberrations in cyclamate treated PHA-stimulated peripheral blood lymphocytes in vitro.

ACKNOWLEDGEMENT

The authors thank Dr. Samuel P. Massie (USNA), Visiting Eminent Scholar, Virginia State University for technical advice; Petersburg General Hospital for plasma supplies; Dr. Earl V. Allgood and Mr. Payton T. Butler for data analyses; Virginia State University biology students Boone, Bibbins, Cheatham, Chambliss and Winbush for technical assistance; and colleagues in the Department of Biology for general assistance. This study was supported by grants from NIH/MBS Program 12-506-AA-08090, for which support we are most grateful.

BIBLIOGRAPHY

1. R.B. Stith, N.L. Bannister and E.W. Jemison, Reaction of Drosophila melanogaster to food substances containing sodium cyclamate, calcium cyclamate and sodium saccharin, Va. J. Sci. 21116, 1970.
2. D. Stone, Cyclamate and human cells, Science 6:132, 1969.
3. D. Stone, E. Lawson and K. Pickering, Cytogenetic effects of cyclamates on human cells in vitro, Science 164:568, 1969.
4. D.R. Stoltz, K.S. Khera, R. Bendall and S. Gunner, Cytogenetics studies with cyclamates and related compounds, Science 167:1501-2, 1970.
5. D.R. Stoltz, B. Stanric, and Klassen, R.E. Bendall Commercial cyclamate mixed with food substances, Environ. Path. Toxicol. 9:11, 1977.
6. G. Obe, B. Beck, C. Dublin, The human leucocyte system: DNA synthesis and mitosis in PHA stimulated 2 days cultures. Mutation Res. 23:279, 1974.
7. D. Merchant, R. Kahn and W. Murphy, "Handbook of Cell and Organ Culture." Burgess Publishing Co., New York (1933).
8. P.S. Moorehead, P. Nowell, W. Mellman, D. Battip and Hungerford, Chromosome preparations of leucocytes culture from human peripheral blood, Exp. Cell Res. 20:613-616, 1960.
9. R.R. Green, Changes in acid hydrolases during lymphocyte stimulation, Exp. Cell Res. 110:215, 1977.
10. W. Heneen, Extensive chromosome breakage occurring spontaneous in a certain individual of Elymus farctus, Hereditas 49:1-32, 1963.
11. E.W. Jemison and B.A. Rivers, Morphological changes in human lymphocytes after calcium cyclamate exposure. Proceedings fo the Fifth Annual Xavier-MBS Biomedical Symposium, Fontainebleau Motor Hotel, New Orleans, Louisiana 70125, 226:127, 1977.

12. M.A. Friedman, V. Aggarwal and E. Goute, Inhibition of epidermal DNA synthesis by cycloheximide and other inhibitors of protein synthesis, J. Nat. Cancer Inst., 1975.

13. S. Wolff and B. Rodin, Saccharin-induced sister chromatid exchanges in Chinese hamster and human cell, Science 200:543.

14. M. Legator, K. Palmov and K. Peterson, Cytogenetic studies in rats of cyclohexlamine, a metabolite of cyclamate, Science 165:1139, 1968.

15. M. Barkin, R. Comisarow, L. Taranger and A. Canada, Three cases of human bladder cancer following high dose cyclamate injection, J. Virol. 118:258, 1977.

16. E.W. Jemison, E. Rivers, C. Boone and R. Knight, Cytogenetic effects of cyclamate, Abstract 32.3:100, Virginia Academy of Science, 1981.

17. F. Ruscetti, D. Morgan and R. Gallo, Functional and morphologic characterization of human T cells continuously grown in vitro, J. Immunol., 119:131, 1977.

MAMMALIAN CELL CULTURE: TECHNOLOGY AND PHYSIOLOGY

Nikos K. Harakas, Charles Lewis, Ronald D. Bartram,
Bernard S. Wildi and Joseph Feder

Monsanto Company
800 North Lindbergh Boulevard
St. Louis, Missouri 63167

ABSTRACT

During the last decade phenomenal advances have taken place in large-scale mammalian cell culture both for microcarriers and suspension methods. The cost of serum and product quality require that such systems be examined in terms of both their physical and chemical parameters. Data are presented as to the method used to more than double the final harvest cell density of 100-liter batch culture reactors by simple temperature measurements and 12-liter reactors by increasing the oxygen potential of the liquid. The results are interpreted in terms of physical transport phenomena of momentum, heat and mass via the theoretical relationships of Navier-Stokes, Fourier and Fick respectively. In addition, data are presented of 3-, 12-, and 100 liter reactors that the spent cell growth media were analyzed for ten different chemistries. The Monod cell growth curve was used to interpret the results. The cost savings and product quality improvements can be enormous by approaching both the technology and physiology of mammalian cell culture from the vantage points of chemistry, thermodynamics and transport phenomena.

INTRODUCTION

The key limiting factor for the economic commercialization of products from mammalian cells has been the very high cost of the growth media required to grow such cells. Mammalian cells, either on microcarriers or in suspension, require serum (fetal bovine, calf, horse, etc.) in addition to amino acids, hormones, vitamins, salts, etc., such as Dulbecco's MEM composition (total of about 40 different chemicals) for growth. Such growth media are costly and serum is available in very limited supply. The major portion of the cost is due to serum which is used at a concentration of about two to ten percent by volume. Hence, a new approach must be taken to reduce the cost of the raw materials per unit volume of cells produced and growth media volume used. The cell culture system must be examined in light of the transport of mass, energy and momentum and the chemical composition of both the growth media and the cell itself. Such results must be analyzed in terms of theoretical relationships, namely: (1) Fick's Law for mass transfer, (2) Fourier's Law for energy transport, (3) Navier-Stokes equation for momentum transfer, (4) Gibbs-Duhem equation for chemical potential, (5) Monod cell growth phase diagram for culture state. Other relationships that could be considered are the Nernst equation and the Onsager reciprocal relations.

The theoretical relationships will be discussed in this paper and examples will be given that indicate marked increase in productivity of cells produced via batch cell culture reactors of

THEORETICAL RELATIONSHIPS

The Monod (4) cell phase diagram (Fig. 1) has been shown to be quantitatively applicable for mammalian cell growth (2,5,6) as for bacterial cells. Cultures that are used for either research, development or production should be defined with respect to one of the possible six phases of the cell growth for the given experiment. This applies for both cells and spent media whichever may be analyzed. The chemical composition of the culture varies widely over this multi-phase growth diagram as expected. Under certain conditions in addition to the cell count, e.g., the dissolved O_2 concentration of the media, may be used to define the culture phase (2,3,6). The phase diagram is probably the most powerful tool available for the physiological analysis of mammalian cultures.

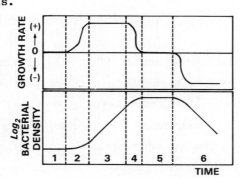

1. lag phase: growth rate null;
2. acceleration phase: growth rate increases;
3. exponential phase: growth rate constant;
4. retardation phase: growth rate decreases;
5. stationary phase: growth rate null;
6. phase of decline: growth rate negative.

Figure 1

Monod Phase Diagram for Cell Growth

Fick's Law: MOLECULAR DIFFUSION

$$(v \cdot \nabla c_A) = \mathscr{D}_{AB} \nabla^2 c_A$$

Diffusion in laminar flow (dilute solutions of A in B)

1. $\mathscr{D}_{AB}, \rho$ = constants
2. Steady state
3. No chemical reactions

Fourier's Law: HEAT CONDUCTION

$$(v \cdot \nabla T) = \alpha \nabla^2 T$$

Heat conduction in laminar incompressible flow

1. k, ρ, c = constants
2. No viscous dissipation
3. Steady state

$$\alpha = \frac{k}{\rho c}$$

Navier-Stokes Equation: MOMENTUM TRANSFER

$$\rho \frac{Dv}{Dt} = -\nabla p + \mu \nabla^2 v + \rho g$$

1. ρ, μ = constant

Gibbs-Duhem Equation: CHEMICAL POTENTIAL

$$-SdT + VdP - \Sigma n_i d\mu_i = 0.$$

Figure 2

Theoretical Relationships for the Cell Culture System

The Gibbs-Duhem (7) equation (Fig. 2) further describes in a generalized form the culture state with respect to the chemical potential (S=entropy, T=temperature, V=volume, P=pressure, n_i=concentration of component i, μ_i=chemical potential of component, i). It can be seen from this equation that it is permissible for the concentration of the various components to vary widely provided that all are present and still satisfy the relationship as has been found, e.g., for normal human blood and its components (8,9). Furthermore, it can be seen very easily from the equation that changes, e.g., in temperature must be compensated by one of its other variables. The direct practical application of the Gibbs-Duhem equation is not difficult, but it requires knowledge of the μ_i and S in reference to some standard state, since chemical thermodynamics deals only with changes of these quantities.

The transport phenomena (10,11) of mass, energy and momentum of the culture system can be handled quantitatively by the respective mathematical relations of Fick, Fourier and Navier-Stokes (Fig. 2). The Laws of Fick and Fourier are mathematically the same (∇=Laplacian operator in three-dimensional space x, y, z). $D_{A,B}$ is the diffusion coefficient for Fick's Law, and α is the thermal diffusivity ($\alpha = \frac{k}{\rho c}$; k=thermal conductivity, ρ=density, c=heat capacity) for Fourier's law. Examples will be given of the practical and simple applicability for both laws in this paper. The equations are solved for specific conditions which describe a given situation. The same is true for the Navier-Stokes equation which describes the transfer of momentum for movement and mixing of cells and of the surrounding extracellular growth fluid.

EXPERIMENTAL

Equipment

   Three-Liter Spinner. The standard Bellco spinner (Cat. No. 1969-03000) is used with the modification of sterile air-CO_2 (95%/5%) overlay gas purging. It has been found that the three-liter spinner is a good model reactor for growing cells in suspension with high cell density and viability.

   Twelve-Liter Spinner. It is made in-house using 3-1/2 gallon capacity glass bottles (Cat. No. 022887B from Fisher Scientific) modified as the three-liter Bellco spinner with air-CO_2 (95%/5%) overlay gas purging and O_2 sparging via a fritted glass about two inches from the bottom of the bottle and about four inches from its center.

One-Hundred Liter Reactor. It is made in-house of 316 stainless steel from a 30-gallon food-grade container and it is stirred via a Vibromixer made by Chemapec. It has been described elsewhere (12).

Hemacytometer. The standard type hemacytometer was used (Cat. No. B3192 from American Scientific Products) to count cells and are reported as number of cells per cm^3 rather than per ml as commonly done because the dimensions are in centimeters (13). Cell viability was determined via Trypan blue exclusion (13).

Materials

Cells. Walker 256 rat carcinoma cells were used in all experiments of the same origin as previously described (14,15) and cell line No. ATCC CCL38-LLC-WRC256 (16).

Growth Media. The cells were grown in 4.5 to 5.5% fetal bovine serum obtained from various suppliers (e.g. K.C. Biological, Gibco, etc.) and Dulbecco's modified Eagle medium (e.g., Gibco Cat. No. 430-2100) with 110 mg/liter sodium pyruvate and $NaHCO_3$ added as the buffer and adjusted to pH 7.4.

Culture Sterility. Cultures and growth media were checked on a periodic basis via plating on the following growth media: blood agar plates obtained from American Scientific Company, thioglycollate broth tubes and Sabouraud's agar slants obtained from Difco Co. (5,17).

Supernatants. For the chemical analyses of the cell supernatants reported in this paper, they were processed as follows: a 20-50 ml sterile sample was removed from each reactor under sterile conditions. The cells were removed quickly via

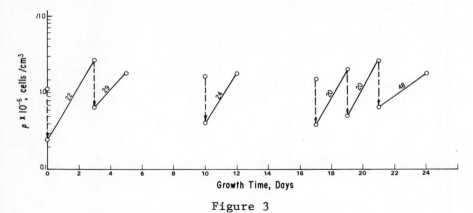

Figure 3

Cell Growth Curves of a 3-Liter Cell Culture Spinner.

centrifugation using an IEC laboratory clinical centrifuge (1000xg for 10 min.); the supernatant was placed in sterile 10 ml tubes and frozen immediately. Then samples were thawed about 2-3 hours before the chemical analyses. All the samples including an unused cell growth medium as a control were analyzed via the ACA-II analyzer (DuPont Co.) at St. Mary's Hospital and Health Center, St. Louis, MO, for ten chemistries (see Results and Discussion) at the same time.

RESULTS AND DISCUSSION

Temperature

Puck (18) clearly demonstrated the enormous sensitivity of mammalian cell growth to small fluctuations in temperatures as it is true with certain types of bacteria, e.g. E. coli (19). In general, below 37°C mammalian cells grow very slowly, if at all, and they cannot tolerate for very long before senescence temperatures just a few degrees above this optimum level. The maximum cell densities achieved with three-liter capacity cell culture reactors were in the range of $2-3 \times 10^6$ cells/cm^3 (Fig. 3). However, for the 100-liter capacity cell culture reactors, maximum

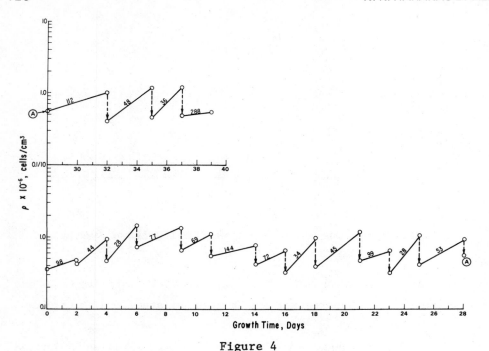

Figure 4

Cell Growth Curves for 100-Liter Cell Culture Reactor.

cell densities were in the range of 1–1.3x10^6 cells/cm^3 (Fig. 4) using the same harvest periods as the 3-liter reactors. The growth medium temperature of an 100-liter reactor was monitored (Fig. 5). It was found that it took about 24 hours before the 100-liter reactor reached 37°C, the optimum growth temperature for mammalian cells. Such a lag in temperature could be predicted from Fourier's Law (Fig. 2) for the boundary conditions of the 100-liter reactor placed in a room which is kept at 37°C.

As a result, the weekly harvest schedule for 100-liter reactors was changed to 3-, 4-days instead of the 2-, 2-, 3-days in order to allow the reactor to reach 37°C and thus achieved a much higher cell density (Fig. 6) of 2–3x10^6 cells/cm^3. The productivity of the operation was nearly doubled with respect to the volume of cells produced per unit time and volume of growth media used. It is noted that for the harvest schedule αεω (Fig.

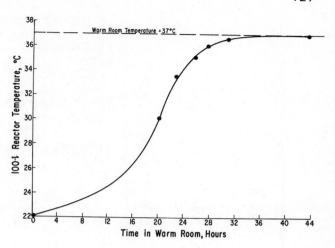

Figure 5
Temperature Change with Time of a 100-Liter Cell Culture Reactor Placed in a Room which is Controlled at 37°C.

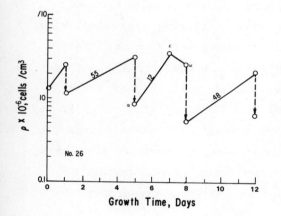

Figure 6
Cell Growth Curves for 100-Liter Cell Culture Reactor.

6) that the reactor should have been harvested sometime in the time interval εω rather than at ω because the cells went into the declining phase (Fig. 1) after time ε. Reactors should always be harvested before reaching the stationary phase (Fig. 1) as has been fully discussed previously (2,3,6). It is a logistics problem which can be managed effectively by proper planning and measurements.

THERMAL PROPERTIES

- **GLASS VESSELS**

$$\alpha_{GLASS} = 0.0058 \; \frac{cm^2}{sec.}$$

Figure 7

- **STAINLESS STEEL VESSELS**

$$\alpha_{ss} \cong 0.060 \; \frac{cm^2}{sec.}$$

- **COOLING RATE RATIO**

$$\frac{\alpha_{ss}}{\alpha_G} \cong 10$$

Effect of Thermal Diffusivity, α, on Relative Cooling Rate of Vessels Made from Glass and Stainless Steel.

Thermal Properties

Up to a few years ago nearly all cell culture was performed using glass vessels. Now metal vessels are also being used, made of 316 stainless steel composition to resist salt corrosion. Extreme cautions should be taken in sterilizing via autoclaving stainless steel vessels with appropriate and adequate sterile filters. Vessels made of stainless steel cool ten times faster than glass vessels (Fig. 7) because of their higher thermal diffusivity (20) as predicted by Fourier's Law (Fig. 2). It is strongly advised not to remove stainless steel vessels from the autoclave while still hot unless extreme precautions have been taken to introduce sterile air into the vessel during its immediate cooling period when placed at room temperature for subsequent use.

Gaseous Diffusion

Purging a stream of mixture of air-CO_2 (95%/5%) above the surface of a cell culture reactor results in establishing a concentration gradient of dissolved O_2 and CO_2 (Fig. 8) with the highest concentration near the interface, i.e. $[C_4] \ggg [C_1]$ as required by Fick's Law (Fig. 2). The extent of the difference between $[C_4]$, $[C_3]$...$[C_1]$ depends on the vessel depth, x, and the diffusivity coefficient, $\mathcal{D}$, of the particular gas into the

MAMMALIAN CELL CULTURE

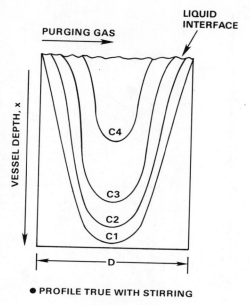

Figure 8

Effect of Vessel Depth on Concentration Profile of Gas, e.g. Air-CO_2 (95%/5%), Purging Over the Top Liquid Interface

aqueous solution, and extent of liquid turnover by stirring, but a concentration profile will be established.

The 12-liter reactors with overlay purging (8-10 inches from bottom of reactor) of air-CO_2 mixture could reach maximum cell densities of only $1-1.3 \times 10^6$ cells/cm^3 whereas the 3- (Fig. 3) and 100-liter (Fig. 6) reactors could reach values of $2-3 \times 10^6$ cells/cm^3. The O_2 potential of the cell growth media of the 12-liter reactors was further increased by sparging O_2 besides the air-CO_2 overlay purge via a fritted glass tube placed about two inches from the bottom of the reactor and about four inches from its center. The result of this simple modification was that the maximum cell densities were increased to $2-3 \times 10^6$ cells/cm^3 (Fig. 9). as expected by Fick's Law (Fig. 2). A similar approach had been used previously by others (5) with a similar size reactor of sparging air-CO_2 instead of O_2 which reduces possible foam

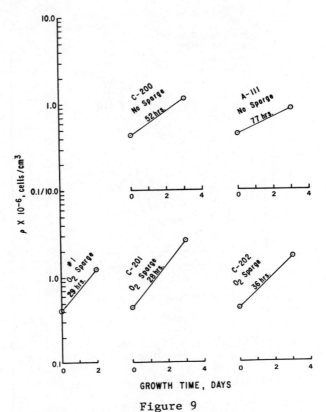

Figure 9

Effect of O_2 Sparging on Harvest Cell Density of 12-Liter Spinners.

problems because of the much smaller volumetric gas rate (~5xless) required to achieve the same O_2 potential.

Supernatant Chemistry

The most complex and the most important aspect of cell culture is the biochemical composition and state of both the surrounding extracellular fluid or supernatant and that of the cell itself (21). As has been discussed in a previous section, there is a wide range in the concentration of a given molecule with normal behavior of the system (8,9). Recent advances indicate that accurate measurement of small, intracellular pH (22) changes may correlate with the turning on of the cell activities. Such results provide excitement and incentive to define the

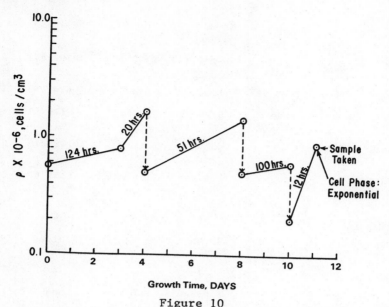

Figure 10
Monod Phase of Cell Growth of a 3-Liter Spinner.

physiology of the cell culture system, and how, for example, such chemical state is related to the Monod growth phase diagram (Fig. 1) of the cell.

An exploratory study was undertaken to analyze the supernatant fluid of nine different cell culture reactors and fresh cell growth media as a control for ten different chemistries, namely albumin, alanine aminotransferase, ammonia, aspartate aminotransferase, glucose, γ-glutamyl transferase, iron, lactic dehydrogenase and liver lactic dehydrogenase. The data and results are tabulated in Table 1. For each reactor the approximate cell phase was established (Table 1) via cell growth plots as, for example, Fig. 10. All the data were analyzed in terms of the Monod phase diagram (Fig. 1) as shown, for example, for glucose, lactic dehydrogenase and asparate aminotransferase and given in Figures 11, 12 and 13 respectively. The results show largely that they fit the model. The points that are circled in Figures 11-13 do not fit the model. In all cases all the

Table 1

	Control Media #99	R-202 #77 100L	R202 #73 100L	R302 #75 100L	A76 12L
Date Sample Colltd.	6/5/79**	6/4/79	6/4/79	6/4/79	6/4/79
Date Reactor Started	Made 6/5/79	5/28/79	5/28/79	5/29/79	6/1/79
Viable Cell Density, Initial, cells/cm^3	-	0.555×10^6	0.5×10^6	0.325×10^6	0.6×10^6
Harvested Cell Density (Sample Taken), Cells/cm^3	-	2.47×10^6	1.65×10^6	6.5×10^5	1.2×10^6
Cell Viability, %	-	90%	94%	97%	80%
Approximate Doubling Time, Hrs.	-	43.2	62.4	201.6	96
Growth Curves, Fig. #	-	-***	-	-	-
Cell Culture Phase (Fig. 1)	-	Stationary	Exponential or Stationary	Long lag, no growth	Stationary very far
pH	(7.4)	-	-	-	6.535
Osmolality, mOsm/kg		-	-	-	351
Albumin (ALB), g/dl	0.67	0.67	0.65	0.68	0.70
Alanine Aminotransferase (GPT), IU/L	A-Code*	2	0	0	1
Ammonia (Amon), μmol/L	320	2908	2792	770	3346
Aspartate Aminotransferase (GOT), IU/L	13	61	19	11	88
Glucose (GLUC), mg/dl	464	151	210	350	85
γ-Glutamyl Transferase (GT), IU/L	0	69	16	0	42
Iron (IRN) μg/dl	11	12	12	6	16
Lactic Acid (LA), mEq/L	0.9	21.1	24.0	5.1	31.5
Lactic Dehydrogenase (LDH), IU/L	26	396	30	24	756
Liver Lactic Dehydrogenase (LLDH), IU/L	0	298	0	0	628

*Absorbance code - too low to read.   **Date fresh media made.

	#12 3L	A50 3L	R201 #76 100L	R201 #68 100L	R201 #68 100L	Data in Fig. No.
	6/4/79	5/24/79	6/4/79	5/25/79	5/29/79	-
	1/29/79	5/10/79	5/31/79	5/21/79	5/21/79	-
	0.18×10^6	0.2×10^6	2.17×10^6	0.518×10^6	0.571×10^6	-
	5.5×10^5	8.7×10^5	1.0×10^5	3.1×10^5	6.8×10^5	-
	74%	95%	N.A.	68%	65%	-
	52.8	12	No vibration	-	-	-
	-	Fig. 10	No plot	-	-	-
	Stationary far	Exponential	Lag	Lag	No growth, cell density declining	-
	6.573	-	-	-	-	-
	372	-	-	-	-	-
	0.67	0.65	0.65	0.63	0.65	-***
	0	1	A-Code	A-code	A-code	-
	3218	3270	2044	2564	2408	-
	64	25	115	38	38	Fig. 13
	120	187	301	209	211	Fig. 11
	27	20	5	10	8	-
	16	18	20	13	36	-
	32.9	26.8	14.1	19.5	21.7	-
	572	36	240	52	60	Fig. 12
	440	0	154	0	18	-

*** Done but not given in paper.   N.A. = not available.

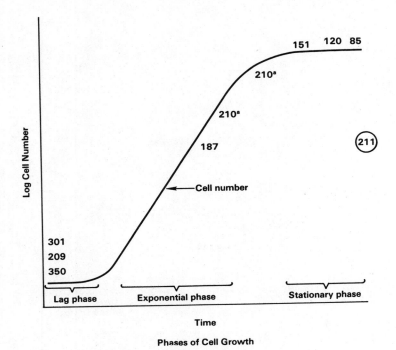

Figure 11

Change of Glucose Concentration of Cell Growth Media with Monod Cell Growth Phases.

Table 2

Physiology	Cell Growth Curve Correlation
Albumin	No correlation—No change
Ammonia	Correlation—increasing type
Glucose	Strong correlation—decreasing type
Lactic Acid	Correlation—increasing type
Lactic Dehydrogenase	Some correlation—increasing type
Liver Lactic Dehydrogenase	Possible correlation for stationary phase—zero at other phases
γ-Glutamyl Transferase	Possible correlation—increasing type
Aspartate Amino Transferase	Correlation—increasing type

chemistries of the control growth media were as expected. Table 2 gives the correlation of the results with respect to the Monod phase diagram for the eight chemistries (Table 1). The results for alanine aminotransferase and iron were not analyzed in terms of the model in view of the low values, it any, found in the

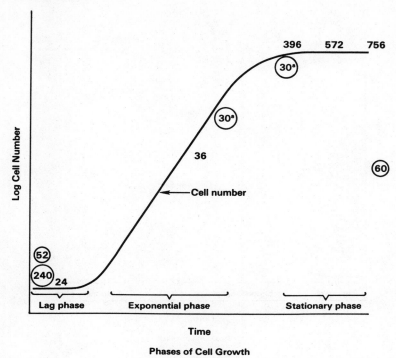

Figure 12

Change of Lactic Dehydrogenase Concentration of Cell Growth Media with Monod Cell Growth Phases.

supernatants. The chemistry results are very much more informative when examined in terms of the Monod diagram as, for example, the lactic acid dehydrogenase (Fig. 12) values in the stationary phase are very large compared to the exponential phase and liver lactic dehydrogenase appears only in the stationary phase (Tables 1,2). Such results have both research and commercial value. Hence, the analysis of the supernatant chemistry of the cell culture system in terms of the Monod diagram proves to be a very powerful and useful tool. The chemistry of the cell itself can be analyzed also in terms of the Monod diagram.

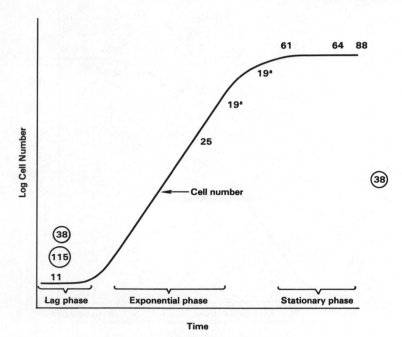

Figure 13

Change of Asparate Aminotransferase Concentration of Cell Growth Media with Monod Cell Growth Phases.

ACKNOWLEDGEMENTS

The senior author wishes to acknowledge gratefully the following persons who carried out the multitude of laboratory experiments in connection with the Monsanto/Harvard Project: M.R. Boyd, G.W. Greenway, C.E. Gross, T.J. Jones, J. Philipp, B. Reddick, R.M. Reynolds, A.L. Robinson and D.S. Tyler. In addition, Monsanto Company and the senior author gratefully acknowledge the generosity of Catherine A. Ashton, Chief Technologist, Chemistry Department, St. Mary's Hospital and Health Center for providing the ACA-II analyzer for the supernatant chemical analyses.

REFERENCES

1. R.T. Acton and J.D. Lynn (Editors), 1977, Cell Culture and its Application, Academic Press, New York, NY.
2. ibid., 1977, Description and Operation of a Large-Scale Mammalian Cell Suspension Culture Facility, in, Advances in Biochemical Engineering, T.K. Ghose et al., (Editors), Springer-Verlag, New York, NY, 7:83-109.
3. N.K. Harakas, 1977, Operation of the Suspension Mammalian Cell Culture Center at the University of Alabama in Birmingham during the Year 1976, Unpublished report: 1-141.
4. J. Monod, 1949, The growth of bacterial cultures, in, Selected Papers in Molecular Biology by Jacques Monod, A. Lwoff and A. Ullman, Editors, Academic Press, New York, NY: 139-162.
5. R.K. Zwerner, C. Runyan, R.M. Cox, J.D. Lynn and R.T. Acton, 1975, An evaluation of suspension culture systems for the growth of murine lymphoblastoid lines expressing TL and Thy-1 alloantigens. Biotech. Bioeng. 17:629-657.
6. N.K. Harakas, J.D. Lynn, R.T. Acton and J.C. Bennett, 1979, Industrial Scale Suspension Mammalian Cell Culture Center: Operation and Performance Dynamics. 72nd Annual Meeting of A.I.Ch.E., San Francisco, CA Abstract 24f:T-28.
7. K. Denbigh, 1955, The Principles of Chemical Equilibrium, Cambridge University Press, New York, NY 90-92.
8. B. Castleman and B.U. McNeely, 1974, Blood, plasma or serum values, New England J. Med. 290:39-49.
9. W.F. Ganong, 1975, Review of Medical Physiology, 7th Edition, Lange Medical Publications, Los Altos, CA.
10. R.B. Bird, W.E. Stewart and E.N. Lightfoot, 1960, Transport Phenomena, John Wiley and Sons, Inc., New York, NY.
11. E.N. Lightfoot, 1974, Transport Phenomena and Living Systems, John Wiley and Sons, New York, NY.

12. W.R. Tolbert, R. Schoenfeld and J. Feder, 1980, Design of 100-liter vessels for storage of medium and growth of mammalian cell suspensions, In Vitro 16(3):227.
13. R.J. Kuchler, 1977, Biochemical Methods in Cell Culture and Virology, Dowden, Hutchinson and Ross, Inc., Strandsburg, PA.
14. J. Folkman, 1975, Tumor angiogenesis, in, Cancer, F.F. Becker, Editor, Plenum Press, New York, NY 3:355-388.
15. J. Folkman and M. Klagsbrun, 1975, Tumor angiogenesis: Effect on tumor growth and immunity, in, Fundamental Aspects of Neoplasia, A.A. Gottlieb et al., Editors, Springer-Verlag, New York, NY:401-412.
16. H.D. Hatt and M.J. Gantt, 1979, The American type culture collection, Catalogue of strains II, 2nd Edition, Rockville, MD:31.
17. R.T. Acton, P.A. Barstad and R.K. Zwerner, 1979, Propagation and scaling-up of suspension cultures, in, Methods of Enzymology, W.B. Jakoby and I.H. Pastan, Editors, Academic Press, NY 58:211-221.
18. T.T. Puck, 1972, The Mammalian Cell as a Microorganism, Holden-Day, Inc., San Francisco, CA.
19. B.D. Davis et al., Editors, 1980, Microbiology, 3rd Edition, Harper and Row Publishers, New York, NY.
20. H.S. Carslaw and J.C. Jaeger, 1959, Conduction of Heat in Solids, Second Edition, Oxford University Press, London, England.
21. A.G. Guyton, 1981, Medical Physiology, 6th Edition, W.B. Saunders Co., Philadelphia, PA.
22. J.L. Marx, 1981, Investigators focus on intracellular pH, Science 213:745-747.

A HIGH EFFICIENCY STIRRER FOR SUSPENSION CELL CULTURE WITH
OR WITHOUT MICROCARRIERS

Norman A. de Bruyne

Techne Inc.
3700 Brunswick Pike
Princeton, NJ 08540

ABSTRACT

The stirrer described in this paper employs "Teaspoon stirring" in which a secondary motion is superimposed on the rotation of the liquid. This secondary motion arises from the viscous drag from the wall and bottom of the flask. The culture medium is rotated by a bulb-ended rod suspended from inside the flask cap: the bulb orbits around in a circular trough formed between a central indent in the base and the rounded periphery of the base. There are no bearings and no stagnant areas. The power used to stir 4 flasks each holding 500 ml of medium is less than 2 watts. The speed is electronically controlled by a tachometer and automatically gives a smooth start and stop as well as "Interval stirring" to assist initial attachment of cells.

INTRODUCTION

We can all expect to stir about 10^5 cups of tea or coffee in a lifetime, but few of us are aware of the subtleties of such stirring. My own enlightenment came from reading the section on "The Madness of Stirring Tea" in Prof. Jearl Walker's entrancing book "The Flying Circus of Physics" which I can strongly recommend.[1]

PURE ROTATION IS INEFFECTIVE

We can see that when we stir we produce rotation of the liquid; but rotation alone will not suspend or disperse particles in the liquid, as is easily demonstrated by rotating a glass of water stood on the center of a "Lazy Susan." The water is dragged around by the inside surface of the glass and will pick up speed until it revolves steadily as though it were a block of ice stuck to the glass. But if you keep the glass stationary and rotate the water with a spoon, a very different kind of motion is produced. The particles not only rotate - they rise up! If the spoon is then removed each dispersed particle will make for the center of the base in apparent defiance of centrifugal force. What is the reason for the difference in behavior of a liquid rotated by a rotating flask and of a liquid rotated in a stationary flask?

SECONDARY MOTION

In a liquid made to rotate in a stationary flask there is a drop in rotational velocity and dynamic pressure due to viscous drag between the liquid and the inside of the glass. Particles at the top of the surface of the liquid will be thrown outwards but as they approach the side of the flask they are slowed down and descend to the bottom, in a helicoidal motion, where they move in to the center and spiral up to the top surface. The existence of this secondary motion was disclosed by James Thomson[2] to a meeting of the British Association for the Advancement of Science, (affectionately known as the British Ass) in 1857. Albert Einstein used this secondary motion to explain the instability of a meandering river and figure 1 showing the secondary motion in a tea cup is adapted from his book[3] "Ideas and Opinions."

ADAPTION OF TEA SPOON STIRRING TO SUSPENSION CELL CULTURE

However maddening in its apparent incomprehensibility "Tea spoon stirring," if properly engineered, is the most effective method of suspending microcarriers and cells at the low Reynolds numbers necessary to avoid damage[4]. The secondary motion has been maximized in the Techne MCS-104 shown in figure 2 by the use of a stirring rod suspended on a flexible connection at the

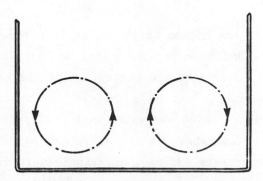

Fig. 1. Einstein's concept of the secondary flow in a stirred tea cup. It is this secondary flow that brings all the tea leaves to the center of the bottom of the cup when the stirring spoon is removed.

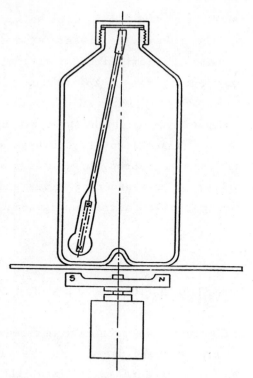

Fig. 2. The MCS-104 stirrer has a stirring rod suspended at the top: the bulb at the bottom contains a bar magnet which is pulled round by a master magnet inside the platform.

top with a bulb at the bottom which orbits in a circular trough formed by the rounded base periphery and an indented center[5]. Figure 3 is from an unretouched photograph of the trace, after 90 seconds of stirring, left by a drop of ink applied to the surface of the liquid which was water with 0.4% polyacrylamide. The picture may be somewhat flattering to the stirrer because polyacrylamide suppresses eddies[6], nevertheless it clearly shows the secondary motion. Note the absence of obstruction to the upward flow: this is in contrast to all stirrers with vertical shafts which have a point of stagnation immediately below the bottom end of the shaft.

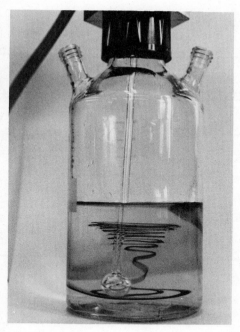

Fig. 3. This shows the trace of an ink drop after 90 minutes of stirring water with 0.4% polyacrylamide. It descended from the surface in a helix close to the side and is now ascending from the center of the base in a coil of increasing radius.

SOME COMPARATIVE TESTS

A paper presented by Hirtenstein, Clark & Gebb[7] to the ESACT Conference at Heidelberg in May 1981 reported results of tests on the four systems represented pictorially in figure 4 which is taken from their paper. A is the traditional spinner vessel developed in the 1950's for use at 50-60 RPM. Although it has given good service it is unsuitable for microcarriers which have a velocity of descent which may be thirty times as fast as cells[4].

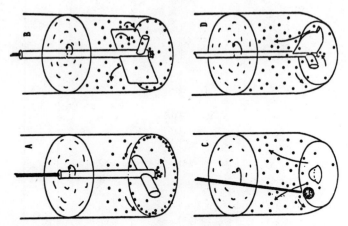

Fig. 4. Four ways of stirring microcarriers. A is traditional, B is A with a paddle, C is the MCS-104, D is A with a scoop (see reference 7).

B is A modified by Bellco Glass Inc. by addition of a paddle. At a stirring speed of 20-40 RPM yields were about 25-30% above those of A but "Stirring motion tended to be erratic with many eddies in the wake of the paddle." D has a plough impeller. Yields at least 50% above A and can be used at lower speeds than B.

C is the MCS-104 which yields 50-70% above A at 20-30 RPM. "Accumulation of microcarriers did not occur." "It appeared that this type of stirring resulted in the least erratic motion of microcarriers of all stirring methods tested. This improved motion of the microcarriers was probably the cause of the improved yields." The cell used in the above tests were MRC5

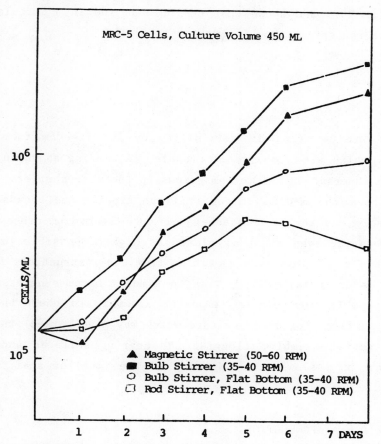

Fig. 5. The ordinates are cells/ml on a logarithmic scale. The curves show that a bulbous stirrer with an indented ("champagne") base is the best combination.

or Vero cells on Cytodex^R microcarriers. The RPM are the minimum values required to suspend the microcarriers: while it is common practice to use RPM as the measure of speed it is incorrect to do so. It is the linear velocity of the tip of the stirrer which is the significant factor and it does not vary with the diameter of the flask as does RPM. Figure 5 is from an investigation made by the Cell Biology Group of Pharmacia at

Uppsala in 1979 and shows that the bulbous rod in a circular trough is the optimum combination.

BEARINGS

The MCS104 gains in reliability by its freedom from bearings which are a source of trouble. A bearing as shown in figure 5 immersed in the medium grinds up the microcarriers: if it is above the medium it can still be troublesome because (presumably due to the Weissenberg effect) the medium creeps up to the bearing even if it is placed as high as possible just inside the cap. This is a quotation from the instructions from Wheaton Instruments; - "From time to time it becomes necessary to disassemble the Celstir^R impeller assembly to check the bearing surfaces for dried culture which may have worked its way up the shaft assembly, clogging the bearing surfaces and resulting in a slowdown or jamming of the impeller shaft."

NEGLIGIBLE HEAT PRODUCTION

The absence of bearings and the low starting torque required in the MCS-104 permit the use of a motor and electronic tachometer control system which together consume 1.5 watts which must be compared with the 50 watts required in other microcarrier stirrers claimed to be "heatless." Its low power requirement permits the MCS-104 to be used in small incubators without upsetting temperature control.

STIRRER FOR SUSPENSION CELL CULTURE

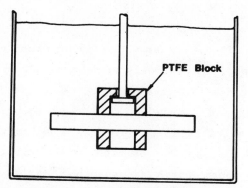

Fig. 6. An immersed bearing, frequently used in early traditional stirrers, is unsuitable for microcarriers.

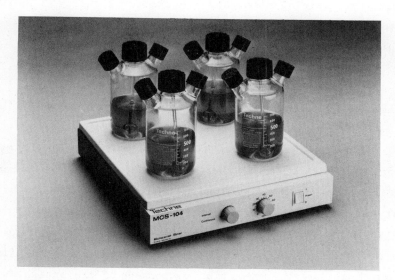

Fig. 7. The MCS-104 stirrer platform with four 1000 ml flasks each containing 500 ml of culture medium operated by a single motor using less than 1.5 watts. Alternatively it will stir a single 10 liter flask containing 5 liters of medium.

SOFT START/STOP AND INTERVAL TIMING

The control circuit in the MCS-104 ensures that the motor accelerates or decelerates slowly when switched on or off. The circuit also provides "Interval timing" which facilitates the initial attachment of the cells to the microcarriers. The stirring is periodically interrupted as long as the interval timing switch is in the ON position. The stirring periods should be just long enough to suspend nearly all the microcarriers.

Figure 7 shows four 1000 ml flasks each containing 500 ml of culture medium standing on the low profile platform containing the driving motor and electronic circuitry. The same platform can be used to drive the stirrer in a 10 liter flask containing 5 liters of culture medium.

ACKNOWLEDGEMENTS

It is a pleasure to give thanks to my wife, and to my colleagues on both sides of the Atlantic, for their forbearance during the past six years in which I have had an "On and off" love affair with the stirring of suspension cell cultures. I must also thank my friend Dr. Leonard Weiss for introducing me to the problem. Finally I wish to make it clear that the final model which is now in production owes much to the insight and ingenuity of Techne (Cambridge) Ltd. particularly to its managing director Mr. James M. Pearson.

REFERENCES

1. Walker, J. 1977 "The Flying Circus of Physics" published by John Wiley (New York & London)
2. Thomson, J. 1857 "On the Grand Currents of Atmosphere Circulation" Trans. of the Sections, Brit. Assoc. Adv. Science 38-39
3. Einstein, A. 1954. "Ideas & Opinions" ("Mein Weltbild") published by Alvin & Redway
4. de Bruyne, N.A. & Morgan, B.J. 1981 "Stirrers for Suspension Cell Culture" American Laboratory 13 52-61
5. U.S. Patents 3913895, 3955802, 39998435, 4204774 and other patents applied for in USA and abroad
6. Walstra, P., 1974. "Turbulence depression by polymers and its effect on disruption of emulsion droplets" Chemical Engineering Science 29 882-885
7. Hirtenstein, M.D., Clark, J.M. & Gebb, Ch. 1981. "A comparison of various laboratory scale culture configurations for microculture of animal cells" to be published in "Development Biol. Standardization" Karger (Basel)

ALTERNATIVE SURFACES FOR MICROCARRIER CULTURE OF ANIMAL CELLS

Christine Gebb, Julian M. Clark, Michael D. Hirtenstein,
Göran E. Lindgren, Björn J. Lundgren, Ulla Lindskog,
and Per A. Vretblad

Cell Biology Group
Pharmacia Fine Chemicals AB
Box 175
S-751 04 Uppsala, Sweden

ABSTRACT

As part of an effort to broaden the applicability and efficiency of microcarrier cell culture various alternative new microcarriers were synthesized. The microcarriers were compared as substrates for the growth of several types of cells and with respect to binding of proteins from the culture medium. Cross-linked dextran has been found to be the most suitable material for a microcarrier matrix and was used as the matrix for the new microcarriers. One type of microcarrier was synthesized so that the charges necessary for cell attachment were present only in the surface layer of the microcarrier in the form of N,N,N-trimethyl-2-hydroxyaminopropyl groups. The resulting microcarriers had a very low capacity to bind proteins from the culture medium (e.g. albumin and IgG) and such proteins could be removed from the cultures more efficiently than when using previous microcarriers. A new principle was used for the development of the other type of microcarrier. A surface layer of cross-linked denatured collagen provided the surface for growth of cells. These microcarriers provided a "natural" substrate for cell growth and allowed improved attachment and spreading of cells with epithelial morphology. Harvesting cells from these microcarriers with proteolytic enzymes was more efficient than with previous microcarriers. The two alternative microcarriers should result in improved process efficiency and have a potential value in the preparation of live vaccines.

INTRODUCTION

Microcarrier culture has proved to be an efficient method for the production of anchorage-dependent cells and their products in high yields (5,14,16,21,22). The most suitable microcarriers produced to date have been those based on a positively-charged cross-linked dextran matrix (10,12). Such microcarriers have been used for the culture of a wide variety of cells in culture systems ranging from a few millilitres to several hundred litres (9,12,14, 16,17,22).

The presence of charged groups on these microcarriers was found to be essential for cell attachment and a specific degree of substitution of the microcarrier matrix with diethylaminoethyl groups was required for optimal cell growth (10,12). This optimum degree of substitution probably reflected a specific density of charges in the surface layer of the microcarriers.

We believe that the charged groups inside the matrix of current dextran-based microcarriers (Fig. 1.1) are not essential for the attachment and growth of cells and these charges provide the microcarriers with an unnecessarily high ion-exchange capacity. To facilitate removal of medium components and cell products it would therefore be advantageous to use a microcarrier

SURFACES FOR MICROCARRIER CULTURE

Figure 1

Schematic representation of the three alternative types of microcarrier.

which only has the charges essential for attachment of cells distributed in the surface layer of the microcarrier.

In this report we describe two alternative types of microcarrier. One of these is surface-charged (Fig. 1.2) and permits growth of cells to at least the same extent as microcarriers with charges distributed throughout the matrix (Fig. 1.1). Removing medium components from surface-charged microcarriers was achieved more easily than from cultures using microcarriers with an evenly charged matrix.

A different principle was used for the development of a second new type of microcarrier. Collagen-coated surfaces have long been used in animal cell culture and have proved important for the growth of many types of epithelial and differentiated cells in culture (1,19). It was therefore this material which we chose for the surface coating of the microcarrier.

The development of a collagen-coated microcarrier should improve the efficiency of the microcarrier technique for the high-yield culture of sensitive types of cells. A further advantage of using a collagen-coated microcarrier for the culture of anchorage-dependent cells is the possibility for improved harvesting of cells with proteolytic enzymes. We describe here the development of a microcarrier coated with covalently bound denatured collagen (Fig. 1.3). This type of microcarrier provides a "natural" substrate for cell growth and broadens the possibilities and efficiency of the microcarrier culture technique.

MATERIALS AND METHODS

All cultures were maintained at 37°C in an atmosphere of 95% air: 5% CO_2. Microcarriers were evaluated in 40-50 ml cultures contained in spinner vessels open to the above-mentioned gas mixture. The culture media did not contain antibiotics and are described in the figure legends. All microcarriers were sterilized by autoclaving (120°C, 15 psi, 20 min) and used at a concentration of 2 mg/ml. Cell number was determined by using the nucleus extrusion method (20). All degrees of substitution are expressed per g dry weight of product.

RESULT AND DISCUSSION

The three types of microcarrier synthesized are represented schematically in Fig. 1. Microcarriers with charged groups distributed throughout the matrix (Cytodex^R 1) were produced by substituting cross-linked dextran (Sephadex^R) with diethylaminoethyl (DEAE) groups (Fig. 1.1). Sephadex beads were also substituted with trimethyl-2-hydroxyaminopropyl (TMHAP) groups in only a surface layer to form the surface-charged microcarriers (Fig. 1.2). A layer of denatured collagen was cross-linked to Sephadex beads to form the third type of microcarrier (Fig. 1.3).

Suface-charged microcarriers

Published procedures for substituting Sephadex with DEAE groups to form microcarriers for cell culture (12,13,15) can lead to the formation of a high proportion (up to 35%) of tandem groups (see 18,8). Such tandem groups are introduced as a result of side reactions and are susceptible to Hoffman-type degradation. The reaction conditions for synthesizing Cytodex 1 microcarriers are

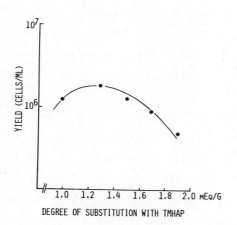

Figure 2

The growth of Vero cells growing on microcarriers having different degrees of substitution with TMHAP groups throughout the matrix. Cultures were inoculated with 10^5 cells/ml. Culture medium was DME supplemented with 4 mM glutamine and 10% foetal calf serum. Cell yield was determined after 113 hours.

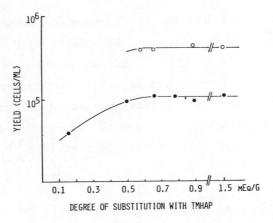

Figure 3

The yield of Vero (- o - o -) and MRC-5 (- • - • -) cells growing on surface-charged microcarriers having different degrees of substitution with TMHAP groups. Cultures were inoculated with 10^5 Vero cells/ml or 1.7×10^5 MRC-5 cells/ml. After a static period of 20 hours the cultures were stirred at 20 rpm. The culture media were DME supplemented with 4 mM glutamine and 10% foetal calf serum for Vero cells and DME supplemented with non-essential amino acids and 10% foetal calf serum for MRC-5 cells. Cell yield was determined after 60 hours.

such that tandem groups comprise only a minor proportion of the charged groups present. In order to eliminate entirely the formation of tandem groups, the quarternary amine TMHAP was used instead of DEAE in the synthesis of surface-charged microcarriers.

The suitability of TMHAP groups for promoting the attachment and growth of cells on microcarriers was first examined with a series of microcarriers substituted to different degrees throughout the entire matrix (Fig. 2). Maximum cell yields were observed when the matrix of Sephadex beads was substituted with TMHAP groups to a total capacity of 1.3 mEq/g. This degree of substitution is slightly less than the optimum value of 1.5 mEq/g observed when using DEAE groups distributed throughout the matrix (10,12).

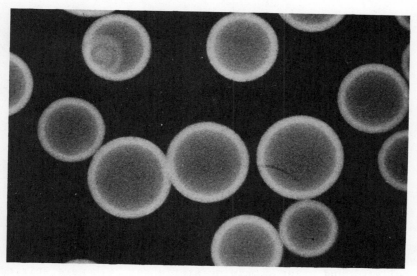

Figure 4

The distribution of TMHAP groups in surface-charged microcarriers with a total capacity of 0.6 mEq/g. Charges are identified by binding of FITC-Dextran T-10 (MW 10,000) which can penetrate the entire matrix of the microcarriers.

The charge capacity residing in the outer layer of microcarriers substituted throughout the matrix with TMHAP groups was calculated and a variety of microcarriers substituted to a similar degree (but only in the outer layer) were synthesized. This series of microcarriers had an identical density of surface charges but the depth of the charged layer was varied. Fig. 3 illustrates that for both Vero and MRC-5 cells a total charge capacity of approximately 0.6 mEq/g distributed only in the surface layer of the microcarrier (Fig. 4) resulted in cell yields equal to those obtained from microcarriers with a capacity of 1.5 mEq/g and charges distributed throughout the matrix. The yield of cells was decreased when lower degrees of substitution in the outer layer of the microcarriers were used (Fig. 3).

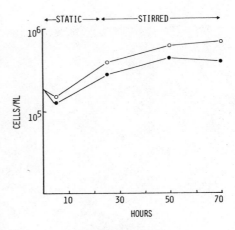

Figure 5

The growth of MRC-5 cells on surface-charged microcarriers (-o-o-) and Cytodex 1 (-•-•-). Cultures were inoculated with 2×10^5 cells/ml and after remaining static for 24 h were stirred at 20 rpm. Culture medium was DME supplemented with non-essential amino acids and 10% foetal calf serum. The surface-charged microcarriers had a total capacity of 0.6 mEq/g.

Fig. 5 illustrates that the growth of MRC-5 cells on the surface-charged microcarriers was slightly better than growth on Cytodex 1 microcarriers. The increase in final cell yields on the surface-charged microcarriers was probably due to an increased plating efficiency and more sustained growth of the late exponential phase culture. The distribution of charges in the surface-charged microcarriers is illustrated in Fig. 4.

Denatured collagen-coated microcarriers

An alternative approach to providing a substrate for microcarrier culture was examined. While most surfaces used in cell culture possess a specific density of small charged molecules to promote attachment of cells, proteins can also provide a surface for the growth of cells in vitro (6,10). Collagen can provide a culture surface for the attachment and growth of a variety of cells. The main advantage of this protein is that it

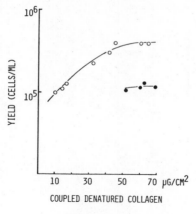

Figure 6

The effect of the amount of denatured collagen coupled to the microcarrier matrix on the yield of cells. Cultures were inoculated with 10^5 Vero cells/ml (-o-o-) or 10^5 BSC-1 cells/ml (-•-•-). After a static period of 24 hours the cultures were stirred at 20 rpm. Vero cells were cultured in DME supplemented with 4 mM glutamine and 10% foetal calf serum. BSC-1 cells were cultured in DME/Medium 199 (75:25) supplemented with 10% foetal calf serum. Cell yield was determined after 40 hours.

permits the growth of many epithelial and/or differentiated cells which are normally difficult to grow in culture (1,19). In addition, sensitive types of cells can be grown at much lower culture densities on collagen-coated surfaces than when grown on artificial culture surfaces (1,19,23). This fact is potentially important for microcarrier culture because it is often necessary to initiate cultures with only a small number of cells.

The process of attachment of cells to collagen surfaces is likely to involve steps in common with attachment to artificially charged surfaces. The glycoprotein fibronectin is one of the main components involved in attachment of cells to both artificial substrates and collagen (4,7,11). Fibronectin binds strongly to collagen (2,3,4) and has an even higher affinity for denatured collagen (2,3).

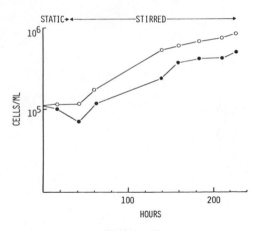

Figure 7

The growth of secondary bovine embryo kidney cells on denatured collagen-coated microcarriers (-o-o-) and Cytodex 1 (-•-•-). Cultures were inoculated with 1.2×10^5 cells/ml. After a static period of 24 hours the cultures were stirred at 20 rpm. Culture medium was DME supplemented with 10% foetal calf serum.

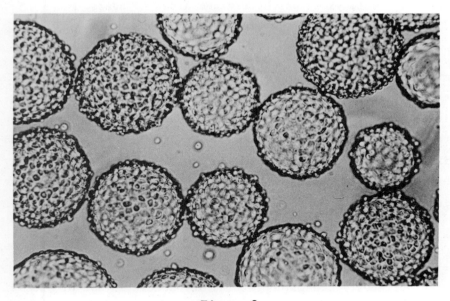

Figure 8

Confluent monolayers of human kidney epithelial cells (Flow 4000/clone 2) growing on denatured collagen-coated microcarriers. Diameter of microcarriers approx. 160 µm.

Denatured collagen from porcine skin (MW 60,000-200,000) was cross-linked to the surface of Sephadex beads. The cross-linking of denatured collagen to the microcarrier matrix avoided the problem of leakage of protein which is encountered with the standard techniques for coating cell culture surfaces. Fig. 6 illustrates the yield of Vero and BSC-1 cells from cultures growing on a series of microcarriers coated with different amounts of denatured collagen. No growth of cells was observed on uncoated Sephadex beads and the yield of cells from microcarrier cultures was greatest when the amount of denatured collagen coupled to the surface of the microcarriers was greater than approximately 50 µg/cm^2.

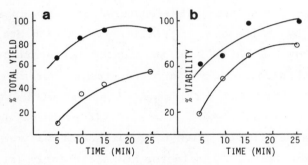

Figure 9

The influence of pretreatment with EDTA on harvesting of human epithelial kidney cells (Flow 4000/Clone 2) from collagen-coated microcarriers with collagenase. The culture medium was DME/Medium 199 (80:20) supplemented with 10% FCS and 2 mM glutamine and cultures were inoculated with 2 x 10^5 cells/ml. For harvesting, the medium was removed by decantation, and the microcarriers were washed once with warmed PBS. The microcarriers were then washed either with EDTA (0.02%w/v) in Ca^{++}, Mg^{++}-free PBS, or with PBS not containing EDTA. These solutions were removed after 2 minutes and collagenase (1 mg/ml) in Krebs II buffer (containing Ca^{++} and Mg^{++}) and the microcarriers were stirred at 37°C for the indicated times. Released cells were counted in a haemocytometer. The live cell proportion was determined by standard dye exclusion techniques.
9A - Yield of cells   o - o   without EDTA treatment
                       • - •   with EDTA treatment
9B - Viability of cells o - o   without EDTA treatment
                        • - •   with EDTA treatment

Fig. 7 compares the growth of secondary bovine embryo kidney cells on denatured collagen-coated microcarriers and Cytodex 1. Cells attached and spread more rapidly on the coated microcarriers and the plating efficiency was clearly greater. As a result of this improved plating of cells on the denatured collagen-coated microcarriers, final cell yield was greater than in cultures using Cytodex 1. The secondary bovine embryo kidney cell culture contained a mixture of fibroblast-like and epithelial-like cells. Fig. 8 illustrates confluent cultures of human epithelial kidney cells growing on the denatured collagen-coated microcarriers.

A denatured collagen coat on the microcarriers is an important advantage when maximum recovery of cells from the microcarriers is required. The denatured collagen is susceptible to digestion by proteases. In preliminary experiments with Vero cells, recovery of cells from the coated microcarriers after treatment with standard trypsin procedures was greater than 95% with 95% viability. Using the same procedures with Cytodex 1 the recovery of cells was 80-85%. The specific protease, collagenase, can provide a method for harvesting cells with maximum retention of membrane integrity. This factor is particularly important when harvesting cells which bind strongly to microcarriers, e.g. human epithelial kidney cells (Flow 4000/clone 2) (see Fig. 8). Some typical results of harvesting this cell type from collagen-coated microcarriers are shown in Fig. 9. When confluent microcarriers were prewashed with EDTA (0.02% w/v) before enzyme treatment (Collagenase, 1 mg/ml, w/v), it was observed that cell yields increased dramatically (Fig. 9A) to 85% in 15 minutes, compared to 55% without EDTA treatment.

The viability of the harvested cells was greater than 90%, and was significantly higher when a prewash with EDTA was included in the harvesting procedure (Fig. 9B). Similar results were obtained with Vero and BSC-1 cells.

Table I: Removal of serum proteins from microcarrier cultures using Cytodex 1 and two alternative microcarriers. (Results kindly supplied by Dr. A.L. van Wezel.)

Sample*	Cytodex 1		Surface-charged microcarriers		Denatured collagen coated microcarrier	
	Albumin µg/ml	IgG µg/ml	Albumin µg/ml	IgG µg/ml	Albumin µg/ml	IgG µg/ml
Culture fluid	16	256	16	256	16	256
First wash	16	64	16	64	16	64
Second wash	16	8	2	2	1	2
Third wash	4	2	0.5	1	0.15	1

*Procedure: Secondary monkey kidney cells were grown for 9 days and cultures were washed with Medium 199 without serum (First wash). The cultures were resuspended in Medium 199 without serum and incubated overnight. The wash fluid was removed (Second wash). The culture was resuspended in Medium 199 (without serum) and infected with polio virus. After 3 days culture fluids were harvested (Third wash). Proteins were determined by counter-current electrophoresis.

Removing medium components from the microcarriers

In many microcarrier cultures and as part of production procedures it is important to be able to wash medium components or cell products from the culture. The alternative microcarriers described here had a much lower ion-exchange capacity than Cytodex 1 microcarriers and it was expected that it would be easier to remove proteins from cultures using these microcarriers. It was observed that only 4.3% of the total protein present in 100% newborn calf serum was adsorbed to Cytodex 1 microcarriers. Less than 2.7% of the protein adsorbed to the surface-charged microcarriers and less than 1.1% absorbed to the denatured collagen-coated microcarriers. Adsorbtion of protein was determined by elution at culture pH and ionic strength. The amount of protein adsorbed to the microcarriers from culture medium supplemented with the usual 10% serum was therefore very small.

Table I shows that it was possible to wash out serum proteins to a much greater extent from cultures using these alternative types of microcarrier than from cultures using Cytodex 1. The third wash in Table I corresponds to the harvest of virus-containing culture fluid when producing vaccines from microcarrier culture.

CONCLUSION

The alternative surfaces described here represent a new approach to the development of microcarriers and provide a way of improving the efficiency and possibilities of this culture technique. These alternative microcarriers are based on a cross-linked dextran matrix. The physical and chemical properties of such a matrix have been found to be optimal for microcarrier culture (10).

While Cytodex 1 microcarriers are suitable for a wide variety of cell types the alternative microcarriers could have advantages in certain situations. The ability to wash out serum proteins more easily than from previous microcarriers indicates that the culture products are likely to be more pure and these microcarriers could have a potential value in the preparation of live vaccines. The improved plating efficiency of cells, particularly on the denatured collagen coated microcarriers suggests that these microcarriers would be useful when growing cells normally difficult to grow in culture and when using very low numbers of cells having a low plating efficiency.

REFERENCES

1. P. Bornstein and H. Sage, 1980, Structurally distinct collagen types, Ann. Rev. Biochem., 49:957-1003.
2. E. Engvall and E. Rouslahti, 1977, Binding of soluble form of fibroblast surface protein, fibronectin to collagen, Int. J. Cancer, 20:1-5.
3. E. Engvall, E. Rouslahti and E.J. Miller, 1978, Affinity of fibronectin to collagens of different genetic types and to fibrinogen, J. Exp. Med., 147:1584-1595.
4. L.I. Gold and E. Pearlstein, 1980, Fibronectin-collagen binding and requirements during cellular adhesion, Biochem. J., 186:551-559.
5. D.J. Giard and R.J. Fleischaker, 1980, Examination of parameters affecting human interferon production with microcarrier-grown fibroblast cells, Antimicrobial Agents Chemotherap., 18:130-136.
6. F. Grinnell, 1978, Cellular adhesiveness and extra-cellular substata, Int. Rev. Cytol., 53:65-144.
7. F. Grinnell, M. Feld and D. Minter, 1980, Fibroblast adhesion to fibronogen and fibrin substrata: Requirement for cold insoluble globulin (plasma fibronectin), Cell, 19:517-525.
8. F. Gubenšek and S. Lapanje, 1968, Potentiometric titration studies of diethylaminoethyl dextran base, J. Macromol. Sci. Chem., A2:1045-1054.
9. M. Hirtenstein and J. Clark, 1980, Attachment and proliferation of animal cells on microcarriers (Cytodex^{TM} 1). In "Tissue Culture in Medical Research" eds. R.J. Richards, K.T. Rajan, pp.97-104, Pergamon: Oxford.
10. M. Hirtenstein, J. Clark, G. Lindren and P. Vretblad, 1980, Microcarriers for animal cell culture: a brief review of theory and practice, Develop. Biol. Standard, 46:109-116.

11. M. Höök, K. Rubin, Å. Oldenberg, B. Öbrink and A. Vaheri, 1977, Cold-insoluble globulin mediates the adhesion of rat liver cells to plastic petri dishes, Biochem. Biophys. Res. Comm., 79:726-733.
12. D.W. Levine, D.I.C. Wang and W.G. Thilly, 1979, Optimization of growth surface parameters in microcarrier culture, Biotechnol. Bioeng., 21:821-845.
13. M. Manousos, M. Ahmed, C. Torchio, J. Wolff, G. Shibley, R. Stephens and S. Mayyasi, 1980, Feasibility studies of oncornavirus production in microcarrier cultures, In Vitro, 16:507-515.
14. B. Meignier, H. Mougeot and H. Favré, 1980, Foot and mouth disease virus production on microcarrier-grown cells, Develop. Biol. Standard, 46:249-256.
15. B. Mered, P. Albrecht and H.E. Hopps, 1980, Cell growth optimization in microcarrier culture, In Vitro, 16:859-865.
16. B.J. Montagnon, B. Fanget and A.J. Nicolas, 1981, The large-scale cultivation of Vero cells in microcarrier culture for virus vaccine production. Preliminary results for killed poliovirus vaccine, Develop. Biol. Standard., 47:55-64.
17. Pharmacia Fine Chemicals, 1980, Microcarrier cell culture, Technical notes, 1, Uppsala, Sweden.
18. Pharmacia Fine Chemicals, 1974, Dextran Fractions, Dextran Sulphate, DEAE-Dextran. Defined polymers for biological research, Technical Book Series, Uppsala, Sweden.
19. L.M. Reid and M. Rojkind, 1979, New techniques for culturing differentiated cells: reconstituted basement membrane rafts, Methods. Enzymol., 58:263-278.
20. K.K. Sanford, W.R. Earle, V.J. Evans, H.K. Waltz and J.E. Shannon, 1951, The measurement of proliferation in tissue cultures by enumeration of cell nuclei, J. Natl. Cancer Res. Inst., 11:773-795.

21. A. von Seefried and J.H. Chun, 1981, Serially subcultivated cells as substrates for poliovirus production for vaccine, Develop. Biol. Standard., 47:25-33.
22. A.L. van Wezel, C.A.M. van der Velden-de Groot and J.A.M. van Herwaarden, 1980, The production of inactivated polio vaccine on serially cultivated kidney cells from captive-bred monkeys, Develop. Biol. Standard., 46:151-158.
23. D. Yaffe, 1973, Rat skeletal muscle cells. In "Tissue Culture: Methods and Applications", eds. P.F. Kruse, M.K. Patterson, pp. 106-114, Academic Press, New York.

INTERFERON PRODUCTION IN MICROCARRIER CULTURE OF HUMAN FIBROBLAST CELLS*

Victor G. Edy

Flow Laboratories Inc.
McLean, VA 22102

ABBREVIATIONS

MEM - Eagle's minimal essential medium, (Flow Labs.); DME - Dulbecco's modified Eagle's medium, (Flow Labs or Gibco); FBS - Fetal bovine serum (Flow Labs.); DEAE - Diethylaminoethyl; poly(I) poly(C) -Polyriboinosinic acid - polyribocytidylic acid homopolymer duplex. (P-L Biochemicals)

ABSTRACT

The microcarrier cell culture technique offers a convenient way of manipulating relatively large amounts of anchorage-dependent cultured mammalian cells in a relatively small volume. Other advantages of this system are ease of direct observation, and of monitoring and control of other culture parameters.

---

*Large portions of the work described in this article were supported by Contract No. NO 1-CM-07370 from the NCI.

The production of human fibroblasts interferon on microcarrier culture on a scale of up to 44 litres (very roughly equivalent to 1100 roller bottles) using a conventional induction with poly(I) poly(C), followed by superinduction by treatment with metalolic inhibitors, is described.

The key to the successful and economical large-scale production of anchorage-dependent cells or cell products is to find a method of conveniently and safely manipulating very large areas of growth substrate. Numerous systems have been developed to accomplish this aim; some relative scale-up possibilities and surface area:volume ratios (a critical determinant of the economic feasibility of a method) were discussed by Levine et al (1) and the use of some methods of large-scale culture in interferon production has been reviewed (Edy, to be published).

The microcarrier concept of growing anchorage-dependent cells on the surface of small beads suspended in growth medium, was first described by Van Wezel (2). Using commercially available ion-exchange materials, he found that only DEAE-Dextran beads (DEAE-Sephadex A50), coated with small amounts of nitrocellulose would support the attachment and growth of cells. Although Van Wezel was successful in using this coated-charged bead approach, and was able to operate on a quite large scale (100 liters or more of monkey kidney culture), considerable problems were experienced, with cells occasionally showing long lag times, significant loss of inoculum viability, and with cell detachment and death after relatively short periods in cultures. These problems could be ascribed either to cell starvation, with the beads sequestering a vital component(s) of the growth medium, or to an ill-defined "toxicity". Nevertheless, some successes were reported (for example see references 3-6).

Thilly and his coworkers examined the causes of this "toxicity" in their studies of optimal conditions for growth of cells on microcarriers. They found that the answer lay in the degree of substitution of the cross-linked dextran substrate with DEAE ion-exchange groups. By altering the reaction conditions, they were able to reproducibly generate microcarrier beads that supported the growth of cells of high density without any of the problems experienced earlier. These beads had a charge density of 1.8 to 2.1 meq. g^{-1} of unsubstituted Sephadex (1, 7-9). The optimal requirements for microcarriers are listed in Table 1. Several commercial companies now make beads meeting most or all of these requirements, (e.g. Biorad "Biocarriers", Flow "Superbeads", Nunc "Biosilon" and Pharmacia "Cytodex") and new further improved microcarriers are promised by various manufacturers. Conditions for microcarrier culture on some of these various beads have been described by their manufactures (10-12).

Table 1. Desirable Properties of Microcarriers

1. Size 100 - 200μm.
2. Narrow Size Range Distribution.
3. Density 1.02 - 1.05
4. Suitable Charge or Other Surface Properties
5. Uniform Shape, Spherical?
6. Steam Sterilisable
7. Transparent
8. Non-Toxic

For interferon production on microcarriers, we have chosen to use the FS-4 cell line, a human diploid fibroblast line of finite lifespan, kindly supplied by Dr. Jan Vilcek of New York University. Because of difficulties in obtaining reproducible growth after subcultivation of microcarrier cultures of these cells by trypsinisation, particularly on a large scale, the cells for interferon production are grown up in roller bottles and only transferred to microcarriers for the final growth phase. Even with this constraint, under these restraints, the use of microcarriers offers significant cost savings, in labor and materials.

The cells in rollers are grown in MEM plus 10% FBS, and can be subcultivated by trypsinisation followed by a 1:2 or 1:4 split, every week or so. When sufficient cells have been amassed to initiate a microcarrier culture, the cells are split one final time, and grown out again in rollers. Before these final rollers are confluent, the cells are again trypsinised and mixed with microcarriers in DME + 5% FBS. Rapidly growing cells seem, at least in our hands, to be essential for good initiation on microcarriers; the reason for this is under investigation. The microcarrier cell mix generally contains $4g.1^{-1}$ of Superbeads (weight calculated as the original unsubsituted Sephadex, i.e. roughly 5g of DEAE-dextran) and 400×10^6 cells. $Litre^{-1}$. The cultures, in 6 litre spinner vessels, are stirred immediately. It has been reported that initial intermittent stirring, with long rest periods, followed by subsequent dilution of the culture, gives more uniform attachment of cells to beads, and an improved final cell yield (12,13). However, in our experience if cultures were set up at final volume, and stirred continuously from initiation, no deleterious effects were seen. If one assumes that a cell-microcarrier suspension is uniformly and continuously mixed and that all beads are eqully attractive to all cells, then before any cell division has occured one would expect the distribution of cells on beads to be Poissonian in form. To determine the actual distribution, samples were taken from microcarrier cultures 3-4 hrs after initiation, when attachment is complete but it is not expected that significant cell division will have occured. These samples were then fixed, stained and photographed, and the cells on each bead counted. A total of 132 beads were scored in this fashion, and the number of cells.bead^{-1} compared with the theoretical distribution, as shown in Figure 1. Statistical comparison of the observed and expected distributions by the χ^2 technique showed that they did not differ significantly.

Cell growth commences within 24-72 hrs after culture initiation, and is monitored both by simple observation of stained samples, as well as by nuclear counts (14).

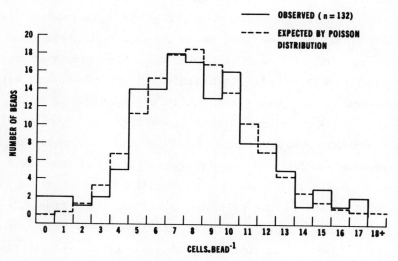

Fig. 1. Observed and expected distribution of cells.bead^{-1} in a freshly inoculated microcarrier cultures. (For details see text).

In dense, rapidly growing microcarrier cultures, pH shifts can be problemmatical, for its seems that the cells rapidly metabolise much of the available glucose and convert it to lactate. This can be avoided by substituting, on a mole-for-mole base, fructose for the glucose generally included in the medium (Imamura et al in press). Growth curves and pH plots for FS-4 cells grown in conventional and fructose-containing DME (HighGEM, Flow Laboratories Inc.) are shown in Figure 2. Preliminary experiments show that substitution of fructose for glucose has no significant effect on interferon yields in microcarrier cultures.

In our current large-scale production, however, the cultures are not held long enough in spinner for pH shifts to become dramatic. Instead, once it is established that the microcarrier cultures are growing well and are sterile, groups of 5 to 12 spinners are pooled, and placed in a fully instrumented fermentor. The units we use ("Bilthoven Novopaljas", Contact-Roestvrijstaal b.v., The Netherlands) monitor and control temperature, pH, pO_2 and agitation. Some slight modifications have been found necessary to make these fermentors suitable for our use. The most significant changes are the replacement of all 304 Stainless Steel components with 316 grade steel, and the substitution of multiple (2 or 3, depending on culture volume) marine propeller-type impellers for the disc turbine normally supplied. No medium reformulation in these fermentor cultures is necessary, as the pH is monitored by an immersed pH electrode, which controls the CO_2 content of the gas mix used to aerate the culture. Aeration is by simple head space gas exchange; sparging of gas is unnecessary with these cells, gives significant foaming problems, and is distinctly deleterious to the appearance and growth of the cultures.

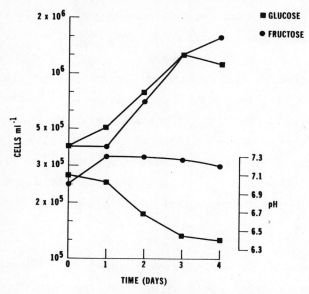

Fig. 2. Growth of FS-4 cells and pH changes in DME containing 25 mH glucose or fructose. (For details see text).

In the fermentors, the cells will grow to about 1.2×10^6 cells ml^{-1} although a density about 80% of this is more commonly achieved. At this point, the cells appear to be confluent, although it is not know whether in fact growth is surface limited.

The induction techniques used are based on the work of Giard et al (15,16), and are described elsewhere in more detail (17). Basically, once cell multiplication has ceased the cultures are primed for 16-20 hrs with 50 I.U. ml^{-1} of interferon, then induced with 50 µg ml^{-1} of poly(I), poly(C) and 10µg ml^{-1} of cycloheximide. After 4 hrs, 1µg ml^{-1} of actinomycin D is added, and the cells incubated for a further 2 hrs. They are then extensively washed and finally refed with DME plus human plasma protein fraction.

Currently, we are evaluating the benefits of holding the cells at 37° throughout the production period, as compared with incubating for 1 hr at 37°, followed by up to 48 hrs at 30°; shown by Fleischacker and Giard (in press) to give significantly better yields. To date, no difference has been observed, although it is difficult to reproduce in the large volume fermentor the exact temperature profile experienced by cells in small spinner flasks.

The bulk of the interferon produced is released into the supernatant; however about 10% more of the interferon remains associated with the beads. This is released slowly into production medium, originally leading us to believe in a "second" or late production. However, it can be rapidly washed off the beads by serum-free medium or by saline, which points rather to some loose binding of IFN by the beads. Significant alteration of pH or ionic strength does not release further interferon.

This system is still not fully controlled, for we experience still quite considerable run-to-run vairation in yields. Peaks in excess of 20,000 I.U. ml^{-1} have occasionally been observed, but a more reasonable mean yield is 5000-6000.

ACKNOWLEDGEMENT

W.G. Thilly, D.I.C. Wang and D.J. Giard, of MIT, and their coworkers, were of considerable assistance in this work.

REFERENCES

1. Levine, D.W., Thilly, W.G. and Wang, D.I.C. 1978. New microcarriers for the large scale production of anchorage-dependent mammalian cells, Adv. Exptl. Med. Biol. 110:15-23.
2. Van Wezel, A.L. 1967. Growth of cell-strains and primary cells on microcarriers in homogenous culture, Nature 216:64-65.
3. Van Hemert, D., Kilburn, D.G. and Van Wezel, A.L. 1969. Homogenous cultivation of animal cells for the production of virus and virus products, Biotechnol Bioeng 11:875-885.
4. Van Wezel, A.L. 1973. Microcarrier culture of animal cells In: Kruse, P.F., Patterson, M.K. (eds.) Tissue Culture, methods and applications. Academic Press, New York p. 372.
5. Horng, C. and McLimans, W. 1975. Primary suspension culture of calf anterior pituitary cells on a microcarrier surface, Biotechnol. Bioeng. 17:713-732.
6. Spier, R.E. and Whiteside, J.P. 1976. The production of foot-and-mouth disease virus from BHK 21 C13 cells grown on the surface of glass spheres, Biotechnol. Bioeng. 18:649-657.
7. Levine, D.W., Wang, D.I.C. and Thilly, W.G. 1977a. Optimising parameters for growth of anchorage dependent mammalian cells in microcarrier cultures. In: Acton, R.T., Lynn, J.D. (eds.) Cell culture and its application. Academic Press, New York, p. 191.
8. Levine, D.W., Wong, J.S., Wang, D.I.C. and Thilly, W.G. 1977b. Microcarrier cell culture: New methods for research-scale application, Somatic Cell Genet. 3:149-155.

9. Levine, D.W., Wang, D.I.C. and Thilly, W.G. 1979. Optimisation of growth surface parameters in microcarrier cell culture. Biotechnol. Bioeng. $\underline{2}$:821-825.
10. Nielsen, V. and Johansson, A. 1980. Biosilon, optimal culture conditions and various research scale culture techniques, Develop Biol. Stand. $\underline{46}$:313-136.
11. Hirtenstein, M., Clark, J., Lindgren, G. and Vretblad, P. 1980. Microcarriers for animal cell culture: A brief review of theory and practice, Develop. Biol. Stand. $\underline{46}$:109-116.
12. Clark, J., Hirtenstein, M. and Gebb, C. 1980. Critical parameters in the microcarrier cultures of animal cells, Develop. Biol. Stand. $\underline{46}$:117-124.
13. Clark, J.J. and Hirtenstein, M.D. 1981. High yield culture of human fibroblasts on microcarriers: A first step in production of fibroblasts-derived interferon (human beta interferon) J. Interferon Res. $\underline{1}$:391-400.
14. Sanford, K.K., Earle, W.R., Evans, V.J., Waltz, K.H. and Shannon, J.E. 1951. Measurement of proliferation in tissue cultures by enumeration of cell nuclei. J. Natl. Cancer Inst. $\underline{11}$:773-795.
15. Giard, D.J., Loeb, D.H., Thilly, W.G., Wang, D.I.C. and Levine, D.W. 1979. Human interferon production with diploid fibroblast cells grown on microcarriers, Biotechol. Bioeng. $\underline{21}$:433-442.
16. Giard, D.J. and Fleischaker, R.J. 1980. Examination of parameters affecting human interferon production with microcarrier-grown fibroblast cells, Antimicrob. Ag Chemother. $\underline{18}$:130-136.
17. Edy, V.G., Augenstein, D.C., Edwards, C.R., Cruttenden, V.F. and Lubiniecki, A.S. (in press). Large scale tissue culture for human IFN-β production. Texas Rep. Med. Biol.

IN VITRO AND IN VIVO EFFECTS OF ENDOTHELIAL CELL-DERIVED GROWTH FACTOR

Corinne M. Gajdusek[1], and Sandra A. Harris-Hooker[2]

[1]Department of Pathology
University of Washington
Seattle, Washington 98195
[2]Mallory Institute of Pathology
Boston, Massachusetts 02118

ABSTRACT

Endothelial cells in culture produce a growth factor with an apparent molecular weight of 10,000 to 30,000 daltons. This material, the endothelial cell-derived growth factor (ECDGF), can replace the requirement of 3T3 and smooth muscle cells for exogenous mitogens. ECDGF, added to plasma-derived serum, a platelet deficient preparation of plasma, supports growth of these cells to confluency. Lyophilized preparations of the ECDGF stimulate migration of smooth muscle cells into an _in vitro_ wound. The same material induces an angiogenic response when implanted on the chick chorioallantoic membrane. Thus two requirements of an angiogenic response, migration and proliferation, are supplied by endothelial cell products.

Abbreviations: PDS-plasma-derived serum; PBS-Phosphate-buffered saline; SMC-smooth muscle cells; BAE-bovine aortic endothelial cells; ECDGF-endothelial cell-derived growth factor; CAM-chorioallantoic membrane; TCA-tricholoacetic acid; ECCM-endothelial cell conditioned medium; SMCCM-smooth muscle cell conditioned medium; SDS-sodium dodecylsulfate

INTRODUCTION

Many species of endothelium are now grown in vitro, including large vessel endothelia derived from aorta of cow (1-3), pig (4), guinea pig (5), and goat (unpublished observations), bovine and human pulmonary artery and vein (6), human umbilical vein (7-9), and capillary endothelium from rat (10), human (11-12), cow (11,13), and cat (14). This paper will deal exclusively with large vessel bovine endothelium. Excellent reviews are available on the current status of capillary (15) and other large vessel endothelium in vitro (15-16).

Most endothelia from large vessels show certain growth characteristics and morphological features which distinguish this cell type from fibroblast-like cells. Growth is characterized by the formation of a non-overlapping monolayer at confluency (8). Although confluent cells cannot be stimulated to divide upon addition of serum (17), endothelium will migrate and replicate when the continuity of the monolayer is disrupted (18-19). In vivo this can occur when tissue is traumatized (20-21), at focal sites of stress in vessels (23), and during inflammation (20,23). Secondly, endothelium proliferates equally as well in whole blood serum and in plasma-derived serum (24-26), which is deficient in the platelet-derived growth factor (27). In contrast, 3T3 cells (28), glial fibroblasts (29), and smooth muscle (30) have requirements for exogenous growth factors and do not proliferate when cells are exposed to plasma. This suggests that factors required by endothelial cells for growth differ from those factors required by these other cell types.

Primary cultures of endothelium are easily obtained from large vessels by enzymatic perfusion of the lumen using either crude collagenase (8) or trypsin-EDTA solutions (31). Many cultures maintained in serum become overgrown by secondary cell types, usually identified as smooth muscle by morphological criteria (17). The smooth muscle cell contamination probably results from prolonged vessel perfusion and digestion of the subendothelial connective tissue. In our laboratory, we observed that primary bovine aortic endothelial cultures, seeded directly into plasma, also showed this secondary overgrowth by smooth muscle. This suggested that the endothelia were producing factors that supported the replication of the smooth muscle (32). This paper presents some of our observations on the biological and biochemical aspects of this material.

MATERIALS AND METHODS

Development of Cell Lines

Culture techniques have been previously described (3) and were adapted from procedures introduced by Gimbrone et al. (8) and Jaffe et al. (7). In brief, segments of bovine thoracic aorta were obtained from a local meat processing plant. The segments were rinsed in phosphate-buffered saline (PBS), transported to the laboratory in a sterile container, and processed within two hours. Collaterals were ligated and the caudal end of the vessel clamped with hemostats. The vessel lumen was filled with a solution of collagenase (1 mg/ml, Worthington Biochemical Corporation, CLSII) dissolved in PBS with calcium (0.15 mg/ml) and magnesium (0.10 mg/ml). After five minutes incubation at 37°C, the solution was centrifuged in sterile conical tubes for 5 minutes at 1,000 rpm in a table top centrifuge. Subsequent incubations produced cultures that showed increasing numbers of smooth muscle. The pellet, containing bovine aortic endothelium (BAE) and some blood cells, was gently aspirated in Waymouth's medium (GIBCO) supplemented with either 10% fetal calf serum (GIBCO) or 10% plasma-derived serum (PDS), 2.5 mmoles glutamine, 100 units/ml pennicilin, 1 ug/ml streptomycin, and sodium bicarbonate. Sera were heat inactivated at 56°C for 30 minutes before use. Falcon plastic tissue culture ware was used throughout these experiments. Cells could be passaged for approximately 50 cell doublings and were used from passages 2 through 10.

Alternately, to avoid smooth muscle contamination and to accelerate tissue processing, the aortic intima was gently scraped with a scalpel. The scrapings were transferred to petris in medium, as above, and placed in a 37°C incubator perfused with 5% CO_2.

Smooth muscle (SMC) were obtained by the explant technique of Ross (33). 1 mm pieces of the intima, previously denuded of endothelium, were placed in a petri and surrounded by a drop of Waymouth's medium containing 10% fetal calf serum. After the tissue adhered to the plastic, sufficient medium was added to cover the explants. When SMC outgrowth occurred, usually by 7 days, cultures were passaged in a 0.05% trypsin- 0.54 mm EDTA, pH 7.4 solution for 5 minutes and transferred to a 75 cm^2 flask. SMC cultures could be transferred for approximately 10 passages and were used from passages 2 through 7.

Swiss 3T3 cells, Balb/c-3T3 cells, and human dermal fibroblasts were maintained as previously described (32).

Preparation of Plasma-derived Serum and Lymph

Blood was obtained immediately after slaughter, collected in polyethelene beakers, and immediately distributed to 250 ml polyethelene centrifuge bottles containing sodium citrate at a final concentration of 0.38%. Blood was chilled in an ice bucket and processed according to the procedures of Vogel et al. (34). CM-Sephadex was used to remove trace amounts of the platelet-derived growth factor. Concentration was done using an Amicon filtration apparatus with PM-10 filters and plasma was dialyzed against PBS in Spectrapor 1 dialysis tubing (6,000-8,000 mw cutoff, Spectrum Industries, CA). The plasma was heat inactivated at 56°C for 30 minutes and sterile filtered. Growth promoting activity was assayed by the ^{3}H-thymidine incorporation into Swiss 3T3 cell DNA (35).

Lymph from goat thoracic duct was provided by Dr. R. Winn, Virginia Mason Research Institute, Seattle, WA. Lymph was collected in the presence of 0.38% sodium citrate, centrifuged, and dialyzed against PBS in 1,000 mw cutoff dialysis tubing. Lymph was heat inactivated, sterile filtered, and stored at -20°C.

Cell Growth and Migration Assays

Trypsin dispersed cells were plated at a density of 5×10^3 cells per cm^2 in multiwells (2.1 cm^2 surface area) in 5% PDS (SMC) or 5-10% PDS or lymph (BAE). 24 hours after cell plating, test media were added and changed every 2 days. Cell number was determined on a Coulter Counter. Migration assays were done by wounding confluent monolayers of SMC with a razor blade. Wounded cultures were refed with test media, irradiated at 2500 rads using a cobalt radiation source, and fixed in buffered formalin 48 hours later. The distance the cells migrated into the wound was measured using an ocular micrometer.

An _in vivo_ assay was used to assess growth promoting activity present in serum-free conditioned medium. Either Swiss 3T3 cells or SMC were plated in multiwells at 5×10^3 cells per cm^2 in 5% fetal calf serum. When cells were quiescent, test media were added. ^{3}H-thymidine incorporation (1 uCi/ml ^{3}H-thymidine, New England Nuclear, NET 027, 6.7 Ci/mM) was measured by activity in acid-precipitable material between 18 to 20 hours after addition of test media. Cells were washed twice in cold 10% TCA for 10 minutes, and solubilized in 0.5N NaOH. This material was counted in a scintillation counter in Aquasol II (New England Nuclear).

Cell Growth on Beads

For determining angiogenic potential of living cells in the chorioallantoic membrane assay, cells were plated and grown on DEAE Sephadex microcarrier beads (Flow Laboratories, Inc., Inglewood, CA). Aliquots of beads were implanted on the membrane and procedures followed as below for examination and quantitation of response.

Chorioallantoic Membrane Assay

The chorioallantoic membrane (CAM) of the chick embryo was used to assess angiogenic potential of cells and cell products. A one cm^2 window was cut into the shell of 10 day old eggs and the CAM dislocated by the false air sac technique (36). Microcarrier beads with cells or lyophilized preparations of cell free material, adsorbed to Sephadex G-200 (Pharmacia), were applied to the CAM using a pipette.

Lyophilized conditioned medium was solubilized in serum free medium, diluted into concentrations of 10 to 80 ug protein, as determined by the Lowry procedure (37), adsorbed to 3 mg Sephadex, and implanted on the CAM. Membranes were examined after 4 days.

Quantitative techniques were developed to measure the degree of angiogenesis induced by cells and cell products (38). Membranes were excised, laid flat, and photographed under standard magnification. The vessel density was determined by covering each photograph with a grid inscribed on a sheet of clear acetate. The grid contained a number of concentric circles (0.5 mm apart) and the number of intersecting vessels was counted as a function of distance from the implant.

Biochemical Determinations

Sensitivity to trypsin and heat was as previously described (32). Acetone precipitation was carried out at -20°C overnight with added albumin as a carrier (50 ug/ml, final concentration). Polyacrilamide gel electrophoresis was performed according to the procedures of Laemli (39). The gel was cut into 1 cm strips, crushed, and protein eluted into 1 mM NH_4HCO_3. Soluble material was acetone precipiated as above. Pure bovine thrombin (2000 units/mg) was obtained from Dr. E. Davies, University of Washington, Seattle, WA. Sephadex G-75 chromatography was done at room temperature, using 1 M formic acid as the solvent system.

RESULTS

Endothelial and Smooth Muscle Proliferation

The endothelium in vivo is normally continuously exposed to plasma or, in the lymphatic system, to lymph. In contrast, smooth muscle and fibroblasts "see" only a filtrate of plasma, the interstitial fluid. It is postulated that, during periods of stress, when the endothelial continuity is disrupted, these cells may be exposed to factors released by platelets (27), other blood cells (23,40-41), and perhaps tissue cells (42).

When the growth promoting activity of serum, plasma-derived serum (PDS), and lymph was assayed <u>in vitro</u>, bovine smooth muscle (SMC) proliferated in fetal calf serum but not in PDS or lymph (Figure 1a). Others have reported similar findings for human SMC exposed to PDS (9,30) and it is suggested that the platelet

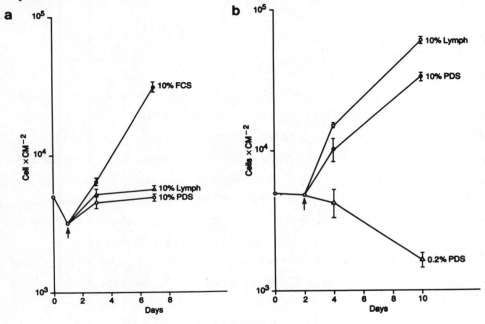

FIGURE 1 a & b

Figure 1a Smooth muscle cell proliferation. Cells were plated at 5×10^3 cells per cm^2 in 5% plasma-derived serum on day zero. Test media, added at arrow, consisted of 10% fetal calf serum (●——●); 10% lymph (△——△); or 10% PDS (○——○). Bars represent standard deviation from the mean. Media were changed every other day. Abscissa: days after plating; ordinate: cell number X cm^{-2}.
Figure 1b Bovine endothelial cell proliferation. Cells were plated at 5×10^3 cells on day zero in 5% PDS. Test media, added at the arrow, consisted of 10% lymph (○——○); 10% PDS (●——●); or 0.2% PDS (△——△). Bars represent standard deviation from the mean. Media were changed every other day. Abscissa: days after plating; ordinate: cell number X cm^{-2}.

derived factors are responsible for conveying competence to progress through the cell cycle (43). In contrast, the bovine aortic endothelial cells (BAE) readily proliferated from sparse densities in both PDS and in lymph (Figure 1b). When lymph was assayed for the presence of inhibitors, by combining lymph with serum, no inhibition of ^{3}H-thymidine incorporation into quiescent 3T3 cells was evident (data not shown).

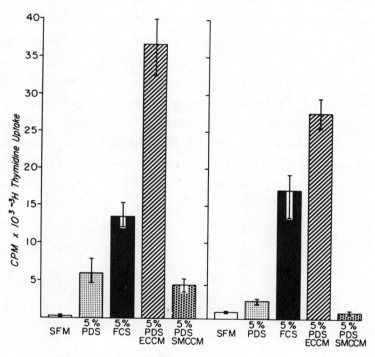

Figure 2

Stimulation of thymidine incorporation by endothelial cell conditioned medium. Left panel: smooth muscle; right panel: 3T3 cells. Quiescent cells were fed with 200 ul test media diluted in 300 ul serum-free medium. Thymidine incorporation was measured at 18 to 20 hours with 1 uCi/ml ^{3}H-thymidine. Bars represent standard deviation from the mean. SFM: serum-free medium; 5% PDS: 5% plasma-derived serum; 5% FCS: 5% fetal calf serum; 5% PDS ECCM: medium containing 5% PDS, preincubated with endothelial cells for 48 hours; 5% PDS SMCCM: medium containing 5% PDS, preincubated with smooth muscle for 48 hours. Ordinate: thymidine incorporation into acid-precipitable DNA.

We then assayed PDS-containing medium, after exposure to confluent cultures of BAE and SMC, for ability to stimulate DNA synthesis in quiescent SMC and 3T3 cell cultures. Figure 2 confirmed in this assay that fetal calf serum, but not PDS, was able to stimulate thymidine incorporation into acid precipitable material. The medium preconditioned by exposure to endothelium stimulated thymidine incorporation at levels 6 to 10 times greater than nonconditioned medium or medium conditioned by SMC cultures. The endothelial cell conditioned medium likewise supported 3T3 and SMC proliferation from subconfluent densities to confluence (32).

If the endothelium was producing a growth stimulating factor, rather than simply altering plasma components or releasing residual platelet-derived growth factor that might have been present in PDS, this effect should also be detected in serum-free conditioned medium. Cultures of both BAE and SMC were exposed to serum-free medium for increasing periods of time. The medium was

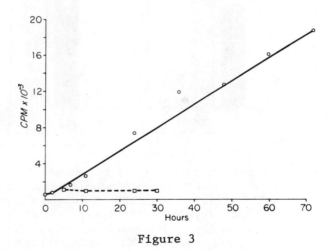

Figure 3

Growth promoting activity in medium as a function of time. Confluent cultures of endothelium and smooth muscle were fed with serum-free medium. Media were collected from replicate cultures at times indicated. Activity was determined by the Swiss 3T3 cell assay. 0: serum-free endothelial cell conditioned medium; ☐ : serum-free smooth muscle cell conditioned medium.

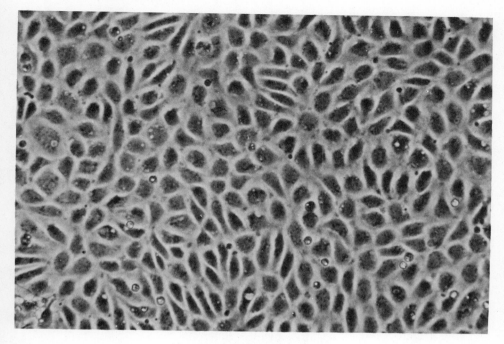

Figure 4

Endothelium after three months in serum-free medium. Cells were maintained in serum-free medium with twice weekly changes in medium. Phase contrast micrograph: 100X.

assayed as before (Figure 3). Growth promoting activity from endothelial cell culture medium was detected by 2 hours and continued to increase linearly with time. Conditioned serum-free medium from SMC cultures, tested as long as 30 hours after exposure to serum-free medium, had no growth promoting activity. Subsequent experiments, which assayed the growth promoting activity in medium from endothelium three months after continuous exposure to serum-free medium, continued to give the same results. Figure 4 shows the morphological appearance of endothelium after prolonged exposure to serum-free medium. Cultures were healthy by morphological criteria and could be replated at subconfluent densities in growth medium with no loss of ability to proliferate.

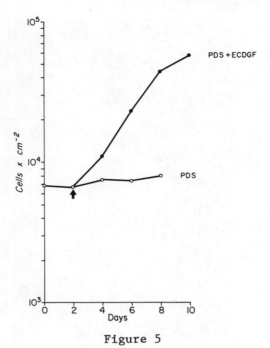

Figure 5

Smooth muscle cell proliferation in endothelial cell-derived growth factor (ECDGF). Cells were plated at 5 X 10^3 cells per cm^{+2} on day zero in 5% plasma-derived serum. Test media, added at arrow, consisted of 5% PDS containing lyophilized preparations of serum-free endothelial cell conditioned medium (●——●) or 5% PDS alone (0——0). ECDGF was reconstituted in 5% PDS to the original volume collected initially from endothelial cell cultures.

Biochemical Characterization

Preliminary data on the characteristics of the growth promoting activity showed that the activity was non-dialyzable (using both 8,000 and 50,000 mw cutoff dialysis tubing) when dialyzed against low ionic strength buffer (1 mM NH_4HCO_3). The non-dialyzable material could also be lyophilized with no loss of activity. Figure 5 indicates the ability of lyophilized material, reconstituted in 5% PDS, to promote SMC proliferation. Further characterization (Table I) indicated the molecule was trypsin sensitive but insensitive to thrombin, suggesting the activity was

TABLE I
Effects of Various Treatments on the Activity of ECDGF

Treatment	Activity (percent of Control)[a]
Trypsin (50 ug for 2 hours)	5.4
Thrombin (1 ug for 24 hours)	100
pH3 through 10 (24 hours)	100
56°C for 30 minutes	100
100°C for 3 minutes	65
100°C for 3 minutes in 0.1% SDS[b]	100
Dithiothreitol (40 uM)	20

[a] Incorporation of ^{3}H-thymidine into Swiss 3T3 cell DNA
[b] SDS removed by acetone precipitation of protein. Recovery of activity is with respect to acetone control. Acetone control is 85% recovery.

a protein or associated with protein. The activity was heat stable at both 56°C and 100°C in the presence of SDS and could be precipitated with acetone. The activity was sensitive to dithiothreitol, indicating the presence of disulfide bonds, and the material was stable to a wide range of pH.

Acid treatment of the material yielded a lower molecular weight activity which was eluted in the retained volume of a G-75 sephadex column. Due to the stability of the activity in SDS, polyacrilamide gel electrophoresis of lyophilized medium, subsequent elution, and precipitation located the activity in the 10,000 to 30,000 mw range of the gel. Further characterization is in progress.

Biological Characterization

In addition to stimulating 3T3 and SMC replication, serum-free endothelial cell conditioned medium and lyophilized preparations were able to stimulate migration of these cells into a wound _in vitro_ in the absence of serum. Other studies (44-45) have indicated requirements for serum components for migration proliferation. Using the _in vitro_ wound technique (45-46), minimal migration was noted in both serum-free medium and in conditioned serum-free medium obtained from SMC cultures (Figure 6). The serum-free endothelial cell conditioned medium and the lyophilized preparations were able to stimulate migration two-fold compared to control cultures, indicating a relationship between stimulation of migration and either the growth promoting activity in endothelial cell conditioned medium or other endothelial cell products.

A variety of cell types, including endothelium, Balb/c-3T3, Swiss 3T3, SMC, and human dermal fibroblasts, and lyophilized preparations of serum-free conditioned medium were assayed for angiogenic potential, using the chick chorioallantoic membrane assay (CAM). Equivalent numbers of confluent cells, grown on microcarrier beads, were transferred to the CAM and the degree of vascularization directed towards the implant quantified. Figure 7

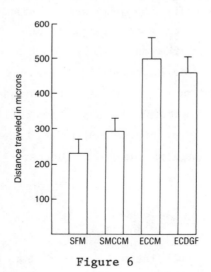

Figure 6

Smooth muscle cell migration. SFM: serum-free medium; SMCCM: serum-free smooth muscle cell conditioned medium after 48 hours exposure to smooth muscle; ECCM: serum-free endothelial cell conditioned medium after 48 hours exposure to endothelial cultures; ECDGF: lyophilized preparations of endothelial cell conditioned medium. Ordinate: distance leading cell traveled into the in vitro wound, measured in microns.

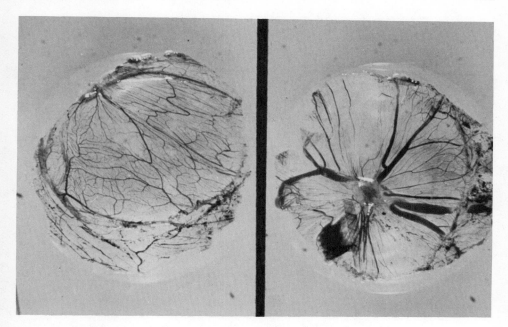

Figure 7

Vascular response in the chorioallantoic membrane. Left panel: control membrane; right panel: vascular response induced by endothelial cells implanted on to the membrane after 4 days. X 120

shows a control membrane with beads compared to a membrane showing positive response induced by endotheial cells.

These responses were quantified by the grid method, described in Materials and Methods. Analysis of the response elicited by the endothelium showed an increased number of vessels intersecting the grid at increasing distance from the implant whereas control values increased only slightly. SMC also invoked a positive response but the values were 50 to 60% less than those induced by equal numbers of endothelial cells. While neither human dermal fibroblasts nor Swiss 3T3 cells induced a response above control values, Balb/c-3T3 cells did, confirming previous reports (47). When varying amounts of the endothelial cell-derived growth factor

TABLE II

Angiogenic Response in the CAM

Distance in mm	Number of Intersecting Vessels ± S.E.		
	Control	Endothelium	Smooth Muscle
0.5	22±2	70±5	42±7
1.0	23±1	87±3	48±7
1.5	25±3	96±5	50±8
2.0	26±1	105±9	66±7
2.5	30±3	110±6	73±5

Both endothelium and smooth muscle were grown to confluence on microcarrier beads and implanted on the chick chorioallantoic membrane (CAM) in equal numbers. The distance represents the number of vessels intersecting the grid at increasing distances from the implant, in increments of 0.5 mm.

TABLE III

Effect of Increasing Amounts of Conditioned Medium on Angiogenesis

Protein (ug)	Number of Intersecting Vessels ± S.E.[a]	
	ECDGF[b]	SMCCM[c]
0	30±3	30±3
10	59±1	33±1
20	66±2	50±4
40	93±2	60±3
80	110±6	59±2

[a] Values are for vessel number at 2.5 mm from the implant.
[b] Lyophilized preparations of endothelial cell conditioned medium. Each value represents 12 assays.
[c] Lyophilized preparations of smooth muscle cell conditioned medium. Each value represents 7 assays.

(ECDGF) were assayed, a linear dose response was found in the number of vessels intersecting the grid at 2.5 mm from the implant (38). Again, SMC conditioned medium (SMCCM) was less effective (TABLE II and III). Histological examination of the membranes showed no evidence of an inflammatory response.

DISCUSSION

Bovine aortic endothelial cells in culture produce a growth promoting activity, the endothelial cell-derived growth factor (ECDGF, 32) that supports migration (38) and replication of 3T3 and smooth muscle cells in the absence of other defined mitogens. The active material is trypsin sensitive and has an apparent molecular weight of 10,000 to 30,000 daltons. Endothelium derived from swine (48) aorta, rabbit and goat aorta, bovine mesenteric vein and pulmonary artery, and human umbilical vein (all unpublished results) also produce this activity. In contrast, conditioned serum-free medium collected from non-endothelial cells, including bovine smooth muscle, Swiss 3T3, human and bovine dermal fibroblasts, do not have similar properties.

Neovascularizaiton requires at least two components that are supplied by endothelial cell products: migration and replication. These and other studies (49-50) suggest that the target cell for our ECDGF preparations is non-endothelial. We have been unable to demonstrate a stimulatory effect of the ECDGF on aortic endothelial cell growth *in vitro*. In another report (49), we have described a low molecular weight material, found in both endothelial cell and smooth muscle cell conditioned medium, that does support endothelial cell proliferation in very low serum concentrations (0.2%). Others have reported on both high and low molecular weight (52-53) angiogenic materials that stimulate endothelial cell proliferation *in vivo* (54-55) and *in vitro* (12,24), although *in vitro* effects have been difficult to demonstrate (56). The activated macrophage, for instance, induces a neovascular response in the guinea pig cornea (54) and also releases a growth stimulating material, the macrophage-derived growth factor, that promotes DNA synthesis in smooth muscle, fibroblasts, and subconfluent endothelium (40). Folkman and his colleagues (57) have postulated that angiogenic materials may act directly, by stimulating endothelium. Indirect induction would include secondary mediators, such as formic acid or silica, which would attract macrophages and subsequently stimulate endothelial cell proliferation.

Our observations suggest an extension of these concepts. Although tumor angiogenic factors directly stimulate endothelial cell proliferation and migration (11-12), in the non-tumor induced response this may be mediated through resident tissue cells. Mitogens, such as the ECDGF and the platelet-derived growth factor, released at an injury site, may stimulate non-inflammatory resident tissue cells to elaborate mediators, similar to the low molecular weight tumor angiogenesis factor or the low molecular weight material we have described in smooth muscle and endothelial cell culture medium (49). These mediators would effect the response by endothelium in neovascularization. ECDGF may also stimulate migration and proliferation of pericytes and other connective tissue cells, providing a pathway for migration of capillary endothelium. A second possibility is that the target cell for neovascularization is a subpopulation of endothelium. Schwartz (3), Cotta-Pereira et al. (58) and McAuslan et al. (59) have described variant endothelial cells that grow over or under confluent endothelial monolayers. These cells show a different synthetic profile than monolayer endothelium, more similar to that of smooth muscle and capillary cells (60). It is feasible that these variant cells respond differently to exogenous mitogens. Indeed it is possible that capillary endothelium also responds to exogenous mitogens.

The factor derived from endothelium therefore could have an important role during the wound healing process, acting as a mediator. Likewise, it may play a role in disease states, such as atherosclerosis. During atherogenesis, both smooth muscle migration and proliferation play a central role in lesion formation. It is postulated in the reaction to injury hypothesis that the central event in lesion formation is endothelial injury with subsequent adhesion and release of platelet components, including the platelet-derived growth factor (27). This mitogen has been shown to be chemotactic for smooth muscle (61). The ECDGF may also contribute to this process.

ACKNOWLEDGEMENTS

We thank Drs. R. Ross, P. DiCorleto, T. Wight and S.M. Schwartz for helpful discussions and support.

REFERENCES

1. F.M. Booyse, B.J. Sedlak and M.E. Rafelson, 1975, Culture of arterial endothelial cells. Characterization and growth of bovine aortic cells, Thrombos. Diathes. Haemorrh. (Stuttg.), 34:825-829.
2. D. Gospodarowicz, J. Moran, D. Braun and C. Birdwell, 1976, Clonal growth of bovine vascular endothelial cells: fibroblast growth factor as a survival agent, Proc. Natl. Acad. Sci. USA, 73:4120-4124.
3. S.M. Schwartz, 1978, Selection and characterization of bovine aortic endothelial cells, In Vitro, 14:966-980.
4. D.N. Slater and J.M. Sloan, 1975, The porcine endothelial cell in tissue culture, Atherosclerosis, 21:259-272.
5. S.H. Blose, M.L. Shelanski and S. Chacko, 1977, Localization of bovine brain filament antibody on intermediate (100 Å) filaments in guinea pig endothelial cells and chick cardiac muscle cells, Proc. Natl. Acad. Sci. USA, 74:662-669.
6. A.R. Johnson, 1980, Human pulmonary endothelial cells in culture. Activities of cells from arteries and cells from

veins, J. Clin. Invest., 65:841-850.
7. E.A. Jaffe, R.L. Nachman, C.G. Becker and R.C. Minick, 1973, Culture of human endothelial cells derived from umbilical veins, J. Clin. Invest., 52:2745-2756.
8. M.A. Gimbrone, R.S. Cotran and J. Folkman, 1974, Human vascular endothelial cells in culture. Growth and DNA synthesis, J. Cell Biol., 60:673-684.
9. G. Thorgeirsson and A.L. Robertson, Jr., 1978, Platelet factors and the human vascular wall. Variations in growth response between endothelial and medial smooth muscle cells, Atherosclerosis, 30:67-78.
10. L.E. DeBault, L.E. Kahn, S.P. Frommes and P.A. Cancilla, 1979, Cerebral microvessels and derived cells in tissue culture: isolation and preliminary characterization, In Vitro, 15:473-487.
11. J. Folkman, C.C. Haudenschild and B.R. Zetter, 1979, Long-term culture of capillary endothelial cells, Proc. Natl. Acad. Sci. USA, 76:5217-5221.
12. G.K. Scherer, T.P. Fitzharris, W.P. Faulk and E.C. LeRoy, 1980, Cultivation of microvascular endothelial cells from human preputial skin, In Vitro, 16:675-684.
13. A.M. Schor, S.L. Schor and S. Kumar, 1979, Importance of a collagen substratum for stimulation of capillary endothelial cell proliferation by tumor angiogenesis factor, Int. J. Canc., 24:225-234.
14. R.N. Frank, V.E. Kinsey, K.W. Frank, K.P. Mickus and A. Randolph, 1979, In vitro proliferation of endothelial cells from kitten retinal capillaries, Invest. Ophth. Vis. Sci., 18:1195-1200.
15. B.R. Zetter, 1981, The endoethelial cells of large and small blood vessels, Diabetes, in press.
16. S.M. Schwartz, C.M. Gajdusek and S.C. Selden, 1981, Vascular wall growth control: the role of the endothelium, Arteriosclerosis, 1:107-126.

17. C.C. Haudenschild, D. Zahniser, J. Folkman and M. Klagsbrun, 1976, Human vascular endothelial cells in culture. Lack of response to serum growth factors, Exp. Cell Res. 98:175-183.
18. S.M. Schwartz, S.C. Selden and P. Bowman, 1979, Growth control in aortic endothelium at wound edges. Cold Spring Harbor 3rd Conference on Cell Proliferation. In: Hormones and Cell Culture. R.Ross and G.Sato, eds., Cold Spring Harbor, N.Y., 6:593-610.
19. U. Delvos, C. Gajdusek, H. Sage, L.A. Harker and S.M. Schwartz, 1982, Interactions of vascular wall cells with extracellular matrix, Lab. Invest., 46:61-72.
20. G.I. Schoefl, 1963, Studies on inflammation. III Growing capillaries: their structure and permeability, Virchows Arch. Abt. A. Pathol. Anat., 337:97-141.
21. W.J. Cliff, 1963, Observations on healing tissue: a combined light and electron microscopic investigation, Phil. Trans. Roy. Soc., 246B:38-83.
22. S.M. Schwartz and E.P. Benditt, 1976, Clustering of replicating cells in aortic endothelium, Proc. Natl. Acad. Sci. USA, 73:651-653.
23. K.K. Thakral, W.H. Goodson and T.K. Hunt, 1979, Stimulation of wound blood vessel growth by wound macrophages, J. Surg. Res., 26:430-436.
24. R.T. Wall, L.A. Harker, L.J. Quadracci and G.E. Striker, 1978, Factors influencing endothelial cell proliferation in vitro, J. Cell Physiol., 96:203-214.
25. G. Thorgeirsson and A.L. Robertson, Jr., 1978, Platelet factors and the human vascular wall. Part 2. Such factors are not required for endothelial cell proliferation and migration, Atherosclerosis, 31:231-238.
26. P.F. Davies and R. Ross, 1978, Mediation of pinocytosis in cultured arterial smooth muscle and endothelial cells by platelet-derived growth factor, J. Cell Biol., 79:663-669.

27. L.A. Harker, R. Ross and J.A. Glomset, 1978, The role of endothelial cell injury and platelet response in atherogenesis, Thrombos. Haemostas. (Stuttg.), 39:312-321.
28. A. Aharonov, R.M. Pruss and H.R. Herschman, 1978, Epidermal growth factor. Relationship between receptor regulation and mitogenesis in 3T3 cells, J. Biol. Chem., 253:3970-3977.
29. C-H. Heldin, A. Wasteson and B. Westermark, 1980, Growth of normal human glial cells in a defined medium containing platelet-derived growth factor, Proc. Natl. Acad. Sci. USA, 77:6611-6615.
30. R.B. Rutherford and R. Ross, 1976, Platelet factors stimulate fibroblasts and smooth muscle cells quiescent in plasma serum to proliferate, J. Cell Biol., 69:196-203.
31. Y. Maruyama, 1963, The human endothelial cell in tissue culture, Z. Zellforsch Mikrosk Anat., 60:69-79.
32. C. Gajdusek, P. Dicorletto, R. Ross and S.M. Schwartz, 1980, An endothelial cell-derived growth factor, J. Cell Biol., 85:467-472.
33. R. Ross, 1971, The smooth muscle cell. II. Growth of smooth muscle cells in culture and the formation of elastic fibers, J. Cell Biol., 52:172-181.
34. A. Vogel, E. Raines, B. Kariya, M-J. Rivest and R. Ross, 1978, Coordinate control of 3T3 cell proliferation by platelet-derived growth factor and plasma components, Proc. Natl. Acad. Sci. USA, 75:2810-2816.
35. R. Pollack and A. Vogel, 1973, Isolation and characterization of revertant cell lines. II. Growth control of a polyploid revertant line derived from SV-40 transformed 3T3 mouse cells. J. Cell. Physiol., 82:93-100.
36. R. Phillips and S. Kumar, 1979, Tumor angiogenesis factor (TAF) and its neutralization by a xenogeneic antiserum, Int. J. Canc., 23:82-88.
37. O.H. Lowry, N.J. Rosebrough, A.L. Farr and R.J. Randall, 1951, Protein measurement with the Folin phenol reagent, J.

Biol. Chem., 241:265-277.

38. S.A. Harris, C.M. Gajdusek, S.M. Schwartz and T.N. Wight, 1979, Role of endothelial cell products in vascular growth responses and neovascularization, J. Cell Biol., 83:CU444.

39. U.K. Laemmli, 1970, Cleavage of structural proteins during the assembly of the head of bacteriophage T_4, Nature, 227:680-685.

40. B.M. Martin, M.A. Gimbrone, Jr., E.R. Unanue and R.S. Cotran, 1980, Stimulation of non-lymphoid mesenchymal cell proliferation by a macrophage-derived growth factor, J. Immunol., 126:1510-1515.

41. S.J. Leibovich and R. Ross, 1976, A macrophage-dependent factor that stimulates the proliferation of fibroblasts in vitro, Am. J. Pathol., 84:501-513.

42. K.E. Paschkis, 1958, Growth-promoting factors in tissues: a review, Canc. Res., 18:981-991.

43. W.J. Pledger, C.D. Stiles, H.N. Antoniades and C.D. Scher, 1978, An ordered sequence of events is required before BALB/c-3T3 cells become committed to DNA synthesis, Proc. Natl. Acad. Sci. USA, 75:2839-2843.

44. A. Lipton, I. Klinger, D. Paul and R.W. Holley, 1971, Migration of mouse 3T3 fibroblasts in response to a serum factor, Proc. Natl. Acad. Sci. USA, 68:2799-2801.

45. R.R. Burk, 1973, A factor from a transformed cell line that affects cell migration, Proc. Natl. Acad. Sci. USA, 70:369-372.

46. G.J. Todaro, G.K. Lazar and H. Green, 1965, The initiation of cell division in a contact-inhibited mammalial cell line, J. Cell Comp. Physiol., 66:325-334.

47. M. Klagsbrun, D. Knighton and J. Folkman, 1976, Tumor angiogenesis activity in cells grown in tissue culture, Canc. Res., 36:110-114.

48. D.N. Fass, M.R. Downing, P. Meyers, E.J.W. Bowie and L.D. Witte, 1978, A mitogenic factor from porcine arterial

endothelial cells. 32nd Annual Meeting, Council on Atherosclerosis, American Heart Assoc. v.11.

49. C.M. Gajdusek and S.M. Schwartz, 1982, Ability of endothelial cells to condition culture medium, J. Cell. Physiol., 110:35-42.

50. S.M. Schwartz, C.M. Gajdusek, M.A. Reidy, S.C. Selden and C.C. Haudenschild, Maintenance of integrity in aortic endothelium, Fed. Proc., 39:2618-2625.

51. B.M. Glaser, P.A. D'Amore, R.G. Michels, A. Patz and A. Fenselau, 1980, Demonstration of vasoproliferative activity from mammalian retina, J. Cell Biol., 84:298-304.

52. A. Fenselau and R.J. Mello, 1976, Growth stimulation of cultured endothelial cells by tumor cell homogenates, Canc. Res., 36:3269-3273.

53. R.A. Brown, J.B. Weiss, I.W. Tomlinson, P. Phillips and S. Kumar, 1980, Angiogenic factor from synovial fluid resembling that from tumors, Lancet, March 29:682-685.

54. P.J. Polverini, R.S. Cotran, M.A. Gimbrone, Jr. and E.R. Unanue, 1977, Activated macrophages induce vascular proliferation, Nature, 269:804-806.

55. D.H. Ausprunk and J. Folkman, 1977, Migration and proliferation of endothelial cells in preformed and newly formed blood vessels during tumor angiogenesis, Microvasc. Res., 14:53-65.

56. J. Folkman and R. Cotran, 1976, Relation of vascular proliferation to tumor growth. In: International Review of Experimental Pathology, G.W.Richter, ed., Academic Press, N.Y., 16:208-245.

57. J. Folkman and C. Haudenschild, 1980, Angiogenesis in vitro, Nature, 288:551-556.

58. G. Cotta-Pereira, H. Sage, P. Bornstein, R. Ross and S.M. Schwartz, 1980, Studies of morphologically atypical ("sprouting") cultures of bovine endothelial cells. Growth

characteristics and connective tissue protein synthesis, J. Cell. Physiol., 102:183-191.

59. B.R. McAuslan and W. Reilly, 1979, A variant vascular endothelial cell line with altered growth characteristics, J. Cell. Physiol., 101:419-430.

60. H. Sage, Collagen synthesis by endothelial cells in culture. In: The Biology of Endothelial Cells, E.Jaffe, ed., Martinus Nijhoff Pub., B.V., the Netherlands, in press.

61. G.R. Grotendorst, H. Seppa, H.K. Kleinman and G.R. Martin, 1981, Attachment of smooth muscle cells to collagen and their migration toward platelet-derived growth factor, Proc. Natl. Acad. Sci. USA, 78:3669-3672.

LYMPHOTOXINS - A MULTICOMPONENT SYSTEM OF GROWTH INHIBITORY AND CELL-LYTIC GLYCOPROTEINS

G.A. Granger, J. Klostergaard, R.S. Yamamoto, J. Devlin, S.L. Orr, D. McGriff and K.M. Miner

Department of Molecular Biology and Biochemistry
University of California, Irvine
Irvine, California

ABSTRACT

Activated lymphocytes from experimental animals and man can release materials, termed lymphotoxins, which cause growth inhibition and cell lysis in vitro. These molecules, from human lymphocytes, are glycoproteins which can be divided into five molecular weight classes. These forms are heterogeneous, for each MW class can be further subdivided into multiple charge subclasses. It is now clear certain MW classes are interrelated and form a system of cell toxins. The larger classes (>140,000 d) are associated with nonclassical antigen-binding receptors (R), which can be of T cell origin. The smaller forms (<90,000 d) do not express R function and are derived from the larger forms, possibly by enzymatic action. Two MW classes, one receptor-associated and one non-receptor associated, have been purified to homogeneity and their peptide composition is being studied. Functional studies reveal the larger MW forms derived from alloimmune cell populations; can induce selective and nonselective destruction of cells in vitro.

Supported by NIH grant 2 R01 AI CA 09460, and Rheumatic Diseases Research Foundation grant 1882.

Antibodies which inhibit the in vitro cell lytic ability of various human LT forms can block different classes of human lymphocyte cell killing reactions in vitro.

Lymphotoxins (LT) are a heterogeneous family of cell-lytic and growth inhibitory glycoproteins released in vitro by stimulated lymphocytes from humans and experimental animals (1). Their release can be specifically stimulated by coculture of immune cells with the immunogen or nonspecifically by incubation of cells from nonimmunized donors with various lectins, i.e., Concanavalin A (Con A) or phytohemagglutinin-P (PHA). In vitro LT release by human lymphocytes is inducible and ceases rapidly in the absence of the inducing agent (2). The rapidity and amount of material released is dependent upon the stage of cellular activation, the type of lymphocyte stimulated, and the nature of the stimulus applied to the cells (3,4,5). Lectin and antigen preactivated human and murine lymphocytes have been shown to contain intracellular LT pools and can release high levels of lymphotoxin in a few hours in response to a secondary in vitro lectin stimulus (4,6). However, non-preactivated cells from the same cell donor require much longer, up to 48 to 72 hrs to release LT when stimulated by identical treatment. Different populations of lymphocytes, i.e., T and B cells can both release these molecules; however as will be discussed later, the kinetics and molecular classes of LT released are different (7,8). In addition, low levels of lymphotoxins of one MW class are spontaneously released in vitro by certain continuous human B cells lines, i.e., RPMI 1788, RPMI 8866, PGLC-33h (9,10). Finally, the level of activity and molecular type of LT detected is greatly influenced by a number of technical variables such as the type of media employed and the stability of the LT form(s) being detected.

Target cells in culture vary in their sensitivity to human lymphotoxins in lymphocyte supernatants (11). The effect of LT on a particular target cell type is concentration dependent for a cell may be lysed at a high concentration and only growth inhibited at a lower level of material (12). A lytic level of lymphotoxins for a sensitive cell may only temporarily growth inhibit or have no effect on an LT resistant cell. It has been reported by Evans and colleagues that neoplastic cell lines are more sensitive in vitro to the effects of the alpha and beta molecular weight class LTs than primary cells (13). More recently these investigators have demonstrated that the presence of these molecules prevents the appearance of transformed cells in cultures of fibroblasts treated with various carcinogenic agents, but has no apparent effect on the nontransformed cells (14). These studies are quite interesting and were repeated with LT from several different animal species. However, others have shown that resistant and sensitive cell subpopulation to human LT forms exist within individual continuous cell lines (15,16). In vitro lysis of LT sensitive cells by LT usually requires from 10-48 hrs. Studies with both human and Guinea pig LT indicates these molecules induce cell lytic effects by action on the cell membrane (11,17-19).

While heterogeneous for both size and charge, biochemical and immunological studies indicate the various LT forms are part of an

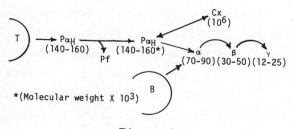

Figure 1

interrelated system of cell toxins (1). A model of the interrelationship of the LT forms released by human T and B cells is shown in Figure 1 (5).

Cultures of T enriched cells appear to first release a form termed <u>precursor alpha heavy</u> ($P\alpha_H$), which upon mild denaturation or upon molecular sieving loses several small MW components termed precursor factors (Pf). These cells can release LT very rapidly, within 24 hrs, upon even primary stimulation with lectins <u>in vitro</u>. Upon disassociation of the Pf, the $P\alpha_H$ becomes the <u>alpha heavy</u> (α_H) 140-160,000 MW form. The α_H may now proceed down one of two possible pathways: a) condense to form the high MW form termed <u>complex</u>; and/or b) dissociate to form the <u>alpha</u> (α) 70-90,000 MW, <u>beta</u> (β) 30-50,000 MW, and <u>gamma</u> (γ) 12-25,000 MW LT classes. The α MW class is further divisible into a number of charge subclasses (α_1, α_2, and α_3) (1,20). Additional studies suggest there are two clearly separable and distinct β classes (21). The β_2 form is quite stable and immunologically related to α class LT forms and the β_1 is unstable and immunologically and biochemically distinct from the α. Immunologic studies indicate the presence of β_2 forms in LT complexsis isolated from lectin stimulated human lymphocytes. The γ MW LT class is quite unstable and to date has not been studied by any laboratory. The dynamics of the system appear to be the following: the larger forms are normally quite short-lived, and the system favors the accumulation of the α and the β_2 LT forms which are quite stable under most culture conditions. The conversion of the larger form into the smaller classes is due to dissociation, protease cleavage or a combination of both. Thus it is not yet quite clear if β and are true subunits of the α class or are cleavage products. Additional studies indicate LT forms released <u>in vitro</u> by several other animal species exhibit similar molecular weight classes that can associate and dissociate in a fashion similar to that observed in

studies of human lymphotoxins (22). In contrast B cells and
continuous B cell lines only release LT of the α MW class and the
α of the individual B cell lines is of limited charge heterogeneity. Moreover B cell enriched cultures release LT very slowly
in response to lectin stimulation, requiring up to 5 days, to
reach levels only 1/3 to 1/2 the level of activity released by T
cells in 24 hrs.

Studies of individual human and murine LT forms reveal they
have different immunological and functional properties (23-25). A
surprising but very interesting finding was that a subpopulation
of human $P\alpha_H$, α_H and Cx but not the smaller LT forms are neutralized with rabbit anti-human F(ab')2 (IgG) antisera but minimally
reactive with anti-H and anti-L chain sera (26,27). In addition
Amino and colleagues had previously observed that human LT forms
were neutralized to some degree by anti-human Ig antibodies in
vitro (28). These findings raised the interesting possibility
that the larger LT forms were associated with or had Ig-like antigenic determinants and could express C and/or V regions. This
latter possibility was supported by showing a subpopulation of LT
forms from lectin stimulated immune cells were capable of binding
to the specific Ag immobilized on beads (29). Moreover their
lytic activity for L-929 cells was also neutralized by the
presence of Ag. Additional studies conducted with LT from human
and murine sources demonstrated the Complex forms are more
lytically active on different target cells in vitro when compared
to the smaller α, β, and γ forms (24). However, the larger forms
are difficult to study, for they are present in small quantities,
and as previously mentioned, quite unstable. Devlin et al (25)
revealed murine Cx forms could be stabilized by introducing
molecular cross-links into the molecule with gluteraldehyde. We
have attempted to examine the lytic activity of both human and
murine Cx forms from alloimmune cytotoxic T cells induced to

release these forms by a short 5 hr lectin restimulation in vitro.
We have found forms from these cells are capable of both specific
and nonspecific lysis of allogenic target cells in vitro measured
in a 10-16 hr ^{51}Cr release assay (30). These results can be
explained in two ways, either the Cx forms were derived from two
separate cell population, i.e., one capable of specific and the
other of nonspecific cell lysis, or there are two Cx forms,
perhaps interconvertible, one capable of specific and one of non-
specific cell lysis. The most commonly observed effect is non-
specific cell lysis.

We have selected two LT MW classes for detailed biochemical
studies. The first, the α class, because: a) it is a form
synthesized by both T and B cells; b) is enzymatically cleaved or
dissociates into the smaller β and γ classes; and c) is a subunit
of the $α_H$ LT forms. These molecules are glycoproteins containing
mannose, N-acetyl glucosamine, N-acetyl galactosamine, and fucose
residues (31). The $α_2$ subclasses from lectin stimulated human
lymphocytes has been purified to homogeneity (32). It is composed
of two noncovalently linked peptides, a heavy chain of 68-70,000,
and a light chain of 25-28,000. Additional studies of this
material are underway. An $α_H$ form from lectin stimulated human
lymphocytes has been purified to apparent homogeneity (33).
Analysis of this material reveals it appears to be composed of two
H chains; with one or more L chains. Studies are underway to
determine the relationship of α and $α_H$ forms and which peptides
possess F(ab')2 reactive and cell-lytic activities.

The above findings must be kept in mind as we consider a role
for LT in lymphocyte mediated cell lytic phenomenon in vitro.
While cell lytic reactions are complex, involving multiple classes
of lymphoid cells with the potential for multiple types of
effector mechanisms, these lytic reactions can be broadly

separated into two steps: a) initial membrane contact which serves as a recognition step(s) and b) delivery of a lethal hit to a target cell which causes actual cell cytolysis. These may be separate or related events at the molecular level. However not a single molecular component(s) of any of these potential molecules has yet been isolated or physically identified. Let us first consider the evidence against the role of LT as an effector in these reactions: the first point is circumstantial, namely the requirement for cellular contact between effector and target cell, which presumably would not be required if "soluble" effector molecules were the causative agents of lysis. The second is the nature of T cell killing which is very rapid, and specific for the sensitizing target cell(s) with little or no lysis of a neighboring or so-called innocent bystander cell(s). These features of cell lytic reactions mediated by intact lymphocytes seem to discount a possible role for LT molecules in these reactions. The arguments against lymphotoxins as effectors in these reactions is especially strong and convincing when one only considers the properties of LT forms which were known at that time namely only the smaller alpha, beta, and gamma forms had been identified. Clearly the putative role of LT molecules in these reactions must now be reconsidered in light of the new information discussed in the previous sections.

Interesting data has been obtained by different investigators studying the effects of xenogeneic anti LT serum directed at the smaller forms of different types of lymphocyte cell lytic reactions *in vitro*. Gately *et al* (34) found an antiserum reactive with Guinea pig β-LT (50,000) inhibited specific target cell lysis mediated by Guinea pig lymphoid cells *in vitro*. The serum, however, only appeared to inhibit lysis when the target cell antigen was projected into the supernatant and would not block lysis when the antigen was in close proximity to the surface of the target cell(s). Sawada and Osawa (19) reported an antiserum to Guinea pig β-LT inhibited nonselective lysis of targets by lectin activated Guinea pig lymphocytes in vitro. Hiserodt et al reported rabbit anti-α_2 LT serum inhibited the actual lethal hit phase of L-929 in vitro cell lysis mediated by lectin activated human cells (3). Rosenau et al reported anti-α LT serum inhibited antibody-dependent cell lysis of nucleated target cells by human lymphocytes in vitro (35).

Our own work examining the effect of rabbit antisera directed at various LT forms on different classes of cell lysis induced by human lymphocytes in vitro has been quite extensive. These studies reveal that the situation is complex; however the data is now quite compelling that these molecules are involved as effectors in different classes of cell lysis mediated by human lymphocytes in vitro. The studies are quite involved because we are dealing with a multicomponent system and an antibody against a single subunit of such a system may have no effect, may inhibit or accelerate the final endpoint event. Moreover the antigenicity of the larger forms change as they assemble and disassemble; thus an antibody may inhibit one form, but have no effect on another, yet they both may share a common subunit. This latter situation has actually been observed with the $P\alpha_H$ and α_H LT forms from human T cells.

Antiserum inhibition studies in our laboratory have been performed only as antibodies which neutralize the various human LT forms have been developed. The first antiserum developed was against the human α_2 MW LT class. As already mentioned, Hiserodt and colleagues demonstrated that this serum was able to inhibit the actual lytic phase of lysis of murine L-929 cells mediated by lectin activated human lymphocytes in vitro (35). Ware et al (26,36) tested a broader spectrum of antisera which neutralized various α, β, and γ classes and subclasses and serum prepared against unfractionated supernatants on cell lysis mediated by human T and anomalous killer cells in vitro. They found antiserum reactive with the smaller LT forms were essentially without effect on these two classes of cell lytic reactions. However anti-whole supernatant was a potent inhibitor of cell lysis and the relevant antigen recognized by this serum was of high molecular weight. Subsequent studies by Weitzan and her collaborators (37) revealed that monospecific anti-alpha LT serum inhibited lysis of K-562 and MOLT-4 targets about 50-60% of the time by human NK effectors and that N-acetyl glucosamine and alpha-methyl mannoside blocks lysis of MOLT-4 target cells mediated by both NK cells and lectin induced supernatants from these same effector cell populations (38). A new antiserum has been developed by Orr and Klostergaard (39) against purified α_H LT from lectin activated normal human lymphocytes (33). This serum has been tested by Devlin and Yamamoto (40) and given quite exciting results for its blocks in vitro cell lysis measured in 4 hr ^{51}Cr release assay mediated by: (A) human T killers sensitized in vitro against xenogeneic and allogeneic target B cells. (B) human NK effectors in cultured peripheral blood tested against MOLT-4 and K-562 targets, and (C) human anomalous killers directed against allogeneic continuous B cell lines. Moreover, while still not monospecific an antiserum from rabbits immunized with the Cx LT form is also inhibitory for the above classes of in vitro cell lysis. These sera do not affect target or aggressor cell viability. It is significant that an antiserum which reacts with the alpha heavy LT form blocks cell lysis for this is the LT form that can assemble into the high molecular weight LT complexsis (see Figure 1). These latest studies are quite exciting and provide strong evidence that these molecules may be involved in rapid cell lytic reactions mediated by human lymphocytes in vitro.

We are attempting to develop monoclonal reagents with the different human LT forms. These will be very powerful probes in further studies of this interesting family of cell lytic molecules. In summary evidence obtained from biochemical, immunochemical, and functional antibody inhibition studies, lend support to the concept that LT molecules form a family of cell lytic glycoproteins. These molecules can be assembled into high molecular weight complexsis perhaps in association with recognition molecules (these complexsis appear to have increased cell lytic capacity). The ability to assemble into complexsis in the soluble phase may be an indications of how these molecules could be employed by intact lymphocytes as effectors of cell lysis. These and other findings suggest lymphotoxin molecules may be assembled on the target cell surface by an effector cell(s) to form and "attack complex", perhaps somewhat similar to that seen when the complement system is activated by antibody bound to a cell surface. Indeed since cellular immunity proceeds humerol immunity in phylogeny this system may actually represent a primitive "complement-like" system. However, the system appears to be programmed to operate locally around the secreting cell for once the large LT components are formed in the soluble phase they rapidly become converted into the smaller less active alpha and beta LT forms. This conversion of active to less active forms may be a control system which ensures that the cell lytic activity is restricted to the immediate area or microenvironment around the effector cell(s). The smaller LT forms may perform secondary roles with other effects on cells besides that associated with cell lysis. Whatever the case further studies of this interesting system of molecules is necessary to understand at the molecular level the mechanism by which the various forms may be assembled and delivered to a target cell by cytotoxic lymphocytes.

LYMPHOTOXINS - A MULTICOMPONENT SYSTEM

REFERENCES

1. J. Klostergaard, R.S. Yamamoto and G.A. Granger, 1979, Molec. Immunol. 17:613.
2. R.A. Daynes and G.A. Granger, 1974, Cell. Immunol. 12:252.
3. J.C. Hiserodt, G.J. Tiangco and G.A. Granger, 1979, J. Immunol. 123:317.
4. M.L. Weitzen, R.S. Yamamoto and G.A. Granger, In: Cell Death (I.D. Bowen, ed.) Chapman and Hall, London (in press).
5. P.C. Harris, R.S. Yamamoto, C. Christensen and G.A. Granger, J. Immunol.
6. J.C. Hiserodt, S.C. Ford, P.C. Harris and G.A. Granger, 1979, Cell. Immunol. 47:32.
7. K. Yano and Z. Lucas, 1974, J. Immunol. 120:385.
8. G.R. Kimple, J.H. Dean, K.D. Day, P.B. Chen and D.O. Lucas, 1977, Cell. Immunol. 32:293.
9. G.A. Granger, G.E. Moore, J.G. White, P. Matzinger, J.S. Sudsmo, S. Shupe, W.O. Kolb, J. Kramer and P.R. Glade, 1970, J. Immunol. 104:1476.
10. D.S. Fair, E.W.B. Jeffes and G.A. Granger, 1979, Molec. Immunol. 16:186.
11. T.W. Williams and G.A. Granger, 1969, J. Immunol. 102:911.
12. E.W.B. Jeffes and G.A. Granger, 1975, J. Immunol. 114:64.
13. J.O. Rundle and C.A. Evans, 1981, Immunopharm. 3:9.
14. C.A. Evans and J.A. DiPaolo, 1981, Nat. J. Canc. 27:45.
15. S.J. Kramer and G.A. Granger, 1975, Cell. Immunol. 15:57.
16. B.A. Lisafeld, J. Minowada, E. Klein and O.A. Holtman, 1980, Int. Arch. All. Immunol. 62:59.
17. S.M. Walker and Z.J. Lucas, 1972, J. Immunol. 109:1233.
18. W. Roseanau, M.L. Goldberg and G.C. Burke, 1973, J. Immunol. 11:1128.
19. J.I. Swada and T. Osawa, 1978, Transplant. 26:319.

20. S.M. Walker, S.C. Lee and Z.J. Lucas, 1976, J. Immunol. 116:807.
21. J.C. Hiserodt, D.S. Fair and G.A. Granger, 1976, J. Immunol. 117:1503.
22. M.W. Ross, G.J. Tiangco, P. Horn, J.C. Hiserodt and G.A. Granger, 1979, J. Immunol. 123:325.
23. R.S. Yamamoto, J.C. Hiserodt, J.E. Lewis, C.E. Carmack and G.A. Granger, 1978, Cell. Immunol. 38:403.
24. R.S. Yamamoto, J.C. Hiserodt and G.A. Granger, 1979, Cell. Immunol. 45:261.
25. J.J. Devlin, R.S. Yamamoto and G.A. Granger, 1981, Cell. Immunol. 61:22.
26. J.C. Hiserodt, R.S. Yamamoto and G.A. Granger, 1978, Cell. Immunol. 38:417.
27. C.F. Ware, P.C. Harris and G.A. Granger, 1978, J. Immunol. 126:1927.
28. N. Amino, S.E. Limm, J.T. Pisher, R. Mier, G.E. Moore and L.J. De Groot, 1974, J. Immunol. 113:1334.
29. J.C. Hiserodt, R.S. Yamamoto and G.A. Granger, 1979, Cell. Immunol. 41:380.
30. P. Harris, R.S. Yamamoto and G.A. Granger, J. Immunol. (submitted).
31. M.K. Toth and G.A. Granger, 1979, Molec. Immunol. 1:671.
32. J. Klostergaard, S. Long and G.A. Granger, Molec. Immunol. 18:1049, 1981.
33. J. Klostergaard and G.A. Granger, Molec. Immunol. 18:455, 1981.
34. M.L. Gately, M.M. Mayer and C.S. Henny, 1976, Cell. Immunol. 23:83.
35. J.C. Hiserodt and G.A. Granger, 1977, J. Immunol. 119:374.
36. C.F. Ware and G.A. Granger, 1981, J. Immunol. 126:1919.
37. M. Weitzen, E. Innins, R.S. Yamamoto and G.A. Granger, Cell. Immunol. (in press).

38. M. Weitzen, R.S. Yamamoto, G.A. Granger, Cell. Immunol. (in press).
39. S.L. Orr and J. Klostergaard, (unpublished results).
40. J. Devlin and R.S. Yamamoto, (unpublished results).

PRODUCTION OF GAMMA (IMMUNE) INTERFERON BY A PERMANENT HUMAN
T-LYMPHOCYTE CELL LINE

Jerome E. Groopman, Ilana Nathan and David W. Golde

Division of Hematology-Oncology
Department of Medicine
UCLA School of Medicine
Los Angeles, California 90024

ABSTRACT

Gamma (immune, type II) interferon is an antiviral and immunoregulatory glycoprotein usually obtained from mitogen- or antigen-induced peripheral blood or splenic mononuclear cells. A human "helper-inducer" T-lymphocyte cell line (Mo) is an inducible source of gamma interferon. Mo gamma interferon has been biochemically characterized and partially purified, and resembles the gamma interferon species derived from normal cell sources.

---

[1] This investigation was supported by PHS grant numbers CA 30388 and CA15688 awarded by the National Cancer Institute, DHHS, and by a grant from the Genetics Institute. JEG is a recipient of an American Cancer Society Junior Faculty Research Award.
[2] Unpublished

Permanent human cell lines may serve as sources of a number
of important bioregulatory molecules including hematopoietic
hormones, growth factors, and alpha (leukocyte) interferon (1-4).
We derived a permanent human T-lymphocyte cell line from the
spleen cells of a patient with a T-cell variant of hairy-cell
leukemia (5). This cell line (Mo) has been in continuous culture
in our laboratory and is a source of a number of T-cell-derived
lymphokines, including colony-stimulating factor (6), erythroid-
potentiating activity (7), T-cell growth factor,[2] neutrophil
migration-inhibitory factor (8), a fibroblast growth factor (9),
and gamma (immune) interferon (10). Monoclonal antibody typing of
Mo cells indicates that they have surface antigenic characteristics of helper or inducer T lymphocytes. Many of the lymphokines, such as colony-stimulating factor, erythroid-potentiating
activity, neutrophil migration-inhibitory factor, and the fibroblast growth factor, are constitutively elaborated by the Mo
cells. Because the Mo cells remain viable under serum-free
conditions for 4 to 7 days, serum-free lymphokine-rich medium is
available and has markedly aided the biochemical characterization
and purification of these factors (6).

The interferons are a family of glycoproteins that confer
virus-nonspecific anti-viral protection on target cells through
modulation of cellular processes. Interferons alpha and beta have
been derived from leukocytes and fibroblasts and much is known
concerning their physicochemical and genetic structures (3,11,12).
Gamma (immune) interferon is derived from T lymphocytes and is
antigenically and physically distinct from interferons alpha and
beta; interferon gamma is acid labile and is not neutralized by
antisera to the other interferons. In preliminary studies,
interferon gamma appears to have more potent anti-neoplastic and
immunoregulatory activities than interferons alpha and beta
(13,14). The only species of interferon produced by Mo cells is

Table I. Induction of Gamma Interferon from Mo Cells*

Inducers	Titer (IU/ml)
Medium (constitutive)	5 - 75
PHA (3%)	500 - 1,000
OKT3 (5 ng/ml)	250 - 1,000
TPA (5 ng/ml)	500 - 1,000
PHA (3%) and TPA (5 ng/ml)	5,000 = 20,000

*10^6 Mo cells/ml are cultured in Iscove's medium (Irvine's Scientific) with the above agents: phytohemagglutinin (PHA) (Burroughs-Wellcome), OKT3 (Ortho), and 12-O-tetradecanoylphorbol diester (TPA).

gamma interferon by criteria of acid lability, cell source, and failure of neutralization with antisera to leukocyte interferon. Small quantities of gamma interferon are constitutively produced by the Mo cells, but upon induction with mitogens such as phytohemagglutinin or the monoclonal antibody OKT3 (Ortho) and chemical agents such as phorbol diesters, relatively high levels of interferon are elaborated (Table I).

We have recently characterized the gamma interferon produced by the Mo cell line (10). It has physical and biochemical characteristics similar to those of gamma interferon obtained from mitogen- or antigen-stimulated peripheral blood mononuclear cells (15-17). Mo gamma interferon is a heat-labile glycoprotein that is readily denatured by urea and guanidine hydrochloride and detergents such as sodium dodecylsulfate and zwitterionic detergents. Interestingly, the Mo gamma interferon is stable in 2-mercaptoethanol (10 mM), indicating that if disulfide bonds are present they are not critical for interferon activity. Mo gamma interferon displays the in vitro properties attributed to both type I and type II interferons in terms of antiviral activity and stimulation of natural killer cell activity.

We have biochemically characterized Mo gamma interferon as a glycoprotein which binds entirely to concanavalin-A-Sepharose and

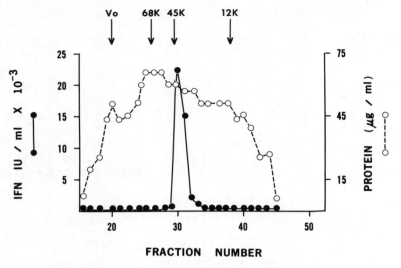

Figure 1

Chromatography of Mo gamma interferon. One liter of serum-free conditioned medium induced with 3% PHA and 5 ng/ml TPA was concentrated using a YM-10 membrane (Amicon) and applied to an Ultrogel AcA 44 column. The column was developed with 20% ethylene glycol 1 M NaCl in 20 pH 7.4. Molecular weight standards are shown.

elutes with 0.2 M α-methyl mannoside. It has a molecular weight of 40,000 to 45,000 daltons (Figure 1). We have been able to partially purify large volumes of induced Mo-conditioned medium by simply swirling the conditioned medium at 4°C for 16 hours with 3.6 g/250 ml of controlled-pore glass beads (Electronucleonics). Because Mo gamma interferon is a cationic hydrophobic molecule, it binds entirely to these glass beads and is eluted with small volumes of 20% ethylene glycol, 1 M NaCl in 20 mM phosphate buffer pH 7.4. An interferon specific activity of 10^5 IU/mg protein with a recovery of 80-100% can be readily obtained using this batch purification method. A further tenfold purification is achieved using gel permeation chromatography on an Ultrogel AcA 44 column (LKB). Active fractions from the gel filtration column are pooled and applied to a poly-U Sepharose 4B (Pharmacia) column which achieves further purification on the basis of affinity binding to the pulyuridine group. Mo gamma interferon of specific activity of about 5×10^7 IU/mg protein is obtained using this simple three-step sequential purification scheme.

Preliminary studies employing semi-purified (10^5 IU/mg protein) Mo gamma interferon indicate that after treatment with endoglycosydase-H (Miles), as well as endoglycosydase-D (Miles) and mixed glycosydases (Sigma) (20 mU/ml) each of neuraminidase, β-acetylglucosiminidase, and β-galactosidase), gamma interferon is still active in vitro. The glycosidase-treated material no longer binds to concanavalin-A-Sepharose but retains full antiviral activity. It is not yet certain that all the sugar residues are cleaved off the Mo gamma interferon after such enzyme treatment, but it is likely that nonglycosylated human gamma interferon will be biologically active.

The production of gamma interferon by the cell line in response to mitogens and chemical agents occurs without the presence of accessory cells such as macrophages. This observation is consistent with reports of production of gamma interferon by cloned murine cell lines in vitro (18). The Mo cell line is a rich source of human gamma interferon and it should prove useful in characterizing this important interferon species.

REFERENCES

1. J.G. DiPersio, J.K. Brennan, M.A. Lichtman, C.N. Abboud and G.H. Kirkpatrick, 1980, The fractionation, characterization, and subcellular localization of colony-stimulating activities released by the human monocyte-like cell line, GCT. Blood 56:717-727.

2. M.-c. Wu and A.A. Yunis, 1980, Common pattern of two distinct types of colony-stimulating factor in human tissues and cultured cells. J. Clin. Invest. 65:772-775.
3. D.V. Goeddel, D.W. Leung, T.J. Dull, M. Gross, R.M. Lawn, R. McCandliss, P.H. Seeburg, A. Ullrich, E. Yelverton and P.W. Gray, 1981, The structure of eight distinct cloned human leukocyte interferon cDNAs. Nature 290:20-29.
4. S. Gillis and J. Watson, 1980, Biochemical and biological characterization of lymphocyte regulatory molecules. V. Identification of an interleukin 2-producing human leukemia T cell line. J. Exp. Med. 152:1709-1719.
5. A. Saxon, R.H. Stevens and D.W. Golde, 1978, T-lymphocyte variant of hairy cell leukemia. Ann Intern. Med. 88:323-326.
6. A.J. Lusis, D.H. Quon and D.W. Golde, 1981, Purification and characterization of a human T-lymphocyte-derived granulocyte-macrophage colony-stimulating factor. Blood 57:13-21.
7. D.W. Golde, N. Bersch, S.G. Quan and A.J. Lusis, 1980, Production of erythroid-potentiating activity by a human T-lymphoblast cell line. Proc. Soc. Natl. Acad. Sci. U.S.A. 77:593-596.
8. R.H. Weisbart, D.W. Golde, L. Spolter, P. Eggena and H. Rinderknecht, 1979, Neutrophil migration inhibition factor from T-lymphocytes (NIF-T): A new lymphokine. Clin. Immunol. Immunopathol. 14:441-448.
9. J.E. Groopman, A.J. Lusis and D.W. Golde, 1981, Characterization of a human fibroblast growth factor derived from a permanent T-lymphocyte cell line. J. Supramol. Struct. (Suppl.) 5:129.
10. I. Nathan, J.E. Groopman, S.G. Quan, N. Bersch and D.W. Golde, 1981, Immune (γ) interferon produced by a human T-lymphoblast cell line. Nature 292:842-844.
11. R. Derynck, J. Content, E. DeClercq, G. Volckaert, J. Tavernier, R. Devos and W. Fiers, 1980, Isolation and

structure of a human fibroblast interferon gene. Nature 285:542-547.

12. S. Nagata, N. Mantei and C. Weissman, 1980, The structure of one of the eight or more distinct chromosomal genes for human interferon-α. Nature 287:401-408.

13. B.Y. Rubin and S.L. Gupta, 1980, Differential efficacies of human type I and type II interferons as antiviral and antiproliferative agents. Proc. Natl. Acad. Sci. U.S.A. 77:5928-5932.

14. J.E. Blalock, J.A. Georgiades, M.P. Langford and H.M. Johnson, 1980, Purified human immune interferon has more potent anticelluar activity than fibroblast or leukocyte interferon. Cell. Immunol. 49:390-394.

15. T. Taniguchi,j R.H.L. Pang, Y.K. Yip, D. Henriksen and J. Vilĉek, 1981, Partial characterization of γ (immune) interferon mRNA extracted from human lymphocytes. Proc. Natl. Acad. Sci. U.S.A. 78:3469-3472.

16. M.P. Langford, J.A. Georgiades, G.J. Stanton, F. Dianzani and H.M. Johnson, 1979, Large-scale production and physico-chemical characterization of human immune interferon. Infect. Immun. 26:36-41.

17. A. Mizrahi, J.A. O'Malley, W.A. Carter, A. Takatsuki, G. Tamura and E. Sulkowski, 1978, Glycosylation of interferons. Effects of tunicamycin on human immune interferon. J. Biol. Chem. 253:7612-7615.

18. F. Marcucci, M. Waller, H. Kirchner and P. Krammer, 1981, Production of immune interferon by murine T-cell clones from long-term cultures. Nature 291:79-81.

THYMIC INHIBITION OF MYELOPOIETIC PROLIFERATION*

Dale F. Gruber and G. David Ledney

Experimental Hematology Department
Armed Forces Radiobiology Research Institute
Bethesda, Maryland 20814

ABSTRACT

Homeostasis may be viewed as the dynamic balance reached between cellular production and loss. The effectors of proliferation are though to be tissue-specific humoral substances. Inhibitory substances appear to negate the proliferate effects of activator substances. We report here that conditioned medium from cultures of murine "nonlymphoid" adherent thymus cells in a source of a potent myelopoietic inhibitor substance(s). Examination of the inhibitor(s) in the soft-agar clonogenic assay show it to abrogate both the 10-day

---

*Supported by the Armed Forces Radiobiology Research Institute, Defense Nuclear Agency, under Research Work Unit MJ 00081. Views presented in this paper are those of the authors; no endoresement by the Defense Nuclear Agency has been given or should be inferred.

Research was conducted according to the principles enunciated in the "Guide for the Care and Use of Laboratory Animals," prepared by the Institute of Laboratory Animal Resources, National Research Council.

granulocyte-macrophage colony-forming cell (GM-CFC) and the 25-day monocyte-macrophage colony-forming cell (M-CFC). The inhibition is most significant when the thymus-conditioned medium (TCM) is present upon culture initiation. TCM added at a time after soft-agar initiation (day 6) also results in signficiant clonogenic inhibition. We have characterized the inhibitor as potent on the basis that volume ratios of inhibitor to L-cell colony-stimulating factor (CSF) as low as 1 part to 40 parts will give near total inhibition of both GM-CFC and M-CFC clonogenic growth. The inhibitor is dialyzable, has a molecular weight of less than 1000, is not significantly cytotoxic, and its effects are reversible with washing.

INTRODUCTION

Hematological modulation may be a morphologic depiction of changes in the quantitative interrelationship(s) between systemic activators and inhibitors. Various organs, or tissues, may have either a direct or indirect effect on the regulation of hematopoiesis. It has been demonstrated that endocrine gland(s), kidneys, and liver play a major role in the production and regulation of stimulating factors controlling erythropoiesis (1). Lymphomyelopoiesis (2,3) has also been shown to be responsive to homeostatic perturbations, and the thymus is suspected as participating in the routine maintenance of homeostatic hemopoiesis. Lymphoidal tumor involvement has been

---

Abbreviations: GM-CFC, granulocyte-macrophage colony-forming cell; M-CFC, monocyte-macrophage colony-forming cell; TCM, thymus-conditioned medium; CSF; colony-stimulating factor; LCM, L-cell-contioned medium; PPD, purified protein derivative; CIF, cloning inhibitory factor; PHA, phytohemagglutinin; PIF, proliferative inhibitory factor.

evident in many instances of hematopoietic dyscrasias (4). There is now compelling ablative-type evidence that the thymus serves an endocrine function and that ectopic thymus allografting is capable of reconstituting some immunodeficient states (5). Generally, thymic epithelial cells have been considered to be the cellular source of thymic humoral factors (6), although lymphocytes also have been implicated. Reduction of Thy 1.2 positive lymphocyte population resulted in a concurrent decrease in both erythroid and granuloid splenic colonies (7). If one accepts the contention that thymic epithelial cells are the source of thymic humoral factors, then these factors should reside in media conditioned by these cells (8). This contention has been supported by findings that supernatants from thymic epithelial cell cultures are capable of inducing Thy-1 antigen expression on splenocytes from athymic mice (9), E-rosette formation in fractionated bone marrow cells (10), and an increase in the mitogenic responsiveness of rat thymocytes (11). In the present study, we present evidence for the in vitro elaboration of humoral inhibitory factors by "nonlymphoid" thymic adherent tissue composed of cells bearing the histological appearances of macrophages and epithelial cells.

MATERIALS AND METHODS

Animals

Male, (C57BL/6 X CBA)F1 Cum BR hybrid (B6CBF1) mice (Cumberland View Farms, Clinton, TN) were used throughout the study. Thymic cultures were prepared from mice 8 to 10 weeks old. All animals were maintained on a 6 AM (light)-6 PM (dark) cycle in filter-covered cages. Wayne Lab-Blox and acidified water (pH 2.5) were available ad libitum. All mice were housed for 2 weeks in quarantine until certified free of histologic lesions and Pseudomonas sp.

Thymic Epithelial and Control Cultures

In each experiment, thumuses from six mice were aseptically removed, pooled in 4°C RPMI-1640 (Flow Laboratories, Rockville, MD), and minced with scissors into small fragments. The organ fragments (15-20) were explanted into 75-cm^2 tissue culture flasks (Corning #25110, Corning Glass Company, Corning, NY) with 5 ml of starter medium. The medium components were RPMI-1640 tissue culture medium (Flow Labs) supplemented with 10% (v/v) heat-inactivated (56°C, 30 min) fetal calf serum (Grand Island Biological Company, Grand Island, NY), 2 mM 1-glutamine (GIBCO), penicillin (100 IU/ml) streptomycin (100 ug/ml), and fungizone (0.25 ug/ml, GIBCO). All cultures were kept at 37°C in a humidified incubator and were gassed with 5% CO_2-95% air. After 24 hours, 3 ml of fresh medium was added to each flask. Attempts were made to not disturb explants during early medium changes. On day 4, 4 ml of fresh medium was added (final culture volume 12 ml). Subsequent medium changes were collected, pooled, and replaced volume for volume with fresh medium. A maximum of 8 ml/flask was replaced as needed. Changing frequency generally increased as the monolayer moved toward confluency. Under these conditions, monolayer confluency was generally established by day 21.

Monolayer Histology

All cultures were regularly observed with an inverted-phase microscope (Opton Model, Zeiss, West Germany). Representative flasks and cytocentrifuge preparations were fixed with 0.2 M phosphate-buffered glutaraldehyde (pH 7.3) and stained with Wrights-Giemsa or α-napthyl acetate to determine the presence and frequency of nonspecific esterase-positive cells.

Ultrafiltration

The TCM after collection was centrifuged at 250 xg for 10 min. The supernatant was filtered through a 0.45-micron filter unit (Nalge Company, Rochester, NY) to remove cells or cellular debris. The filtrate was fractionated by molecular ultrafiltration by subjecting it to an Immersible-CS Molecular Separator-type PTGC (Millipore Corporation, Bedfore, MA) with nominal cutoff at 10,000 daltons. The ultrafiltrate was then passed through an Amicon Molecular Ultrafiltration Membrane UM-2 (Amicon Corporation, Lexington, MA) with nominal cutoff at 1000 daltons. The ultrafiltrate was adjusted to a pH of 7.3 and had an osmolarity of 300 m Osm/kg.

GM-CFC and M-CFC Assays

The soft-agar clonogenic assay was used to assess the stimulating activities of TCM and L-cell-conditioned medium (LCM) alone or in conjunction with each other on the granulocyte-macrophage and monocyte-macrophage progenitor cell populations. The bottom of the two-layered system was 0.5% nutrient supplemented agar (Bacto-agar, Difco Laboratories, Detroit, MI) containing known aliquots of CSF and/or TCM. The top layer was nutrient-supplemented 0.33% agar containing 2.2×10^4 bone marrow cells. All plates were incubated in a humified incubator (Model 3016, Forma Scientific, Marietta, OH) at 37°C and were gassed with a 5% CO_2-95% air mixture. Plates were counted for GM-CFC colonies (>50 cells) at 10 days and M-CFC colonies (>50 cells) at 25 days, using an inverted-phase microscope (Opton Model, Carl Zeiss Incorporated, Oberkochen, West Germany) at a magnification rate of 25X.

Reversibility of Inhibitor

Bone marrow cells were incubated (37°C) with the RPMI-1640 and 10% inactivated fetal calf serum previously described, and divided into four aliquots. One aliquot received no additional treatment for 90 min. The other three aliquots received 100 ul each (5% volume) TCM. One of these three treated aliquots received no further treatment, one was washed once with medium after 90 min and centrifuged at 250 xg for 10 min, and the third aliquot was washed twice after incubation and treatment. The cells from all four aliquots were resuspended in culture medium, counted, and plated in the usual manner in soft agar to determine proliferative competency.

RESULTS

Monolayer Growth

Outgrowths from thymus explants were visible on day 2 of culture and were confluent by 21 days. The monolayers exhibited heterogenous morphology composed mainly of what appeared to be epitheloid polygonal cells and dendritic type cells with cytoplasmic extensions. Microscopic examination of cytocentrifuge preparations of aspirated cells showed the cultures to be lymphocyte-free after approximately 12 days.

The monolayer constituents appeared to consist preponderantly (80%) of nonspecific esterase positive cells resembling macrophages. The other cell type is nonspecific esterase negative epitheloid type, which exists within the monolayer in small aggregates or colonies.

Reversibility of TCM Inhibition

Table 1 demonstrates the reversibility of the TCM inhibitor as demonstrated by the soft-agar clonogenic assay for GM-CFC and M-CFC. Baseline control values were obtained of GM-CFC and M-CFC in the presence of LCM. Addition of 5% by volume of TCM inhibits almost all GM-CFC and M-CFC growth. Washing the TCM-treated bone marrow cells once resulted in clonogenic mean values that were 77% (GM-CFC) and 78% (M-CFC) of control values. A second washing resulted in further improvement, giving 93% (GM-CFC) and 95% (M-CFC) of baseline control values.

TCM Inhibition

Table 2 shows TCM inhibition, the results of which have been seen in three successive replications. TCM appears to be a potent inhibitor of both GM-CFC and M-CFC when present initially in the bottom layer. Volumes as small as 1 part to 39 parts CSF are capable of significant inhibition. In an attempt to ascertain whether there was colony stunting, TCM was added in varying amounts at 6 days after initiation. Results (Table 3) show a direct relationship between the volume of TCM added and the degree of GM-CFC and M-CFC inhibition. There was no noticeable morphometric difference between colonies, normal colonies, and those receiving TCM on day 6.

Cytotoxicity

The cytotoxic effects of TCM on murine bone marrow cells was followed through 48 hours (see Table 4). Counts were

Table 1. TCM Inhibition and Reversibility Demonstrated by GM-CFC[1] and M-CFC[2]

Treatment	GM-CFC (x±SEM)	% of Control	M-CFC (x±SEM)	% of Control
L-cell medium	18.8 ± 0.9	100	10.2 ± 0.8	100
L-cell medium, TCM[3]	0	0	0	0
L-cell medium, TCM, 1 wash[4]	14.5 ± 0.6	77	15.2 ± 0.9	78
L-cell medium, TCM, 1 washes	17.4 ± 0.5	93	18.2 ± 0.9	95

[1] 10-day granulocyte-macrophage colony (>50 cells) count per 2.2×10^4 live nucleated cells plated
[2] 25-day monocyte-macrophage colony (>50 cells) count per 2.2×10^4 live nucleated cells plated
[3] Thymus-conditioned medium (2.5% by volume L-cell medium)
[4] Washed in medium and centrifuged at 250 xg for 10 min.

Table 2. Graded Dose Inhibition of GM-CFC[1] and M-CFC[2] by TCM

Treatment	GM-CFC (x ± SEM)	M-CFC (x ± SEM)
L-cell medium[3]	17.8 ± 1.4	18.2 ± 0.9
L-cell medium + 2.5 vol % TCM	1.0 ± 0	1.0 ± 0
L-cell medium + 5.0 vol % TCM	0	0
L-cell medium + 10.0 vol % TCM	0	0
L-cell medium + 25.0 vol % TCM	0	0

[1] See footnote 1 in Table 1.
[2] See footnote 2 in Table 1.
[3] L-cell-conditioned medium, 200 ul/plate

Table 3. GM-CFC[1] and M-CFC[2] Response to Graded Doses of TCM Added Six Days After Initiation by L-Cell Medium[3]

Volume % TCM Added	GM-CFC (x ± SEM)	% of Control	M-CFC (x ± SEM)	% of Control
0	17.8 ± 1.4	100	18.2 ± 0.9	100
2.5	12.8 ± 1.7	72	13.5 ± 0.6	74
5.0	10.8 ± 1.5	60	4.3 ± 0.5	23
12.5	4.8 ± 0.6	27	4.5 ± 0.5	25
25.0	4.7 ± 0.9	26	4.2 ± 0.3	23

[1]See footnote 1 in Table 1.
[2]See footnote 2 in Table 1.
[3]L-cell-conditioned medium, 200 ul

Table 4. Percent Viability of Murine Bone Marrow in Graded Quantities of Thymus-Conditioned Medium

Volume % TCM[1] Added to Control Medium	Percent Viability[2] Tested at:			
	6 hr	24 hr	30 hr	48 hr
0	99	97	95	95
2	87	86	85	83
5	83	82	80	80
10	82	82	80	79
25	83	81	81	79

[1]TCM, thymus-conditioned medium
[2]Determined by hemocytometer and trypan blue exclusion

established using standard hemocytometer and trypan blue exclusion techniques. Over the full range of amounts (2%-25% TCM by volume) and times tested, TCM reduced bone marrow viability by 10%-17% (mean 14.4%).

DISCUSSION

Hematopoietic homeostasis is a dynamic equilibrium reached when proliferative and inhibitor substances have exacted their influences on target cell populations. The thymus is known to play a significant role in the maintenance of homeostatic hemopoiesis. Lymphocyte-derived factors are known regulators of hemopoiesis. Lebowitz and Lawrence (12) reported that tissue culture medium conditioned by PPD-stimulated lymphocytes possess a cytostatic property that inhibits HeLa cell clonogenic formations. The term applied to that particular cytostatic property was "cloning inhibitory factor-CIF" (13). Further it was found that medium conditioned by phytohemagglutinin (PHA)-stimulated lymphocytes contained factors that inhibited DNA synthesis in HeLa cells. The capacity was termed "proliferative inhibitory factor - PIF' (14). The factors in both these instances are lymphocyte-derived. In the present work, lymphokine-like factors cannot be ruled out completely as there is a decreasing lymphocyte presence through day 12 of culture. In all likelihood, however, it is most probably a member of the adherent "nonlymphoid" cell population that is responsible for the elaboration of the GM-CFC/M-CFC inhibitor. It appears that the _in vitro_ cells elaborate the inhibitor in the absence of either antigenic or mitogenic stimulation, in contrast to the CIF/PIF factors previously mentioned.

A number of investigators have implied that thymic humoral factors are elaborated by epithelial-type cells (8,10). On the basis of histologic and morphologic criteria, it appears that epitheloid cells represent only a small fraction (20%) of the adherent cell culture population. The majority (80%) of the cells appear to be of macrophagic lineage based on their demonstration of positive esterase staining and ingestion of IgG-coated bovine red blood cells. Quantitatively, these observations are in agreement with the morphological descriptions of thymic microenvironmental cells reported by Jordan and Crouse (15) and Jordan et al. (16).

Preliminary work (not reported here) on the physicochemical characteristics of the inhibitor show the inhibitor to be different from both CIF and PIF. It does not possess the thermolability shown by CIF (17) and, in contrast to PIF (18), the inhibitor is dialyzable. Further work will be conducted in an attempt to determine if the inhibitory property is of monokine origin.

Data presented here corroborate recent observations of Maschler and Maurer (19), who reported that ultrafiltrate fractions obtained from calf thymus inhibited both lymphocyte formation and granulocyte soft-agar colony formation in capillary tubes.

Conclusions that may be drawn from these preliminary data are:

1. When present initially, TCM is a strong inhibitor of both GM-CFC and M-CFC.
2. When added at a later date (day 6), TCM exhibits a growth-restricting effect.
3. The inhibitor is elaborated *in vitro* without antigenic or mitogenic treatment.

4. The molecular weight of the TCM inhibitor appears to be less than 1000 daltons.

REFERENCES

1. Fisher, J.W. and Busuttil, R.W. 1977. Sites of production of erythropoietin. In Kidney Hormones, Vol. 2 (Fisher, J.W., editor). Academic Press, New York, p. 165.
2. Craddock, C.G. 1972. Production, distribution and fate of granulocytes. In Hematology (Williams, W.J., Beutler, E., Erslev, A.J. and Rundles, R.W., eds). McGraw-Hill, New York, p. 607.
3. Bentwich, Z. and Kunkel, H.G. 1973. Specific properties of human B and T lymphocytes and alterations in disease. Transplant. Rev. 16:29.
4. Fisher, E.R. 1964. Pathology of the thymus and its relation to human disease. In The Thymus in Immunobiology (Good, R.A. and Gabrielsen, A.E., eds.). Hoeber-Harper, New York, P. 676.
5. Bach, J. and Carnaud, C. 1976. Thymic factors. In Progress in Allergy (Kallos, P., Waksman, B.H. and Weck, A., eds.). S. Karger A.G., Basel, Switzerland, p. 342.
6. Oosterom, R., Kater, L. and Oosterom, J. 1979. Effects of humam thymic epithelial-conditioned medium on mitogen responsiveness of human and mouse lymphoctes. Clin. Immunol. Immunopathol. 12:460.
7. Goodman, J.W., Basford, N.L. and Shinpock, S.G. 1978. On the role of thymus in hemopoietic differentiation. Blood Cells 4:53.

8. Kruisbeek, A.M., Astaldi, G.C.B., Blankwater, M.J., Zijlstra, J.J., Levert, L.A. and Astaldi, A. 1978. The in vitro effect of a thymic epithelial culture supernatant on mixed lymphocyte reactivity and intracellular cAMP levels of thymocytes and on antibody production to SRBC by Nu/Nu spleen cells. Cell Immunol. 35:134.

9. Papiernik, M., Nabarra, B. and Bach, J. 1975. In vitro culture of functional human thymic epithelium. Clin. Exp. Immunol. 19:281.

10. Pyke, K.W. and Gelfand, E.W. 1974. Morphological and functional maturation of human thymic epithelium in culture. Nature 251:421.

11. Kruisbeck, A.M., Krose, T.C. and Zylstra, J.J. 1977. Increase in T cell mitogen responsiveness in rat thymocytes by thymic epithelial culture supernatant. Eur. J. Immunol. 7:375.

12. Lebowitz, A. and Larence, H.S. 1969. Target cell destrubtion by antigen stimulated lymphoctes. Fed. Proc. 28:630.

13. Lebowitz, A.S. and Lawrence, H.S. 1971. The technique of clonal inhibition: A quantitative assay for human lymphotoxin activity. In In Vitro Methods in Cell Mediated Immunity (Bloom, B.R. and Glade, P.R., eds.). Academic Press, New York, p. 375.

14. Cooperband, S.R. and Green, J.A. 1971. Production and assay of a lymphocyte derived "proliferation inhibitory factor - PIF." In In Vitro Methods in Cell Mediation Immunity (Bloom, B.R. and Glade, P.R., eds.). Academic Press, New York, P. 381.

15. Jordan, R.K. and Crouse, D.A. 1979. Studies on the thymic microenvironment: Morphologic and functional characterization of thymic nonlymphoid cells grown in tissue culture. J. Reticuloendothel. Soc. 26:385.

16. Jordan, R.K., Crouse, D.A. and Own, J.J.T. 1979. Studies on the thymic microenvironment: Nonlymphoid cells responsible for transferring the microenvironment. J. Reticuloendothel. Soc. 26:373.
17. Holzman, R.S., Lebowitz, A.S., Valentine, F.T. and Lawrence, H.S. 1973. Preparation and properties of cloning inhibitory factor. Cell. Immunol. 8:240.
18. Yoshida, T. 1979. Purification and characterization of lymphokines. In *Biology of the Lymphokines* (Cohen, S., Pkc, E. and Oppenheim, J.J., eds.). Academic, Press, New York, p. 278.
19. Maschler, R. and Maurer, H.R. 1979. Screening for specific calf thymus inhibitors (chalones) of T-lymphocyte proliferation. Hoppe Seylers Z. Physiol. Chem. 360:735.

PROBLEMS IN THE BIOASSAY OF PRODUCTS FROM CULTURED HEK CELLS:
PLASMINOGEN ACTIVATOR

Marian L. Lewis[1,3], Dennis R. Morrison[2],
Bernard J. Mieszkuc[2], and Diane L. Fessler[1]

ABSTRACT

For research and commercial purposes, the production of materials from cultured cells is progressively becoming routine. Measuring products in cell culture medium, however, is not always straightforward even when the most appropriate product assays are utilized.

A researcher interested in a particular cell product does not intend to spend time developing or refining existing assays. Instead, he would prefer using techniques and standards reported in the literature. Evaluations of enzyme activity using purified standards or control products may inadvertently lead to the

---

[1] Technology Incorporated, NASA-Johnson Space Center, Houston, Texas 77058

[2] NASA-Johnson Space Center, Houston, Texas 77058

[3] To whom correspondence should be addressed

Abbreviations: PA - plasminogen activator, HEK - human embryonic kidney, WHO - World Health Organization, CTA - Committee on Thrombolytic Agents, HMW - high molecular weight, LMW - low molecular weight, CFES - continuous flow electrophoresis system

assumption that the same procedure will give reproducible results when cell culture medium is assayed for the product.

This report describes the selection of commercially available lots of human embryonic kidney cells and the activity of the plasminogen activator (PA) produced by these cells. PA activity was measured by the fibrin plate assay. The problems of comparing activity in conditioned culture medium with that of purified standard preparations are presented. Factors contributing to non-linearity in dose response curves, inconsistencies in activity in replicate flask cultures, and variations in repeated assays after sample storage are considered. Sample handling procedures and alternate assay systems are discussed.

INTRODUCTION

The use of normal human diploid cells as factories for the production of therapeutic or commercially important natural products is appealing. For most products there are no non-human sources available which are safe or as effective as the natural cell product.[1] Genetic engineering products are not yet standardized. Even if future gene products can be purified, the natural product from the cell will serve as the "model molecule." Also it may be some years before the glycosylated products are available from genetically programmed bacterial cells.[2] Thus, production of many natural products from cultured human cells has application far into the future.

Prior to developing full production capabilities, assays to measure the activity of the desired product must be developed. Generally, the assays reported in the literature are selected. This paper illustrates that measuring activity of a product in tissue culture medium may not be as simple or straightforward as one would assume. We will discuss several of the difficulties encountered in the assay of plasminogen activator (PA) present in human embryonic kidney (HEK) cell culture medium.

Since the report by Bernik and Kwaan in 1967[3] of the production of plasminogen activators by HEK cells in culture, various other cell types have been shown to produce the enzyme.[4,5,6] Numerous studies to standardize the assays and reagents,[7] to elucidate the presence of different molecular weight forms[8,9] and to describe inhibitors[4,10] and activators[11] have been reported. The most common assays employed include timed clot lysis,[12] radioactive assays using ^{125}I labeled fibrinogen,[13] fibrinolysis in fibrin plates,[14] and colorimetric assays using synthetic substrates.[15,16]

Bernik and Kwaan[3] have reported that only 5-10% of the cells in a heterogeneous HEK cell population produce plasminogen activators. Enhanced production could be achieved if producer cells could be selectively separated from non-producers. This would also serve as a marker of efficiency for separation techniques. It was our objective to isolate producer cells by sophisticated electrophoretic separation techniques. We chose the fibrin plate assay because of the availability of standard reagents, simplicity of the procedure, ability to assay a large number of samples at once, and good assay sensitivity. Some of the problems we encountered in assaying for specific PA production in separated cell fractions were due to assumptions that the stability of the enzyme in conditioned culture medium and in solutions of purified standard were the same under similar conditions of testing and handling.

In order to obtain valid reproducible assays of products from cell cultures, the parameters of the dose response, as well as sample storage and handling must be carefully controlled. Often project objectives or production plans do not provide enough time or resources to dwell on assay refinements. However, the impact of research or operational decisions which ultimately are based on the accuracy of the assays warrants meticulous attention to details. Adequate proof must be gathered to clearly show that the assay is accurate and reproducible in the presence of unknown factors such as those in conditioned culture media.

MATERIALS AND METHODS

Cells and Medium

Human embryonic kidney (HEK) cells were obtained from M. A. Bioproducts, Walkersville, MD. The cells were grown in M-199 with Earle's salts (Gibco) supplemented with 10% calf serum (Flow Laboratories), 100 units/ml of penicillin and 100 µg/ml of streptomycin sulfate (Gibco). Cells were detached from culture vessel surfaces by treatment with a solution of trypsin (Gibco) and ethylenediaminetetraacetic acid (EDTA) (Sigma Chemical Co.), each at a final concentration of 0.05% in Ca^{++} Mg^{++} free phosphate buffered saline (PBS). Serum free PA production medium prepared by the method of Barlow,[17] consisted basically of balanced salt solution supplemented with lactalbumin hydrolysate, glycine, human serum albumin, glucose and glutamine.

Evaluation of HEK Cell Lots

For comparison of the growth characteristics of different lots of HEK cells, ampules containing frozen cells were immersed in a 37°C water bath and cells were thawed quickly. The cells were resuspended in warm growth medium and counted (Coulter Electronics). The viability was evaluated by trypan blue exclusion.

Culture vessels were seeded at $5-7 \times 10^4$ viable cells per cm^2 of growth surface and incubated at 37°C in 5% CO_2 95% air in a humidified incubator. After 24 hours, duplicate cultures were evaluated for attachment of viable input cells by trypsinization and counting. For the remaining cultures, the medium was changed after 48 hours and cells reached confluent density at seven to ten days. Cells in primary culture and after three sequential passages were karyotyped[18] and chromosome morphology was noted.

Production of PA by HEK Cells

For evaluating PA production, the HEK cells were grown either in Nunclon Delta Multidishes (Vangard International) or in 25 and 75 cm^2 growth surface flasks (Corning). The cells, seeded at approximately 4×10^4 cells/cm^2, were grown to approximately 80-90% confluence (subconfluent), the medium was discarded, and cells were washed twice with Ca^{++} Mg^{++} free PBS. A volume of serum free production medium was added to culture vessels at a ratio of 0.4 ml per cm^2 of growth surface. Cultures were incubated and sampled over a two to four week period.

Assay of Plasminogen Activator

The activity of PA produced by HEK cells was assayed by a modification of the method of Brackman.[14] Activity was expressed as CTA (Committee on Thrombolytic Agents) units/ml. Fibrin plates were prepared as follows. Bovine fibrinogen, 75% clottable (Miles), was made up to a final concentration of 0.5% in Tris HCl buffer, pH 8.8. Ten ml of the solution was added to 100 mm petri plates (Falcon) and mixed with 0.2 ml of bovine thrombin (Miles) at 100 units/ml in distilled water. Clotting was allowed to occur at room temperature and plates were placed in a refrigerator and used on the day of preparation. PA activity in samples of conditioned serum-free medium was evaluated by adding 20 μl of sample to the clot. Four or five spots were placed on each plate, and the lysis zone diameters were read orthogonally on a calibrated viewer (Transidyne General) 15 to 17 hours later. A standard preparation of urokinase, World Health Organization (WHO) International Laboratory for Biological Standards, England was run with each assay of conditioned medium. The dose response curve of the WHO standard diluted in pH 8.0 buffer (0.5 M Tris, 0.25 M NaCl, 1 mg/ml gelatin) was linear in the range of 10-60 CTA units/ml. Conditioned medium samples were diluted in the same buffer to measure within the range of 10-60 CTA units/ml. All lysis zone data were subjected to statistical analysis using a modified linear regression program (the UCLA Biostatistics BMDP package) on a Digital VAX-VMS computer.

RESULTS

Characterization of HEK Cell Lots

In our laboratory the qualification of HEK cell lots for selection as candidates for the study of PA production depended on their characteristic viability, outgrowth, and karyology. The characterization of eight cell lots is shown in Table I. Immediately after thawing cell suspensions of tissue digests received from the vendor, the cell viabilities ranged from 45% to 74%. At 24 hours of incubation after plating, 12% to 18% of input viable cells had attached. One lot, 8507, had low viability and only 2% of the cells attached after 24 hours of incubation. This lot did not reach confluence by day 10 and was eliminated from the study. Normally, cells reached confluency within 10 days. As an added indicator of cell growth, proliferative lifespans were also determined. Cells of most of the lots phased out by passage level five. The cells in several lots were predominately of small size which may account for the large number of population doublings for some lots (see lots 4347 and 5807). All of the cell lots tested showed normal karyology in primary culture; however, at passage three, instability was often manifested as high aneuploidy levels or chromosome gaps.

Biostatistical Analysis - PA Activity

In addition to careful screening for acceptable HEK cell lots, application of adequate biostatistical analysis of assay data is mandatory. Dose response curves, in the linear range of 10 to 60 CTA units/ml of WHO standard urokinase were analyzed by standard regression analysis. An example is shown in Table II. For each WHO standard dilution the diameters of four zones were

Table I

Cell Lot No.	Viability		Growth		Karyology					
	TBE[1] %	Attach. 24-hour (%)	Confl. by Day 10	Prolif. Life-span[2]	Primary			Pass 3		
					Count	Ploidy	Morph	Count	Ploidy	Morph
4347	74	15	+	11.2 (8)	46	Dip	Norm	46	Dip	Gap #1,#9
5807	67	18	+	27.3 (10)	46	Dip	Gap #7	46-45	Aneu 4%	Norm
8922	45	ND	+	3.03 (3)	46	Aneu 4%	Norm	46-45-44	Aneu 15%	Gaps
0085	48	15	+	4.4 (4)	46	Dip	Norm	46	Dip	Norm
8507	25	2	-	ND	ND	ND	ND	ND	ND	ND
8514	70	12	+	8.3 (5)	46	Dip	Norm	46-45	Aneu 8%	Norm
5360	58	ND	+	6.1 (3)	46	Dip	Norm	46	Dip	Norm
9064	60	14	+	5.5 (5)	46	Aneu 5%	Norm	46-45-43	Aneu 21%	Norm

(1) Trypan Blue Exclusion
(2) Proliferative Lifespan - Population Doublings (Phase-out Passage Level)
ND, indicates not done

Table II

WHO STANDARD (CTA UNITS/ML)	SAMPLE MEAN (MM) (1)	COMPUTED CTA UNITS/ML	% ERROR (2)
10	19.1125	10.5959	5.9594
20	20.4875	18.7851	-6.0745
30	21.5125	28.0906	-6.3647
40	22.5250	41.0380	2.5950
50	23.0750	50.0654	0.1308
60	23.7125	62.6741	4.4568

CORRELATION COEFFICIENT .997

(1) MEAN OF EIGHT VALUES, 4 ZONES EACH READ ORTHOGONALLY.
(2) CALCULATED AS THE PERCENT DIFFERENCE BETWEEN STANDARD CONCENTRATION AND ESTIMATED CONCENTRATION CALCULATED FROM THE STANDARD CURVE AT THAT PARTICULAR DILUTION.

read orthogonally, thus generating eight values. The mean of these values was used to calculate the computed activity at each dilution. Figure 1 shows a typical computer generated regression line plot of a WHO standard curve. Typically the correlation coefficients were quite acceptable (0.95-0.99) and the individual points were near the regression line. We found, however, that sometimes these were not satisfactory indicators of acceptable variability for certain specific sample dilutions within the normal dilution range. When actual PA concentrations were compared with estimated concentrations obtained from the regression equations, an odd dilution value could result in 20-25% error in the calculated concentration of PA. Operationally, we then chose to use only those standard curves whose individual measurements varied by 10% or less from the calculated PA concentration using the regression line equation. As sources of experimental variation were eliminated, we achieved routine standard dilutions with a maximum of 6-7% error for any one dilution as shown in Table II.

Reproducibility of Standard Curve Data

During the test period, 35 WHO standard curves were generated and three lots of fibrinogen were qualified as acceptable. Fibrin plates were made using each of the fibrinogen lots and dilutions

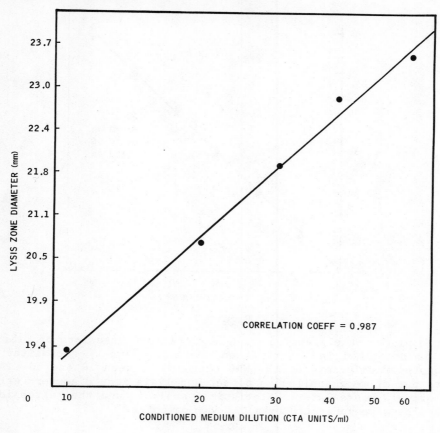

Figure 1

A typical computer generated regression line plot of a WHO standard curve. The mean of eight lysis zone diameter values for each dilution of the WHO standard from 10 to 60 CTA units/ml were used in the linear regression program on a VAX-VMS computer.

of the WHO standard were applied to the fibrin layer as previously described. The fibrinogen lots gave almost comparable results (Figure 2); however, there were differences in the lysis zone size. In selecting a lot of fibrinogen, attention should be given to zone size since the diameters must be within the range of the measuring device. Zones must not be larger than 24.8 mm for accurate measurement with our device.

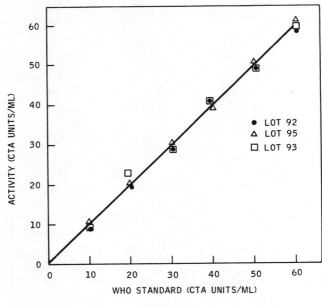

Figure 2

Fibrin plates were prepared using each of the three lots of fibrinogen as described in Materials and Methods. Twenty microliter aliquots of dilutions of the WHO standard between 10 and 60 CTA units per ml were applied to plates and lysis zone diameters were measured 17 hours later. Statistical analysis and the regression line plot were generated by a VAX-VMS computer.

To ascertain the reproducibility of our standard over a prolonged period of time the mean and standard deviations for 35 assays were plotted (Figure 3). The WHO standard was routinely diluted to contain 10 to 60 CTA units/ml, and 0.1 ml aliquots were stored frozen at −20°C until used. Standard curves with dilutions having errors greater than 10% (Table II) were not included in the plot, and the aberrant dilutions were discarded. These data show the consistent linear dose response curve of the purified urokinase standard stored at −20°C. The reproducibility of these tests using the WHO standard led us to believe that serum free conditioned medium from cell cultures could be handled and stored in the same manner. As we will discuss later this was misleading.

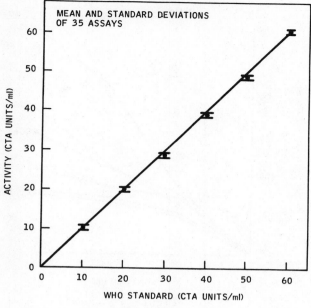

Figure 3

Computer generated regression line plot of the WHO urokinase standard tested by the fibrin plate method. Data were accumulated for assays performed during an 18 month period. All WHO standard dilutions were made in 0.05 M tris, 0.25 M NaCl, gelatin (1 g/l) buffer at pH 8.0. Dilutions were stored as 0.1 ml aliquots at $-20°C$.

Production of PA by Different HEK Cell Lots

Having shown that the WHO standard gave reliable linear dose response curves, the PA activity in serum-free conditioned medium from several lots of HEK cells was measured. Figure 4 shows some cell lots to be superior producers of PA. For instance, lot 8514 produced approximately four times higher levels of activity than the lowest producers. This lot also had a high viability and grew well in culture (Table I).

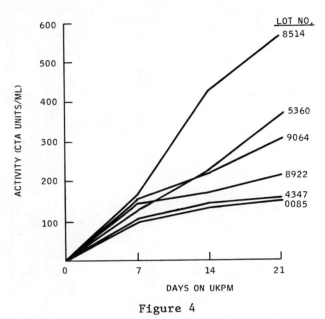

Figure 4

The fibrinolytic activity in serum-free conditioned medium from cultures of various HEK cell lots were measured by the fibrin plate assay. Cells were grown to approximately 80-90% confluency, washed twice with Ca^{++} and Mg^{++} free phosphate buffered saline and serum free medium was added at a ratio of 0.4 ml/cm^2 of growth surface area. Samples were taken at weekly intervals for evaluation of PA activity. Aliquots were stored at -20°C until tested in fibrin plate assays. UKPM = Uronikase Production Medium.

Production of PA by Fractions of Separated Cells

Since a part of our objective was to isolate target PA producer cells, we examined several ways of separating the heterogeneous subpopulations in HEK cell cultures. The method of choice proved to be the McDonnell Douglas Continuous Flow Electrophoresis System (CFES). This is a new state-of-the-art device constructed at the McDonnell Douglas Astronautics Corp. in St. Louis. Other methods included sedimentation at unit gravity in 10% serum-medium, sucrose and ficol gradients.

Figure 5 shows PA production by fractions of CFES separated HEK cells. It appeared that fraction numbers 94, 95 and 103 were the highest PA producers; however, some inconsistencies in activity of diluted samples were apparent (see Figure 6).

Since many samples were tested simultaneously, only three dilutions were initially used hoping to bracket the linear activity range of 10-60 CTA units/ml as previously established for the WHO standard. The extent to which some diluted medium samples showed a non-linear dose response is illustrated (Figure 6). Some undiluted fractions had activity higher than 60 CTA units/ml. Many times both the 1:2 and 1:5 dilution values were within test range but did not result in comparable values when multiplied by dilution factors. Sometimes higher dilution of sample resulted in higher activity, and at other times the 1:5 dilution values were lower than those of the 1:2 dilution.

Sources of Inconsistencies: Non-linear Dose Response Curves

Since the measurement of cell production trends is a preliminary step toward increasing production or selecting high producer cell cultures, it is important to be aware of factors, however trivial they may seem, which could result in inconsistent assays. Inconsistencies in the observed activity levels in culture medium (Figure 6), resulted in a re-examination of our operational assumptions concerning sample dilutions. We designed a set of experiments to determine the proper dilution ranges of culture medium samples. The computer generated linear regression plot of PA activity in serum free culture medium (Figure 7) appeared linear with an acceptable correlation coefficient (0.987) and a reasonable amount of sample variation. However, when the activity was expressed as a function of the dilution factor (Figure 8) the range of PA activity was constant only between dilutions of

1:5 to 1:8. This was in contrast to dilutions of the WHO standard which were constant up to dilutions of 1:10.

These data showed that "bracketing" or short cut dilutions may not provide sufficient information for selection of actual activity values. A series of incrementally close dilutions should be tested in order to find the constant dilution range for each sample even though this is often laborious and time consuming.

Inconsistencies in PA Activity after Storage of Samples at -20°C

Since we had shown that the WHO standard was stable for over 18 months when stored at -20°C and earlier reports indicated that

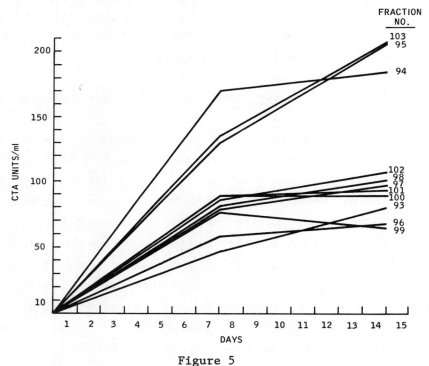

Figure 5

HEK cells in heterogeneous populations were separated on the Continuous Flow Electrophoresis System (McDonnell Douglas). The cells in the separated fractions were grown to 80-90% confluency, maintained on serum-free medium and sampled for PA activity as described in Figure 4.

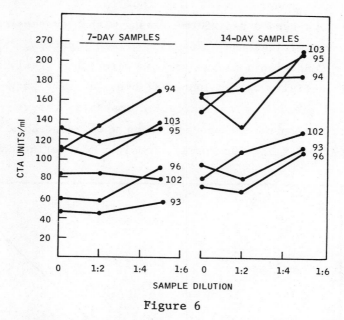

Figure 6

Data represented in Figure 5 plotted for selected fractions showing non-linearity of resulting activity when values at 1:2 and 1:5 were multiplied by dilution factors. All samples were diluted in buffer (0.05 M Tris, 0.25 M NaCl, 1 mg/ml gelatin).

PA is stable in culture medium at 37°C,[19] we routinely stored our conditioned culture medium samples at -20°C. However, when we retested samples, we found that activity from test to test varied (Table III). In some cases the activity was higher in retests (108 and 115) and in other samples the activity dropped. The activity was approximately the same after retesting in less than a third of the samples.

PA Activity in Replicate Flasks Cultures

Since one of our objectives was to compare the absolute activity among fractions of separated cells, we were concerned that replicate flask cultures might also show PA activity variation. To test this, we seeded ten 25 cm^2 flasks with a suspension of HEK cells mixed by swirling the suspension before pipetting cells into each flask. Conditioned medium samples were taken weekly for four weeks and PA activity was evaluated (Table IV). The large standard deviation and flask to flask variation emphasize the requirement for standardization of techniques. Even when reasonable efforts were made to seed flasks uniformly, considerable variation was

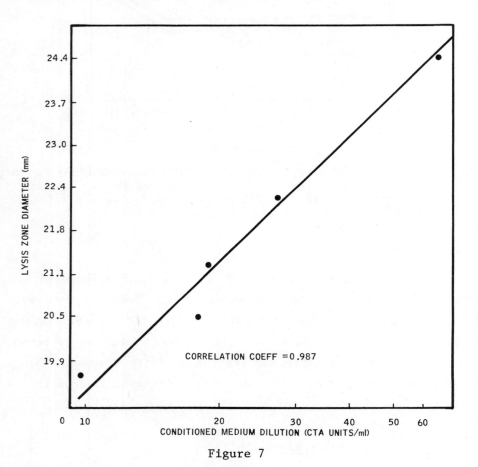

Figure 7

Conditioned medium was sampled from a representative 14 day HEK cell culture. The medium was diluted with buffer and PA activity was measured as described in Materials and Methods. Lysis zone diameter data were subjected to statistical analysis and a plotted by the VAX-VMS computer.

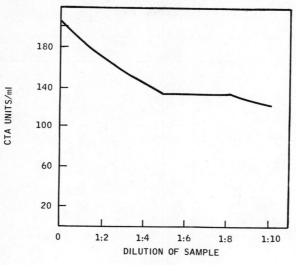

Figure 8

When PA activity in CTA units/ml for the same data described in Figure 7 was plotted against sample dilution, a plateau of activity in the dilution range of 1:5 to 1:8 was obtained. The plateau shows the appropriate dilution range values for assessing the activity for this sample.

Table III

Sample No. Day	CTA Units/ml		
	Test 1	Test 2	Test 3
107 (7)	32.5	32.4	ND
115 (7)	35.6	93.8	ND
111 (7)	145.0	156.5	ND
104 (16)	91.8	80.0	56.4
105 (16)	96.8	81.0	50.3
106 (16)	84.8	53.9	40.2
108 (16)	114.8	291.5	267.8

Samples were tested for PA activity as described and stored at $-20°C$ for one to three months. Subsequently, the samples were thawed and retested as shown.

still apparent. Consequently, extra care must be taken to ensure the dispensing of consistent cell numbers and representative cell types into each flask. Unequal distribution of cell functional types could produce large PA activity differences among replicate flasks. Early reports of Bernik and Kwaan[20] indicated that some

Table IV

FLASK NO	CTA UNITS/ML ON DAY:			
	7	14	21	28
1	66.9	73.3	71.0	59.6
2	56.6	76.2	60.4	48.6
3	68.0	61.6	61.1	47.3
4	73.8	78.0	65.9	48.2
5	51.6	48.2	40.9	35.4
6	96.0	110.0	67.2	52.5
7	60.0	50.0	59.6	45.7
8	51.1	45.7	41.5	28.1
9	56.2	58.1	63.1	42.3
10	68.0	75.0	68.0	46.1
MEAN	64.8	67.6	59.8	45.4
S.D.±	13.4	19.3	10.5	8.7

Replicate flasks (25 cm^2 growth surface) were seeded with cell suspensions by pipetting 10 ml volumes of growth medium containing 1 x 10^5 cells/ml into the flasks. Before dispensing cells into each flask the suspension was swirled to ensure a homogeneous suspension. When cells reached 80-90% confluence, the growth medium was replaced with serum-free medium and samples were assayed as described in Materials and Methods.

cell cultures produce PA inhibitors concomitant with PA. Consequently culture flasks should be seeded with cells from continuously stirred suspensions.

Expression of Activity - A Common Denominator Unit

We wanted to monitor PA production over a period of weeks, but with a limited number of cells available we did not want to routinely sacrifice cultures to evaluate cell-associated PA. However, we found that it may be necessary to periodically compare units/ml with units/cell to determine actual production levels in continuing culture. As shown in Figure 9, cell numbers decline with time in culture; whereas PA activity levels increase. If the decrease in cell numbers with time is not linear and predictable,

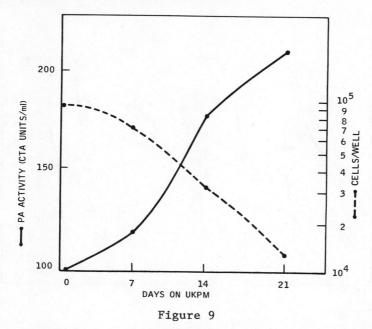

Figure 9

HEK cells were grown in multidishes (Nunclon) and maintained on serum-free medium as described. At 7, 14 and 21 days the conditioned medium was sampled for PA activity and the cells were trypsinized and counted. The graph represents the mean cell counts and PA activity for four cultures evaluated at each sampling interval.

it becomes very difficult to accurately compare the accumulated PA production among different cell cultures.

Figure 10 shows PA activity expressed as units/ml and units/100 cells for four replicate culture groups. When the activity is expressed as CTA units/ml, considerable variation is apparent among replicates and it would appear that groups 2 and 3 produced much less PA than cultures 1 and 4 after 14 and 21 days of culture. However, when the PA activity is normalized to CTA units per 100 cells the production appears to be more consistent among all four replicate groups. Obviously, the decrease in cell number must be relatively linear and consistent from replicate to replicate to make this expression of PA activity meaningful.

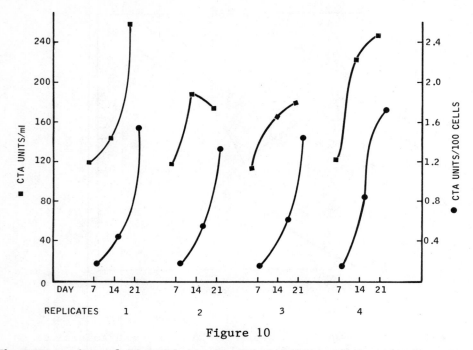

Figure 10

The expression of PA activity in CTA units per ml compared to CTA units per 100 cells for values described for Figure 9.

Even though a larger number of samples must be evaluated, it is apparent that enough replicate cultures should be seeded initially to allow sacrifice and counting of cells at each sampling interval. Early evaluations of PA activity/ml of media may be quite useful especially within the first week of production. However, the expression of PA units/ml should be compared with the measured PA units/cell to ensure reasonable consistency in accumulated production levels, especially if cultures are maintained for several weeks.

DISCUSSION

Research involving natural products from mammalian cells in culture usually requires the bioassay of the products in the conditioned culture medium. The selection of cells to be used for production requires preliminary studies to determine viability, outgrowth, and stability of karyotype after subculture.

Experiments should also be conducted to verify the usefulness of the assays chosen to measure the levels of the target product. Such assays are generally performed using a purified standard or known control material. Results of bioassays using standards or control products can lead a researcher to assume that the assay is reproducible under the same conditions when testing unknown samples. This may not always be true, especially when one is assaying samples of conditioned medium from cell cultures. This may lead to inconsistent results and difficulties in demonstrating clear cut effects of alterations to the culture medium or test conditions. Initial data should be evaluated for reproducibility of the assays used and should accurately determine the product concentration in the culture medium. It is be necessary to identify the sources of variability and eliminate them. This requires biostatistical analysis of data.

In the assay of PA in HEK cell culture medium, the WHO standard urokinase preparation produced consistent results while assays of the product in serum free culture medium did not. The following inconsistencies were noted: non-linearity in dose response curves, failure to obtain reproducible results in tests repeated after storage of medium samples at $-20°C$, and variability in activity among replicate flask cultures. Solutions to these problems included making additional dilutions to ensure accurate calculation of activity from a constant dilution range, storage of samples in aliquots for retesting, storage of samples at a colder temperature (-70°C), and refining culture procedures to reduce inconsistencies among replicate flasks.

Special mention should be made regarding the storage of conditioned serum-free medium samples at -20°C. Increases in apparent PA activity after storage were surprising. If activity had been consistently lower in retests, one might assume that the enzyme was being inactivated. This may be true for some samples. However, other factors appeared to be operating. For instance, freeze-thaw cycles and sample handling may cause the disassociation of inhibitors from PA or activation of PA by other proteolytic enzymes in the medium.[11] Other investigators[21,22] have reported variation in absolute activity of PA in different experiments. Since we had shown our fibrin plate assay to be quite reproducible using the WHO standard within the range of 10-60 CTA units/ml, we considered that other factors were responsible for variations in culture medium samples stored at -20°C. Therefore, we adopted the policy of storing five to 10 aliquots of each culture medium sample at -70°C. No retesting was done on sample aliquots which had been previously thawed.

Standardization of these test parameters may not totally eradicate the problem of assay inconsistencies. Other factors also influence the assay. For example, the fibrin plate assay based on measurement of lysis zones is subject to variation caused by different concentrations of high and low molecular weight forms of the enzyme. It has been reported that the low molecular weight (LMW) form diffuses faster, thus producing larger lysis zones.[23] In samples containing both forms, the continuous increase in volume of fluid in the lysis zone due to rapid lysis by the LMW form mass dilute the slower diffusing high molecular weight (HMW) form,[14] resulting in predominant expression of only the LMW form.

Assessing the true concentration of PA may also be complicated by the autocatalytic conversion of HMW urokinase to the LMW form at 37°C.[24] The use of a chromogenic assay using the synthetic substrate S-2444 (Kabi Diagnostics), though not as sensitive as the fibrin plate assay, is not affected by the differential presence of the molecular weight forms.[25] Therefore, multiple assays may be necessary especially when attempting to detect subtle changes in product levels secreted into culture medium. However laborious, multiple assays often provide unforeseen advantages. For example, when using both the clot lysis and chromogenic assays for urokinase, results from assays of known mixtures of HMW and LMW forms may be used to determine the ratios of HMW to LMW in culture media samples.[25]

Bernik and Kwaan[20] and Roblin et al[26] reported the concomitant production of PA and PA inhibitors by cells in culture. For maintenance of hemeostasis in the body, the simultaneous presence of activators and inhibitors seems desirable. However, in culture medium the presence of inhibitors may mask PA production by the cells. Loskutoff and Edgington[27] reported that the action of PA inhibitors at the cell surface is immediate and irreversible. This is not clear, however, since Aoki and Kawano[28] report that the inhibition is reversible. The accuracy of different fibrin lysis assays may be compromised by the presence of PA inhibitors in test samples.[29] Consequently, the accurate determination of PA in culture medium may necessitate a separate assay for presence of inhibitors. Culture techniques are also important. Not only the seeding density, but also the type of cells cultured may affect PA production. Careful and continuous mixing of the seeding suspension during seeding of replicate cultures is important if, as Bernik and Kwaan have reported,[20] the degree of inhibitor activity partially depends on preferential survival and growth of inhibitor producing cells.

For purposes of comparing PA production in fractions of separated cells it is necessary to standardize the expression of the PA activity. Activity expressed as units per ml and units per cell (Fig. 10) should be compared to determine actual production levels in continuing cultures especially when cell numbers are decreasing. Also, cell death may result in release not only of PA, but factors which inactivate PA.[11]

Consistent with the theme of this Congress, when the emphasis is placed on the production of commercial or therapeutic products from cells in culture, one must consider rigorous control of sample handling and test parameters, use of adequate biostatistical analyses and possible use of multiple product assays. Often "off-the-shelf" assays are not useable unless modified. The true operational limitations of assays of target cell products in culture medium should be determined and project plans should include adequate provisions for these studies. Much time can be saved if product assays are suitably verified and refined early in the project.

ACKNOWLEDGEMENT

The authors gratefully acknowledge the technical assistance of Ms. Charlotte Stelly.

REFERENCES

1. J.M. Price, G.H. Barlow and J.E. Donahoe, Mammalian cell culture as a source of biological materials. In: "Principles and Techniques of Human Research and Therapeutics, A Series of Monographs," (F. Gilbert McMahon, ed.), Futura Publishing company, Inc., Mt. Kisko, NY, Vol. XV:37-54, (1978).

2. R. Davis, Technical issues of recombinant DNA in medical products. Presented at Symposium on Medicinals, Chemicals, Diagnostic Products, and Evaluation of New Biotechnology, Stanford Res. Inst., Palo Alto, CA, 1980.

3. M.B. Bernik and H.C. Kwaan, Origin of fibrinolytic activity in cultures of human kidney. J. Lab. Clin. Med. 70:650-661, 1967.

4. M.B. Bernik and H.C. Kwaan, Plasminogen activator activity in cultures from human tissues. An immunological and histochemical study, J. Clin. Invest. 48:1740-1753, 1969.

5. L.J. Lewis, Plasminogen activator (urokinase) from cultured cells. Thrombo. and Haem. 42:895-900, 1979.

6. E.F. Howard, C.Y. Cheng and J.S. Howard, Mitogenic activity and plasminogen activator in harvest fluid concentrates from mammary cells in culture. Cancer Res. 40:4385-4389, 1980.

7. A.J. Johnson, D.L. Kline and N. Alkjaersig, Assay methods and standard preparation for plasmin, plasminogen and urokinase in purified systems, in: "Recommendations of the NHI Committee on Thrombolytic Agents, Subcommittee for Standardization, (F.K. Schallauer, ed.), Verlag., Stuttgart, NY Vol. XXI:259-272 (1969).

8. W.F. White, G.H. Barlow and M.M. Mozen, The isolation and characterization of plasminogen activators (urokinase) from human urine, Biochem. 5:2160-2169, 1966.

9. L. Holmberg, B. Bladh and B. Astedt, Purification of urokinase by affinity chromatography, Biochim. Biophys. Acta. 445:215, 1976.

10. M.E. Soberano, E.F. Ong and A.J. Johnson, The effects of inhibitors on the catalytic conversions of urokinase, Thromb. Res. 9:675, 1976.

11. M.B. Bernik and E.P. Oller, Increased plasminogen activator (urokinase) in tissue culture after fibrin deposition. J. Clin. Invest. 52:823-834, 1973.

12. Plough and N.D. Kjeldgaard, Urokinase; an activator of plasminogen from human urine. I. Isolation and properties. Biochem. Biophys. Acta. 24:278-282, 1957.

13. J. Unkless, K. Dano, G.M. Kellerman and E. Reich, Fibrinolysis associated with oncogenic transformation. Partial purification and characterization of the cell factor, plasminogen activators. J. Biol. Chem. 249:4295-4305, 1974.

14. P. Brackman, ed., A standardized fibrin plate method and a fibrinolytic assay of plasminogen, in: "Fibrinolysis," Scheltema and Holkema, Amsterdam, p. 124 (1967).

15. W.L. Bigbee, H.B. Weintraub and R.H. Jensen, Sensitive fluorescence assay for urokinase using synthetic peptide 4-methoxy-B-napthylamide substrates. Analyt. Biochem. 88:114-122, 1978.

16. G. Claeson, P. Friberger, M. Knos and E. Eriksson, Methods for determination of prekallikrenin in plasma, glandular kallikrenin and urokinase, Haemostasis. 7:76-78, 1978.

17. G.H. Barlow, A. Reuter and I. Tribby, In: "Proteases in Biological Control," (e. Reich, D.B. Rifkin and E. Shaw, eds.) Cold Spring Harbor Laboratories, NY, pp. 325-331 (1975).

18. P.S. Moorhead, Harvesting human leucocyte cultures for chromosome cytology. In: "Tissue Culture Methods and Applications," (P.F. Kruse, Jr. and M.K. Patterson, Jr., eds.) Academic Press, NY, pp. 768-773 (1973).

19. R.N. Hull, W.R. Cherry, R.M. Huseby and S.A. Clavin, Studies on tissue culture produced plasminogen activator. I. Preliminary observations and the enhancing effect of colchicine and other antimitotic agents, Thromb. Res. 10:669-677, 1977.

20. M.B. Bernik and H.C. Kwaan, Inhibitors of fibrinolysis in human tissues in culture, Amer. J. Physiol. 221:916-921, 1971.

21. D.B. Rifkin and R. Pollack, Production of plasminogen activator by established cell lines of mouse origin, J. Cell Biol. 73:47-55, 1977.
22. I.N. Chou, R.O. Roblin and P.H. Black, Calcium stimulation of plasminogen activator secretion/production by Swiss 3T3 cells, J. Biol. Chem. 252:6256-6259, 1977.
23. G.H. Barlow, C.W. Francis and V.J. Marder, On the conversion of high molecular weight urokinase to the low molecular weight by plasmin, Submitted, 1981.
24. M. Nobuhara, M. Sakamaki, H. Ohnishi and Y. Sasuki, A comparative study of high molecular weight urokinase and low molecular weight urokinase, J. Biochem. 90:225-232, 1981.
25. R.D. Philo and P.J. Gaffney, Assay methodology for urokinase: Its use in assessing the composition of mixtures of high- and low-molecular weight urokinase. Thromb. Res. 21:81-88, 1981.
26. R.O. Roblin, P.L. Young and T.E. Bell, Concomitant secretion by transformed SVW138-VA13-2RA cells of plasminogen activator(s) and substance(s) which prevent their detection. Biochem. Biophys. Res. Com. 82:165-172, 1978.
27. D.J. Luskutoff and T.S. Edgington, An inhibitor of plasminogen activator in rabbit endothelial cells, J. Biol. Chem. 256: 4142-4145, 1981.
28. N. Aoki and T. Kawano, Inhibition of plasminogen activators by naturally occurring inhibitors in man. Amer. J. Physiol. 223:1334-1337, 1972.
29. G. Wijngaard, Interfering factors in the assay of plasminogen activators by the fibrin plate method. Occurrence of different inhibitors against tissue plasminogen activator and urokinase. Thrombos. Haemostas. 41:590-600, 1979.

A SIMPLE METHODOLOGY FOR THE ROUTINE PRODUCTION AND PARTIAL PURIFICATION OF HUMAN LYMPHOBLASTOID INTERFERON

P.J. Neame[1,3] and R.T. Acton[1,2,3]

Departments of Microbiology[1], Epidemiology[2] and
Diabetes Research and Training Center[3]
University of Alabama in Birmingham
University Station, Birmingham, Alabama 35294

ABSTRACT

We report a methodology suitable for the large scale production and partial purification of human lymphoblastoid interferon with a minimum expense and reasonable (10%) degree of purity of product. The cells used were Namalwa lymphoblastoid cells, which have the advantage of being both easy to grow and well characterized. They were grown in RPMI 1640 containing 10% fetal calf serum to a density of $1.5 - 2 \times 10^6$ cells/ml and then diluted 50% by the addition of serum free medium containing 2mM sodium butyrate. After 48 hours, the medium was removed and the cells induced to produce interferon by Sendai virus. Typical initial interferon titres were in the region of $4 - 4.8 \log_{10}$ units/ml, while specific activities were in the region of 10^6 units/mg protein. The initial stage in the purification involved batchwise adsorption of the interferon to Procion red-HE7B Sepharose CL6B. The Sepharose

---

* Neame and Acton; unpublished results.
  This work was supported by NCI grant # CA09128
  <u>Abbreviations</u>: IFN: Interferon; SDS-PAGE: Sodium dodecyl sulfate polyacrylamide gel electrophoresis

was then packed into a column, washed with 0.5M KCl and the interferon eluted with 2M KCl. The interferon was further purified by gel filtration to give an activity of approximately 10^7 units/mg protein. Yields were between 30-50% of the initial interferon in a volume of 25ml. Further separation of the components of the heterogeneous alpha interferon could be obtained on Procion blue-HER Sepharose, or by utilizing reverse phase HPLC.

INTRODUCTION

The purification of human lymphoblastoid interferon (Hu IFN-alpha) to homogeneity has been achieved in a number of laboratories (1,2), as has the conceptually similar purification of leucocyte interferon (3). Major problems encountered by most workers are in the regions of cost and in the purification required for an acceptable product. Our work has been concerned with reduction of cost and, at the same time, improvement of purification protocols to achieve at least 10% pure IFNalpha on a routine basis, preferably using methods which would be applicable to the transfer of production to pilot plant scale.

It has been demonstrated that human leucocyte and lymphoblastoid interferons are complex mixtures of a variety of proteins of similar molecular weight but rather different sequences (4,5), but, so far, little effort has been made to resolve these species in production methodologies which do not involve recombinant DNA technology. Rubinstein et al. (3) have used HPLC for isolation of multiple species of leucocyte IFN, while other groups have concentrated on obtaining one single type of IFN from a complex mixture (2). We have used the versatile methodology of dye-ligand chromatography to select a column which would resolve lymphoblastoid IFN into at best four components, but invariably two, in a three step procedure suitable for use on a relatively large scale. Previously obtained data (6) indicated that the HE series of Procion dyes [bis-(monochlorotriazine) polysulfonated dyes] immobilized to Sepharose were useful for IFN purification. Screening of a selection of these dyes* indicated that red-HE7B and blue-HER were the most useful dyes for IFN purification.

MATERIALS

Sendai virus and cell culture medium (RPMI 1640) were purchased from Flow Laboratories, tissue culture plasticware was purchased from Costar and fetal bovine serum from either Flow Labs or Dutchland Labs. Procion dyes were free samples generously supplied by Imperial Chemical Industries, Wilmington, Delaware. Sepharose CL-6B was purchased from Pharmacia and Ultrogel AcA 44 was purchased from LKB.

METHODS

Interferon Assay

Interferon was assayed by a method essentially similar to that described by Stewart (7). The cytopathic effect of Semliki forest virus on VERO cells was scored in 96 well microtiter plates using the neutral red dye uptake assay described by Finter (8). NIH reference standard #G-023-901-527 was included on each plate and all assays were performed in duplicate. In cases where specific activity is quoted, the results are an average of a variety of determinations on separate plates.

Interferon Production

Interferon was produced by a modification of the method of Johnston (9). Namalwa cells were maintained in log phase growth to a density of about $1 - 2 \times 10^6$ cells/ml in the desired volume of RPMI 1640 supplemented with 10% fetal calf serum and penicillin/streptomycin. At this point, the volume was doubled with serum free medium supplemented with sodium butyrate to give a final concentration of 1mM. The cells were incubated for 48 hours and then the medium was removed by centrifugation at 600 x g for 10 minutes in 250 ml conical tubes (Corning). The cells were resuspended in approximately 1/10 of the previous volume, Sendai virus was added at 2.5 HAI units/10^5 cells and the cells incubated for 80 minutes. The cells were again centrifuged at 600 x g for 10 minutes and resuspended in fresh serum free medium at a final density of $2 - 3 \times 10^6$ cells/ml and incubated for 21 - 23 hours prior to removal of the cell supernatant by two spins: the first at 1200 x g for 10 minutes and the second at 14,000 x g for 20 minutes. Higher cell concentrations than this resulted in significantly lower final cell viability.

Interferon Purification

The cell supernatant was stirred overnight with approximately 100ml of red-HE7B-Sepharose CL-6B for each 2 liter volume at 4° C. The slurry was then filtered on a sintered glass funnel and washed with 10mM Tris HCl buffer, pH 7.4, before being packed into a chromatography column (2.6cm diameter) and washed with one column volume of buffer. Initial elution of some of the contaminating protein was achieved by a wash with one volume of buffered 0.5M KCl and the IFN was eluted with 2M KCl. It was found that the IFN eluted in the latter part of the protein peak and was collected in a volume of approximately 20ml. This was applied directly to a

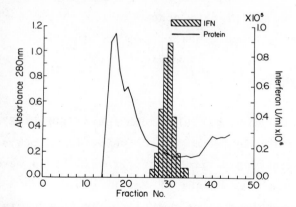

Figure 1. <u>Elution of Interferon from Ultrogel AcA 44</u>.
   IFN eluted from red-HE7B Sepharose by 2M KCL was applied to a 300ml column of Ultrogel AcA 44 equilibrated with 100mM ammonium bicarbonate. Fractions (7.5ml) were collected and assayed for protein (O.D. 280nm) and interferon.

column of AcA 44 (2.6 x 60cm) to remove higher molecular weight contaminants and the KCl.

The IFN peak eluted from the Ultrogel column was bulked with the products of other purifications having similar specific activity, the pH adjusted to 7.2 and the conductivity to 1,000 µMHOS and applied to a 20ml column (1cm diameter) of blue-HER-Sepharose CL-6B. The column was then washed with two volumes of Tris-HCl buffer (10mM, pH 7.4). Elution was achieved by a KCl gradient (0 - 2M, 40ml) in Tris buffer.

Protein assay

The coomassie blue binding assay of Bearden (10) was employed for assay of protein where mentioned. Otherwise the absorption of samples at 280, 214 or 206nm was used to obtain a close approximation to the protein concentration, using bovine serum albumin and bovine gamma globulin as standards.

Gel-electrophoresis

SDS-Polyacrylamide gel electrophoresis (SDS-PAGE) was performed as described by Laemmli (11) using 12.5% polyacrylamide gels in a non-reducing environment. Protein bands were detected by the silver stain technique of Merril et al. (12). IFN was detected by homogenization of gel slices in 100mM Tris buffer pH 7.4, followed by assay of the buffer as described above.

Dye-column synthesis

Dye columns were synthesized as described elsewhere (6) by base mediated nucleophilic substitution of the reactive chlorine on the dye by the hydroxyl groups on the agarose at 60 - 70°C.

RESULTS AND DISCUSSION

The average results of five separate production runs and the partial purification of the IFN at the four to six liter scale are outlined in table 1. In addition to these results we have also performed production runs at the seventy liter scale. While the latter obviously produce more IFN, we have concentrated on small scale runs as removal of serum proteins is crucial to obtaining the required high specific activity of IFN. Larger scale production requires the use of a continuous flow basket centrifuge and this leaves rather more medium behind than is desirable. By using 250ml conical centrifuge tubes, approximately half to a third of the volume is left with the pellet as compared to the basket centrifuge. As serum proteins probably consist of about 90% of the contamination, this is a major factor in obtaining high final purity. We have not found any serious problems associated with cell viability in serum free medium over these short incubations. The most critical step appears to be the butyrate treatment: cell viability should be at least 95% at the end of this stage.

Initial cell supernatants containing IFN usually had a protein concentration of 50 - 100 µg/ml and an IFN concentration of $1 - 6 \times 10^4$ units/ml. After batchwise adsorption of the crude IFN to Procion red-HE7B Sepharose, 40 - 50% of the protein bound, while 60 - 70% of the IFN was removed from solution. By adsorbing the IFN in a column system, rather higher retention of the IFN (80 - 90%) was obtained, with the disadvantage of low speed and considerable problems in handling large volumes. The unbound IFN does not bind to a fresh column of red-HE7B Sepharose, and is presumably a different sub-class of IFNalpha. Elution was achieved by either a salt gradient, which gives good resolution and a partial resolution of the eluted IFN into two peaks(6), or by stepped elution as described above with 75 - 100% recovery. The latter procedure provides slightly lower specific activity, but the IFN eluted has a very high titre per ml which is a great advantage when applying the peak to a gel filtration column. In addition, the results are very predictable and enable the eluate to be loaded onto the AcA column in a tandem system resulting in slightly higher specific activity in the peak eluted from the second column (fig. 1). As shown in table 1, the IFN eluted from the Ultrogel column can be as much as 10% pure though more usually it is approximately 1 - 5% pure. Furthermore, as shown in figure 2, the IFN bands can be readily identified in the fractions eluted from the gel filtration column. This to a certain extent obviates the need to wait for results of assays before further processing of the IFN. Recoveries were also in the region of 75 - 100% on this step, probably due to the high protein and salt concentration in the initial sample preventing excessive non-specific adsorption of the sample.

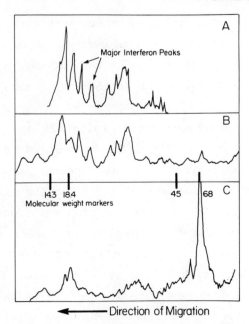

Figure 2. SDS-PAGE Analysis of Crude and Partially Purified Interferon Fractions.
Samples (0.1ml) were treated with SDS and applied to 12.5% acrylamide gel (0.75mm thickness) and stained with the silver stain of Merril et al. IFN activity was detected as described in the text. Gels were analysed using a scanning laser densitometer.
(A) IFN peak from Ultrogel AcA44 gel filtration
(B) IFN peak from red-HE7B-Sepharose
(C) Crude interferon: Namalwa cell supernatant

Further fractionation on Procion blue-HER Sepharose (fig. 3.) does not actually result in greater purification for the first IFN peak eluted, but the last peak (25 - 30% of the adsorbed IFN) is usually of quite high purity(better than 30% and possibly close to 80%), based on specific activity calculations and SDS-PAGE. Both major peaks, when re-chromatographed on blue-HER-Sepharose separately, migrated as single peaks and in approximately the same position on the gradient. Interestingly, all the protein which has been eluted from red-HE7B Sepharose that is applied to the HER column is bound. The capacity is also very high and whereas a 100ml column is required for the first stage of the purification,

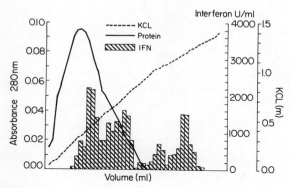

Figure 3. <u>Separation of Interferon Sub-types on Blue-HER Sepharose CL6B.</u>
Partially purified interferon eluted from Ultrogel AcA 44 was applied to a 120ml column of blue-HER agarose (2.6cm diameter). No protein or interferon activity was detected in the void. Elution was by a linear gradient of 0 - 2M KCl in 10mM TrisHCl pH 7.4 and the fractions were assayed for protein (O.D. 280nm), interferon and conductivity.

a 20ml column (1cm diameter) appears to be adequate for the final separation, albeit with the loss of some resolution. The elution profile shown in fig. 3 is typical of that obtained on a 100ml column. Recovery of IFN is between 50 and 100% in both cases.

Initial data from high performance liquid chromatography (HPLC) indicate that this may be the method of choice for preparative resolution of sub-species of alpha IFN from this partially purified material.

In conclusion, the purification and separation of IFN sub-types is not technically difficult, provided that protein and salt

Table 1

	Interferon Activity U/ml	Total IFN	Protein mg/ml	Volume ml	Specific Activity U/mg
Crude interferon	10,000 – 60,000	2×10^7 – 12×10^7	0.05 – 0.1	2,000	10^5 – 10^6
2M KCl eluate from red-HE7B Sepharose	2×10^6	6×10^7	1.0	30	2×10^6
IFN peak from AcA 44	2×10^6	5×10^7	0.06	25	3×10^7
Applied to blue-HER Sepharose (20ml)	0.8×10^6	38×10^6	0.06	50	1.3×10^7
First peak from blue-HER Sepharose	4×10^6	23×10^6	0.18	6	2×10^7
Last peak from blue-HER Sepharose	2×10^6	12×10^6	0.01	6	2×10^8

An outline of the stages of IFN concentration and purification. The upper section is typical of the results obtained during work-up of crude IFN and is an average obtained over five production lots. The lower portion of the table outlines the results from a typical separation of IFN components on a 20ml column of blue-HER Sepharose.

concentrations are kept as high as possible, the major difficulties which occur being due to the variable yields of IFN/cell. This purification protocol would probably also be applicable to leucocyte IFN.

We are currently in the process of attempting to elucidate the nature of the structural variations in terms of hydrophobicity and charge responsible for the different elution properties of alpha interferons on blue-HER sepharose.

REFERENCES

1. Yonehara, Y. Yanase, T. Sano, M. Imai, S. Nakasawa and H. Mori, 1981, J. Biol. Chem., 256, 3770-3775.
2. K.C. Zoon, M.E. Smith, P.J. Bridgen, D. zur Nedden and C.B. Anfinsen, 1979, Proc. Nat. Acad. Sci., 76, 5601-5605.
3. M. Rubinstein, S. Rubinstein, P.C. Familleti, R.S. Miller, A.A. Waldman and S. Pestka, 1979, Proc. Nat. Acad. Sci., 79, 640-644.
4. G. Allen and K. Fantes, 1980, Nature, 287, 408-411.
5. Goeddel,D.V., Leung,D.W., Dull,T.J., Gross,M., Lawn,R.M., McChandliss,R., Seeburg,P.H., Ullrich,A., Yelverton,E. and Gray,P.W. (1981) Nature, 290, 20-26.
6. P.J. Neame and I. Parikh, 1981, J. Applied Biochemistry and Biotechnology 7, 295-305
7. W.E. Stewart, II, 1979, The Interferon System, 18. (Springer, New York).
8. N.B. Finter, 1969, J. Gen. Virol., 5, 419-425.
9. M. Johnson, 1980, J. Gen. Virol., 50, 191-194.
10. J.C. Bearden, Jr, 1978, Biochim. Biophys. Acta, 533, 525-529.
11. U.K. Laemmli, 1970, Nature, 277, 680-685.
12. C.R. Merrill, D. Goldman, S.A. Sedman and M.H. Ebert, 1980, Science, 211, 1437-1438.

THE CLONING, ISOLATION AND CHARACTERIZATION OF A BIOLOGICALLY
ACTIVE HUMAN ENZYME, UROKINASE, IN E. COLI*

P.P. Hung

Genetics Division
Bethesda Research Laboratories, Inc.
Gaithersburg, Maryland 20760

ABSTRACT

We have cloned double-stranded cDNA copies of plasminogen activator messenger RNA isolated from human fetal kidney cells. Some of the clones express protein of discrete sizes ranging from 32,000 to 150,000 daltons. These products possess antigenic determinants related to human plasminogen activator from kidney cells, bind to an affinity column specific for serine protease and activate human plasminogen to dissolve fibrin clots.

INTRODUCTION

Acute thromboembolic events, venous and arterial thrombosis, pulmonary embolism, intracardiac thrombosis and systematic embolism are important medical problems. In all of these pathologies, the plasminogen activator urokinase has attractive potential applications, yet it is difficult to isolate in large quantities. We have designed and constructed, therefore,

---
*This work was carried out while the author was at Abbott Laboratories, North Chicago, Illinois

bacterial plasmids which instruct the synthesis of biologically active plasminogen activators in a microbial cell. Moreover, such clone facilitate elucidation of urokinase structures and mode of biosynthesis as well as of urokinase gene regulation and expression.

A plasminogen activator, urokinase, was first observed in urine in 1951 by Williams (1). More recently, fibrinolytic activity in cultures of human kidney cells was demonstrated (2), and it was found that this activity was immunologically indistinguishable from urokinase from urine. Based on their sizes, two major types of urokinase have been described. These are the high molecular weight species (HMW) with molecular weight of 54,000 and the low molecular weight species (LMW) with 32,000.

RESULTS

Relationship between two species of Urokinase. Since there are two species of urokinase, it is important to know whether they are derived from one gene or from two genes. If they are coded by the same gene, we need only to clone that particular gene. Therefore we studied biochemical relationship between these two species.

To show the close relatedness of the two species, they were digested by trypsin and their fingerprints were compared. The fingerprints from the low molecular weight species had a distinctive distribution pattern with 16 ninhydrin positive spots (Figure 1A). Fifteen of these were also located from the high molecular weight urokinase (Figure 1B). However, the high molecular weight urokinase had a few extra spots not found with the low molecular weight urokinase, and these spots could be derived from the third peptide chain. From these observations, it was very clear that only one gene was involved in coding for the two species of urokinase.

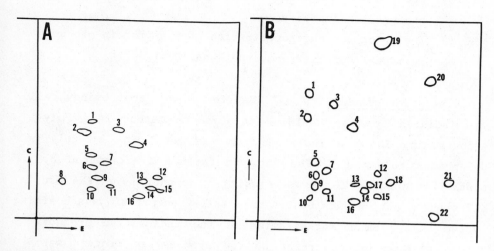

Figure 1. Tryptic fingerprints of urokinase from tissue culture. (A) Low molecular weight species; (B) High molecular weight species.

Approaches in Cloning. Experimental strategies for cloning the human urokinase gene and isolating its protein products took the following steps:

1) Isolation of RNA from human fetal kidney cells at the time which produce a large amount of urokinase.
2) Enrichment of urokinase mRNA by fractionation with sucrose density centrifugation and demonstration of urokinase mRNA in cell-free protein synthesis. Alternatively, immunoprecipitation of polysomes with anti-urokinase antibody and the demonstration of urokinase mRNA in cell-free protein synthesis.
3) cDNA synthesis from the mRNA and insertion of the cDNA at the Pst I site of pBR322.
4) Transformation of E. coli X1776 with the recombinant molecules followed by colony selection with tetracycline.

5) Detection by radioimmunoassays of E. coli colonies that produce urokinase.

6) Isolation and characterization of the protein products derived from the recombinant DNA.

Isolation of mRNA. Total RNA from human fetal kidney cells was isolated by the guanidine thiocyanate method (4). Poly A-containing RNA was isolated from the total RNA by Poly U-Sephadex column (5). The RNA was further fractionated by centrifugation in a sucrose-density gradient. The RNA in different fractions was tested in a cell-free protein synthesizing system (6). Immunoprecipitation of the radioactive products showed that about 7 percent of the synthesized protein was precipitated when mRNA before fractionation was used (Table I). However, only 1 percent of radioactivity was precipitated by anti-urokinase antibody when rabbit globin mRNA was used. After fractionation, the highest urokinase mRNA activity was located in the fraction corresponding to sedimentation values of greater than 28.

mRNA isolated by immunoprecipitation (7) was also studied. Some preparations of mRNA thus isolated were electrophoresed in urea gel (Fig. 2). The electrophoretic mobilities corresponded to about 4,900 nucleotides which can code for a protein of approximately 160,000 daltons.

In cell-free protein synthesis, one preparation gave 50 percent and the other, 90 percent of immunoprecipitable radioactivity which indicated a high degree of purity of this mRNA species (Table I).

The immunoprecipitated radioactive products were analyzed by SDS-polyacrylamide gel electrophoresis. A major peak of radioactivity was observed approximately corresponding to a

Table I. Cell-free protein synthesis using urokinase mRNA

mRNA Preparation	Radioactivity precipitated by anti-urokinase
	%
Globin mRNA (control)	1
mRNA before fractionation	7
mRNA after fractionation	
Greater than 28S	39
28S	5
18 to 28S	14
18S	4
Immunoprecipitated mRNA	
Sample No. 1	50
Sample No. 2	90

Figure 2. Urea-gel electrophoresis of immunoprecipitated mRNA.

molecular weight of 50,000. Other minor protein peaks corresponding to 30,000 daltons as well as to higher molecular weight materials were also detected (Fig. 3).

<u>Recombinant DNA Synthesis</u>. Single-stranded cDNA was synthesized from poly A-containing mRNA with reverse transcriptase. The second strand of cDNA was synthesized by using the large fragment of DNA polymerase I. The double-stranded cDNA was treated with S_1 nuclease and then the DNA was centrifuged in a sucrose-density gradient. DNA of size greater than 2000 base pairs was obtained and was used to attach poly dC tracts. The circular pBR322 DNA was digested with <u>Pst</u> I and tailed with poly dG tracts. The double-stranded cDNA with poly dC tail was annealed to pBR322 DNA with poly dG tail to form recombinant DNA.

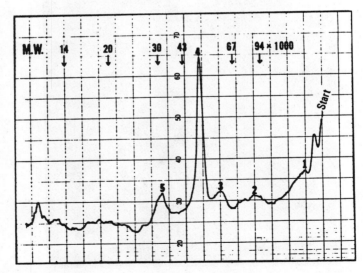

Figure 3. Polypeptides synthesized by <u>in vitro</u> translation mRNA.

<u>Initial Detection of Urokinase-like Material</u>. After transformation of E. coli X1776 with the recombinant DNA, we observed a total of 32 tetracycline resistant colonies (Table II). Of these, four were ampicillin-sensitive and contained inserts in their plasmids. Analyses of <u>Pst</u> I digests of the plasmids by gel electrophoresis revealed that three inserts were about 4.2 kilobase pairs whereas the fourth was about 900 base pairs. Two of these transformants, colony No. 19 and 26 and the negative control, X1776 transformed by pBR322, were grown and cell lysates were prepared. Antigens in the lysates were spotted on and covalently bound to cyanogen bromide-activated paper (8). Detection of urokinase-like materials was carried out by reaction with ^{125}I-labelled anti-urokinase antibody (9). As shown in Fig. 4, the lysates from both transformants showed weak but positive immunoreactivity as compared to the control, X1776 transformed by pBR322. Known amounts (0.08, 0.3, 1.25 and 5 ng) of urokinase (SMW) were also spotted on the paper as the positive control. The weak positive reaction of the transformants suggested that they produced small amounts of urokinase. Therefore, the clones were examined more closely by affinity chromatography.

<u>Isolation and Purification of Urokinase-like Material</u>. Benzamidine is an inhibitor of urokinase and other serine proteases and has been used to purify these enzymes by coupling it

Table II. Selection of Transformants

CLone No.	Tetracycline	Ampicillin	Size of Insert (base of pairs)
27-19	Resistant	Sensitive	4,200
27-20	Resistant	Sensitive	900
27-26	Resistant	Sensitive	4,200
27-29	Resistant	Sensitive	4,200
Other 28 clones	Resistant	Resistant	None

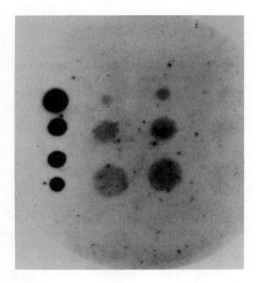

Figure 4. Detection of urokinase-like material in the lysates of transformants.

to Sepharose for affinity chromatography. Aliquots of the cell lysate were loaded on a benzamidine-Sepharose affinity column which was then eluted to collect fractions. Aliquots from each fraction were assayed by radioimmunoassay in plastic-well microtiter plates (8). Figure 5 shows the results of such an experiment. A single positive peak in radioimmunoassays was observed for Transformant X1776 (pABB26), which clearly indicates the presence of urokinase-like material. Transformant X1776 (pABB19) showed a similar result (data not shown). As expected, a parallel experiment with X1776 transformed by pBR322 did not show any material which reacted with antibody against the human protein.

Plasminogen Activator Activity of Products from Transformants. To test the products from recombinant DNA expression for functional activity as a human plasminogen activator, the material was subjected to a plasminogen-dependent radioactive fibrinolysis assay (9). Since crude bacterial lysates

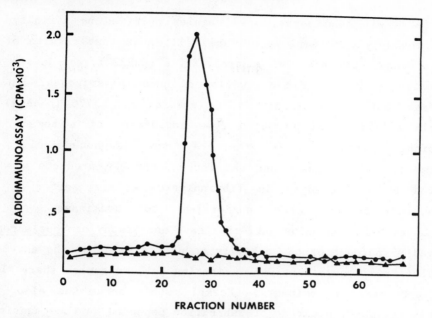

Figure 5. Benzamidine affinity column chromatography of urokinase-like material.

Table III. Plasminogen Activator Activity in Bacterial Transformants

Sample	Antisera	CPM	Units $(\times 10^{-3})$	% Activity Remaining
Background	None	798	---	---
Urokinase Std.	None	7,564	35.0	100
Urokinase Std.	Anti-urokinase	1,630	2.0	7
Urokinase Std.	Normal Rabbit Serum	3,458	12.0	35
X1776(pBR322)	None	952	(0)	---
X1776(pABB19)	None	18,560	175.0	100
X1776(pABB19)	Anti-urokinase	1,886	6.0	3.4
X1776(pABB19)	Normal Rabbit Serum	17.651	156.0	90
X1776(pABB26)	None	10,887	50.0	100
X1776(pABB26)	Anti-urokinase	1,479	2.0	4.6
X1776(pABB26)	Normal Rabbit Serum	5,127	23.0	46

interfered with the assay, only samples purified by affinity chromatography were employed. Table III shows that results of such a study. The negative control, X1776 (pBR322), was devoid of any activity in specific activation of human plasminogen. In contrast, both Transformants X1776 (pABB26) and X1776 (pABB19) clearly exhibited functional enzyme activities of a human plasminogen activator. Furthermore, the immunochemical relatedness of urokinase and the recombinant DNA products was studied by the changes in fibrinolytic activity after immunoprecipitation with anti-serum to urokinase. Immunoprecipitation using anti-urokinase and Staphylococcus aureus removed 95 percent of the activity of urokinase or of the enzyme produced from recombinant DNA indicating both molecules share the same antigenic determinants. Normal rabbit serum was also inhibitory, but the extent of inhibition depended on the quantity of plasminogen activator present in the assay. Plasminogen activator activity derived from urokinase or from the product expressed by the recombinant DNA behaved identically in this respect. Serum is known to be inhibitory in this type of assay.

Molecular Species of the Product Expressed by Recombinant DNA. To study the molecular size of the products, samples eluted from the affinity column were subjected to SDS polyacrylamide electrophoresis and then transferred onto cyanogen bromide-activated paper. After reaction with ^{125}I-anti-urokinase antibody and autoradiography, the products were identified and their molecular weights estimated by comparing their mobilities with known protein standards in the same gel. Figure 6 shows that X1776(pABB26) in Lane 1 produced five bands. Because of diffusion, protein bands were not very sharp. The positions of the observed bands corresponded to materials of molecular weight of about 150,000; 125,000; 87,000; 52,000 and 32,000 daltons. The predominant band was that of the 52,000 species. The smallest two of the products resemble HMW and LMW urokinase in size (Lanes 2

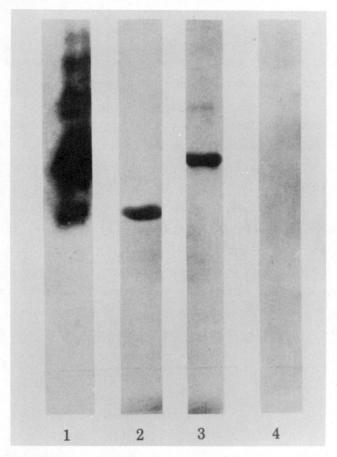

Figure 6. Electrophoresis and filter affinity transfer analyses of urokinase-like material.

and 3; visualized with Coomassie blue). The negative control, X1776(pBR322), in Lane 4 does not show any immunoreactive band against urokinase antibody.

Characterization of Recombinant Plasmid pABB26. This recombinant plasmid which carries the coding sequence for plasminogen activator in the Pst I site of pBR322, was characterized by restriction endonuclease digestion. Figure 7 shows the restriction map of pABB26. The total DNA length of pABB26 is about 8,550 base pairs, with an insertion (thick line of the circle in Fig. 10) of about 4,200 base pairs.

DISCUSSION

The evidence we have presented in this communication indicates that we have designed and constructed plasmids containing a DNA sequence that specifies a plasminogen activator related to human urokinase, and that E. coli transformed by these plasmids produces a biologically active enzyme. This conclusion is based on the comparisons of the products expressed by the recombinant DNA with urokinase in the following properties: (i) sharing of antigenic determinants as detected by anti-urokinase antibody; (ii) binding to benzamidine, an inhibitor to the active site of urokinase; (iii) similarity in molecular sizes; and (iv) functional enzyme activity as a human plasminogen activator in fibrinolysis. This report represents the first human enzyme gene to be cloned by recombinant DNA techniques, and the products are the largest human proteins engineered and expressed to date.

REFERENCES

1. J.R.B. Williams, Brit. J. Exptl. Pathol. 32:530-535, 1951.
2. M.B. Bernik and H.C. Kwaan, J. Lab. Clin. Med. 70:650-661, 1967.
3. R.O. Studer, G. Roncari and W. Lergier, In: Thrombosis and Urokinase, Paoletti, R. and Sherry S. (eds), Academic Press, p. 89-90, 1977.
4. A. Ulrich, J. Shine, J. Chirgwin, R. Pictet, E. Tirscher, W.J. Rutter and H.M. Goodman, Science 196:1313-1318, 1977.
5. R.G. Deeley, J.I. Gordon, A. Burns, K.P. Mullinix, M. Bina-Stein and R.F. Goldberger, J. Biol. Chem. 252:8310-8319, 1977.
6. H.R. Pelham and R.J. Jackson, Eur. J. Biochem. 67:247-256, 1976.

7. R.E. Rhodes, G.S. McKnight and R.T. Schmike, Biol. Chem. 248:2031-2039, 1973.

8. L. Clarke, R. Hitzeman and J. Carbon, In: Methods in Enzymology, Vol. 68, pp. 436-442, Academic Press, Inc., New York, 1979.

9. J. Unkeless, K. Dano, G.M. Kellerman and E. Keick, J. Biol. Chem., 249:4295-4305, 1974.

CLONING OF THE RAT ENDOGENOUS HELPER LEUKEMIA VIRUS DNA SEQUENCE AND EXPRESSION OF THE HELPER ACTIVITY ENCODED BY THE CLONED DNA SEQUENCE IN NORMAL RAT KIDNEY CELLS BY MICROINJECTION

Stringer S. Yang[1], Rama Modali[1] and Edwin Murphy, Jr.[2]
[1]Laboratory of Cell Biology
National Cancer Institute
Bethesda, Maryland 20205
[2]Department of Tumor Virology
University of Texas System
Houston, Texas 77030

ABSTRACT

By the use of recombinant DNA technology and microinjection in cultured cells, the molecular genetic elements involved in the evolution of a retrovirus with the multipotential to infect, transform and replicate in host cell, have been critically examined in this investigation.

Recently we have identified and purified the integrated and proviral DNA sequences specific for two rat endogenous helper leukemia viruses, WR-RaLV[*], originated from a chemically induced wild rat fibrosarcoma, and RHHV[*], isolated from a chemically induced rat hepatoma, HTC-H1[*] (1). By using a multidisciplinary

---

[*]WR-RaLV: wild rat leukemia virus, RHHV: rat hepatoma helper virus, HTC-H1: hepatoma tissue culture cell, line H1, LTR: long terminal repeat polynucleotide sequence, Src: sarcomagenic DNA sequence, K-NRK: Kirsten sarcoma virus transformed non-producer, FCS: fetal calf serum, kb: kilobase, bp: base pair, KSV(RHHV): Kirsten sarcoma virus rat hepatoma helper virus pseudotype, TEN: Tris-HCl, 0.01 M, pH 7.5, EDTA, 0.001 M, NaCl, 0.15 M.

approach combining restriction endonuclease analysis, reverse phase V-column chromatography, agarose gel electrophoresis, Southern blot transfer and filter nucleic acid hybridization, we were able to demonstrate that the rat helper leukemia viral DNA sequence was approximately 8.4-8.8 kb*. The 8.8 kb RHHV DNA was molecularly cloned via the EK-1 certified vector pBR 322 plasmid into E. coli RRI cells. A successful recombinant clone, 8/32, that carried one entire RHHV 8.8 kb DNA sequence was mapped by restriction endonuclease analyses. Restricted DNA fragments of various sizes throughout the complete RHHV genome were isolated and purified for intranuclear microinjection into normal rat kidney cells. Release of type C infectious helper virus in these microinjected cells was investigated by superinfection on K-NRK, Kirsten sarcoma transformed non-producer cells. Recombination of the helper viral DNA sequence, en toto or of subgenomic sizes, carried in microinjected cells, with the sarcomagenic DNA sequence, carried in K-NRK cells, was also studied by genome-rescue and cell-transformation experiments. Our observations led to the conclusion that all critical genetic elements including the 5' LTR* helper DNA sequence, gag, pol, and env genes, encoded for the biological activity of the type C helper virus resided within the 6.0 kb proximal to the 5' terminus of the endogenous rat type C helper virus DNA. They proved vitally essential for the recombination with the Src* sequence during the evolution of an infectious, transforming and replication-competent retrovirus.

INTRODUCTION

The advent of technology in recombinant DNA and microinjection in cultured cells has opened a new horizon for analyzing precisely the biological function of specific DNA sequences encoded within a viral or cellular genome. It also provides powerful tools for dissecting the molecular elements involved in the evolution of a retrovirus with the multipotential to infect, transform, and replicate in host cells. It has been

proposed on the basis of accumulated evidence that type C helper
leukemia viruses of low pathogenicity had the capability to
recombine with certain cellular DNA sequence, designated the Src
gene, when passing through the host cell to give rise to
sarcomagenic viruses. The genetic constitution of a sarcoma virus
may therefore be considered a recombinant of helper leukemia virus
DNA and the cellular Src gene (2-5). In earlier reports we
documented the isolation, morphological, immunological and
biochemical characterizations of a rat endogenous type C helper
virus, RHHV, released from a chemically induced Morris hepatoma
adapted to tissue culture, HTC-H1 cell line (7-9). Upon
co-cultivating K-NRK[*] cells with HTC-H1 cells, a pseudotype virus,
KSV(RHHV)[*] the apparent recombinant product, with the
multipotential to infect and transform cells, was released at high
titer (9). When KSV(RHHV) was cultured in Fisher rat embryonic
cells (FREC) the ratio of physical particle counts, estimated by
electron microscopy to cell transformation units, estimated by
focus-formation assay, was between 5:1 and 10:1. This implicated
either an excess of helper type C viruses or transformation-
defective particles over the sarcomagenic viruses (1,5,6). This
was reflected in the predominance of an 8.8 kb DNA over a 6.4 kb
DNA at a ratio of 8:1, generated by Eco R1 digestion of cell DNA,
both of which demonstrated homology with viral cDNA in Southern
blot-filter hybridization analysis (1). By terminal dilution we
obtained an RHHV infected FREC culture, RHHV/FREC, that replicated
primarily the helper virus. Proviral DNA preparation from
infected cells 40 hours post infection showed primarily the 8.8 kb
species (1). We have molecularly cloned both the 8.8 kb and 6.4
kb DNAs (5,6). In this communication we described the molecular
cloning of RHHV 8.8 kb proviral DNA. We have also localized the
DNA sequence coding for the helper biological activity essential
for the recombination with the Src sequence in the synthesis of an
infectious, transforming and replication-competent virus by intranuclear microinjection studies.

MATERIALS AND METHODS

Viruses and cells. RHHV/FREC was cultured in Dulbecco's modified Eagle's medium supplemented with 5% heat-inactivated (56°C, 30 min) FCS* plus penicillin (50 units/ml) and streptomycin (25 µg/ml) in a 5% CO_2 atmosphere at 37°C. NRK^{153} and NRK^{B101}, normal rat kidney cell lines derived from the original NRK line (10) were supplied by Biotech, Rockville, MD. NRK cells were used from passages 11 to 14. NRK and K-NRK were cultured in medium supplemented with 5% FCS, whereas HTC-H1 was kept in medium supplemented with 3% FCS.

Preparation of proviral DNA and cDNA. The RHHV proviral DNA was extracted from the cells by the Hirt procedure (11). The DNA was analyzed by 1% agarose gel electrophoresis and the 8.8 kb DNA was identified by Southern blot transfer-hybridization to ^{32}P-cDNA. DNA samples run simultaneously was extracted from the gel and used for molecular cloning.

RHHV was concentrated and purified from 20 liters of RHHV/FREC culture medium by ammonium sulfate precipitation and repeated banding in sucrose gradient centrifugation. RHHV specific cDNA was synthesized by the endogenous reverse transcriptase reaction described earlier (12) using ^{32}P-labeled all four deoxynucleoside triphosphates (Amersham Corporation, IL, 3000-3500 Ci/mmole). The ^{32}P-cDNA was further purified by chemical extractions and sulfopropyl sephadex column chromatography (12); it was then heated prior to its use in Southern blot filter hybridization.

Molecular cloning. The EK1-certified vector pBR322 was utilized for molecular cloning of the RHHV 8.8 kb DNA under conditions as prescribed in the revised National Institutes of

Health Guidelines for Recombinant DNA Research. All ligation reactions, propagation of the recombinant plasmid and DNA isolation were carried out in P2 containment conditions. RHHV provirus DNA was digested with Bam Hl and ligated to previously Bam Hl digested pBR 322 DNA at 5° for 20 hours. The digested and the ligated products were verified by electrophoresis in 0.9% agarose gel before the ligated DNA was used in transformation of calcium shocked E. coli RRI cells as described by Curtiss et al. (13). The transformed cells were first plated in ampicillin supplemented L-broth agar. After 20 hrs at 37°C, resistant clones were then screened for tetracycline sensitivity since the Bam Hl site is located at the tetracycline sensitivity locus. Colony hybridization assay (14) with RHHV ^{32}P-labeled cDNA was carried out on ampicillin resistant-tetracycline sensitive clones cultured on nitrocellulose filter. Ampicillin resistant tetracycline sensitive clones that demonstrated positive hybridization with the RHHV ^{32}P-cDNA were then selected for detail restriction endonuclease mapping of the RHHV genomic DNA.

Hybridization reactions, end-labeling and nick-translating of DNA. Southern blot-nitrocellulose filter hybridization was carried our using 2×10^6 cpm of ^{32}P-cDNA per filter in Deinhar solution (1,15). Liquid scintillation was carried out in 50 μl volume with approximately 50,000 cpm per reaction as described earlier (16).

End-labeling of 3' and 5' termini of the DNA were described in detail by Maxam and Gilbert (17). The 3' termini of the digested DNA fragments were labeled by the T_4 DNA polymerase (Boehringer-Mannheim, IN) reaction with only ^{32}P-dATP and ^{32}P-TTP (Amersham Corp., 3000 Ci/mmole) at the staggered end of the Eco Rl digested site. Eco Rl generated 5' ends of the recombinant DNA were labeled by the reverse reaction of the polynucleotide kinase using an extremely high specific activity of γ^{32}P-ATP (Amersham Corp., 5700 Ci/mmole).

Nick-translating of DNA fragments was carried out using all four ^{32}P-deoxynucleotide triphosphates (Amersham Corp., IL) with specific activities ranging 3000-3500 /mmole (18). Generally 10^9 cpm per μg of DNA was obtained.

<u>Preparation of cloned DNA</u>. DNA was prepared from the recombinant clone as described previously (5,6). RHHV DNA was excised from pBR 322 DNA with either Eco R1 or Bam H1 digestion followed by electrophoresis of the DNA in 0.85% low-melting agarose gels (Biorad Laboratories, CA). The DNA was then localized by ethidium bromide staining under UV illumination and recovered by the borosilicate adsorption technique (5,6). The DNA was then eluted from the borosilicate powder with sterile distilled water and adjusted to TEN*.

<u>Restriction enzyme assays</u>. All restriction enzymes were purchased from either Boehringer-Mannheim Co., IN, or Bethesda Research Laboratory, MD or New England Biolabs, MA. Unless where specified, all digestion conditions were according to the instructions provided by the suppliers.

<u>Microinjection</u>. For microinjection, the target cells, NRK^{153} or NRK^{B101} from the same culture, were trypsinized and seeded onto individual coverslips and cultured in 35 mm plate under the conditions described above until 30-50% confluent. Intranuclear injections were carried out on NRK^{153} and NRK^{B101} cells on the coverslip inverted for the purpose of microinjection on a Leitz Laborlux II fixed-stage microscope at 400X magnification. Microneedles were forged on a modified DeFonbrune microforge (19-20) to achieve an orifice of approximately 0.1-0.5 um. DNA samples dissolved in 0.1% KCl were concentrated for microinjection by centrifugation in a sealed 20 μl capillary pipet at 40,000 rpm for 40 min. One μl of each sample was loaded under pressure into

the tip of a water-filled microneedle by displacement of water (20). Movement of the microneedle was controlled by a DeFonbrune micromanipulator. An estimated 200-700 DNA molecules in 10 femtoliters dependent on the size and concentrations of the DNA fragments were microinjected. This efficiency was found compatible with documented microinjection reports (19-20). One hundred cells per coverslip were microinjected; the coverslip was then cultured in a petri plate as described above until ready for trypsinization.

Focus-formation assays, superinfection and genome-rescue experiments. Focus-formation assay was carried out on 150,000 NRK^{153} cells cultured overnight in 2 µg/ml of polybrene supplemented medium in 60 x 15 mm petri plate as described above. After the medium was withdrawn, the cells were infected with 0.5 ml of cell-free culture media prepared from either the superinfected cultures or genome-rescue cocultures. The cultures were gently rocked at 37° for 30 min. Fresh media with 5% FCS were then added to the cultures. Scoring of transformed foci was carried out on day 6-7.

Superinfection experiment was carried out with K-NRK cells, infected with cell-free culture media prepared from the various microinjected NRK^{153}_m clones. After five days of cultivation release of virus in the various superinfected K-NRK culture media was determined by the focus-formation assay.

Genome-rescue experiment was carried out by cocultivating the various microinjected cells with various control and transformed non-producer cells such as NRK^{B101}, HT-1, HTC-H1, and K-NRK at a ratio of 2:1 in cell number. The selected transformed, non-producer cell lines were known to carry differing numbers of copies of Src sequence (1,21) for recombination with the rat helper leukemia viral DNA sequence carried in the microinjected NRK_m^{153} cells. Cell-free media were prepared from each coculture at intervals for focus-formation assays on NRK^{153} cells in order to determine any release of transforming viruses. A flow scheme of the genome rescue experiments with the microinjected NRK_m^{153} cells is shown in Figure 1.

RESULTS

<u>Positive recombinant clones that carried the RHHV genomic DNA, en toto or of subgenomic size.</u> Plasmid pBR 322 carried both the ampicillin resistance and the tetracycline resistance genes. Since the ligation site of the RHHV DNA onto the pBR 322 DNA was at the Bam Hl site located within the tetracycline resistance locus, the <u>E. coli</u> RRI cells transformed by the recombinant DNA were therefore first tested for ampicillin resistance and tetracycline sensitivity. After screening approximately 5,750

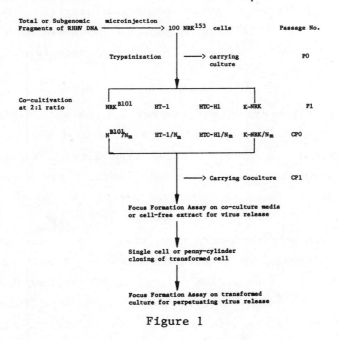

Figure 1

Flow scheme of genome rescue experiment.

ampicillin resistant clones, 38 were found persistently tetracycline sensitive. These ampicillin resistant, tetracycline sensitive clones were then subjected to colony-hybridization using high specific activity ^{32}P-cDNA probe synthesized in the endogenous reverse transcriptase reaction of purified RHHV as described in Methods. Nine clones showed varying degrees of hybridization to the ^{32}P-cDNA of RHHV. Four of these positive clones demonstrated persistent strong hybridization. And among them two clones contained the complete 8.8 kb genomic DNA of RHHV as shown in Figure 2A and 2B. As shown in Figure 2B short-termed (15 min digestion) of the circular form of the recombinant pBR322/RHHV DNA by Bam H1 restriction endonuclease resulted in the appearance of a 13.2 kb, 8.8 kb, 5.8-6.2 kb, and 4.3 kb bands. Prolonged digestion by Bam H1 resulted in the cleavage at the ligation site which bore the Bam H1 specificity and thus increased the concentrations of 8.8 kb and 4.3 kb (Fig. 2C). DNA within the "cir" band was considered the supercoil form of the recombinant

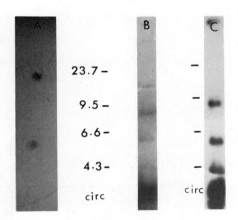

Figure 2

(A) Colony-hybridization of two positive recombinant clones using ^{32}P-cDNA synthesized in an endogenous reverse transcriptase reaction of purified RHHV. (B) Bam H1 digestion of recombinant DNA prepared from clone 8/32 at 37° for 15 mins. and (C) for 180 mins.

pBR 322/RHHV DNA since it generated the linear 8.8 kb, 5.8-6.2 kb, 4.3 kb bands and a 3.20-2.8 kb band when it was subjected to a second digestion by Bam H1 endonuclease. Recombinant DNA of one of these clones, 8/32 was subjected to detailed restriction endonuclease analysis and had been reported elsewhere (5,6). A restriction endonuclease map of the cloned RHHV DNA in this particular isolate is discussed below.

Six other positive recombinant clones were found to carry subgenomic fragments of RHHV DNA when their DNAs were analyzed by Bam H1 digestion. DNAs of these clones were consisted of 5.8-6.0 kb and 3.2-2.8 kb fragments derived from RHHV and 4.0-4.3 kb fragment derived from pBR 322. Considering the Bam H1 site situated at 3.2 kb proximal to the 5' terminus of the RHHV DNA (fig. 3), recombinant clones carrying subgenomic DNA fragments were expected. Three of these clones were extremely unstable and were eventually lost. After laborious investigation on their recombinant DNAs we found that the primary cause underlying their instability was due to "wobbling ligation". Apparently Bam H1 tended to wobble and would attack an Eco R1 site in an environment of excess concentrations of any one of the following factors: cation, hydrophobic reagents such as glycerol and ethanol (22-23).

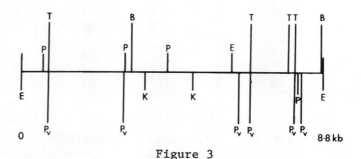

Figure 3

Restriction endonuclease map of RHHV DNA.

All these chemicals were used for the preparation, or during the Bam H1 digestion of the DNA prior to the ligation reaction. Since both the Eco R1 and Bam H1 generated ends were "sticky", ligation occurred between the two termini readily. Recombinant clones consisted of DNA resulted from "wobbling ligation" however tended to be extremely unstable.

Restriction endonuclease map of RHHV 8.8 kb DNA. The molecularly cloned RHHV 8.8 kb DNA (fig. 3) was oriented with respect to the 3' end of the viral RNA by hybridization to the 3' end specific poly(A^+) mRNA (5,6). Having determined the orientation of the RHHV DNA, the various restriction endonuclease sites, illucidted by single, multiple or sequential digestions on either 3' end or 5' end labeled 8.8 kb DNA (5,6), were then assigned as shown in figure 3. A total of fourteen restriction endonucleases were employed in this study. The following including Eco RI, Bam HI, Pst I, Kpn I, Taq I, Pvu II, and Sma I yielded consistent cleavage patterns, and proved to be instrumental in deducing the restriction map of RHHV DNA. A striking feature associated with the RHHV restriction map is the repeated pattern of Taq I, Pvu II, and Pst specificities at both 3' and 5' termini of the 8.8 kb DNA (Fig. 3). The polynucleotide sequence flanked by these restriction endonuclease recognition sites measured 600-800 nucleotides in length, and may be considered the most plausible candidate for LTR^*. The length of LTR for retroviruses among various species varied greatly, ranging from 400 bp to 1,500 bp (24-27). At least one to three copies of LTR had been reported for each viral genomic DNA. In this rat helper leukemia viral genomic DNA, at least two copies of the 600-800 bp LTR was detected per 8.8 kb DNA. Currently, the nucleotide sequence of this candidate LTR is in the process of being resolved so that its length and characteristic may be more precisely defined.

<u>Morphology of the Microinjected Cells</u>. With the deduction of the detailed restriction map for the RHHV genomic DNA, it became feasible to isolate defined subgenomic DNA fragments by precise restriction enzyme cleavages for microinjection studies. This might prove informative for resolving the molecular basis and the relationship between the helper leukemia viral DNA sequence and the Src sequence during the elicitation of cell transformation. The DNAs chosen for microinjection studies are the 8.8 kb total genomic DNA, and four subgenomic fragments of 5.8-6.2 kb, 4.0 kb, 2.65-2.85 kb, 2.5 kb as well as the supercoil recombinant DNA as shown in Fig. 4 A,B. Since the subgenomic DNA fragments used in this study spanned the entire 8.8 kb, biological activities transferred by any of the fragment should provide some insight on both the gene function and order.

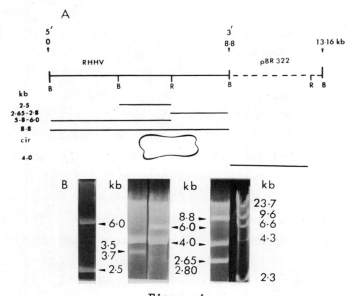

Figure 4

(A) Schematic restriction map of RHHV depicting the locations of the subgenomic DNA fragments used in microinjection experiments. (B) Agarose gel electrophoresis of total and subgenomic DNA fragments of RHHV, concentrated and recycled, prepared by Eco Rl and/or Bam Hl digestions. All experimental details were described in Methods.

No significant change in cell morphology occurred during the initial growth of some 24 microinjected cell clones (Fig. 5A). During the second passage two particular clones microinjected with the total of 8.8 kb DNA, $NRK^{153}_{m8.8}$, or the supercoil RHHV DNA, $NRK^{153}_{m3.5/cir}$, demonstrated morphological changes such as a disoriented and overlapping growth pattern, a loss of contact inhibition and high growth density (fig. 5B). Morphology of $NRK^{153}_{m3.5/cir}$ culture especially resembled that of chemically or spontaneously transformed cultures. This morphology persisted with the $NRK^{153}_{m3.5/cir}$ clone. The transformed culture morphology of the $NRK^{153}_{m8.8}$ however was transient and the cells reverted back to contact-inhibited pattern.

Immediate Release of type-C helper leukemia Viruses. Immediate release of helper leukemia viruses by the microinjected cells into the culture medium, was determined by superinfection on K-NRK cells. Table 1 summerizes the results of the superinfection experiments. Immediate release of helper C-type viruses was

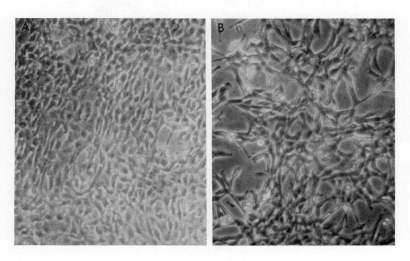

Figure 5

Morphology of (A) normal NRK^{153} culture and (B) $NRK^{153}_{m3.5/cir}$ culture that departed from normal characteristics.

observed only in the $NRK^{153}_{m3.5/cir}$ clone microinjected with the supercoil form of the RHHV DNA. Superinfection of the K-NRK cells with cell-free medium from the $NRK^{153}_{m3/5/cir}$ culture resulted in the rescue of a pseudotype transforming virus even though the number observed in the assay was low (5 FFU/ml). Morphology of one transforming focus is presented in fig. 6. It had the characteristic of a KSV(RHHV) transformed focus. This observation suggested that the circular form of the viral DNA might have expedited integration and expression of viral particle synthesis. No immediate release of C-type helper viruses was observed by all other microinjected clones carrying linear RHHV DNA, en toto or of subgenomic sizes (Table 1).

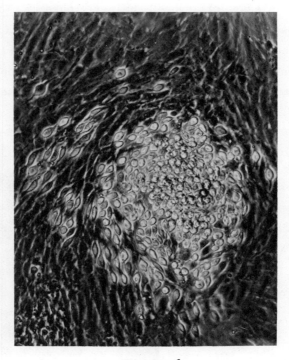

Figure 6

Morphology of one transformed NRK^{153} focus post infection with cell-free medium from K-NRK cells superinfected with $NRK^{153}_{m3.5/cir}$ culture medium.

Table 1

Focus-Formation Assays of Transforming Viruses in Culture Media of K-NRK Cells Superinfected with Various Microinjected NRK_m^{153} Culture Media

RHHV/pBR Recombinant DNA Fragments Microinjected into the NRK^{153} Nucleus*						
kb	8.8	5.8–6.0	4.0	2.65–2.85	2.3	3.5–3.7 Cir DNA
approximate no. DNA molecules**	205	260	500	700	400	100
FFU/ml of Superinfected K-NRK Culture Media†						
Experiment 1	0	0	0	0	0	3 (110)
Experiment 2	0	0	0	0	0	5 (350)

* All microinjected samples were done in duplicate
** The no. of DNA molecules was estimated on the basis of size and concentration of the particular DNA molecule or fragment microinjected (19-20).
† All experimental details were described in Methods. All samples were done in triplicate. The no. of FFU/ml represented averaged values. The number in the bracket represented averaged FFU per culture.

<u>Rat Type C Helper Leukemia Viral DNA Sequence Essential in The Recombination with Src Sequence for The Production of Transforming Viruses</u>. The possible helper leukemia viral biological activity associated with the various microinjected DNA fragments was further ascertained by cocultivating the various microinjected cells at the first passage with various cell lines carrying differing copies of the Src sequence in their cellular DNA in a genome rescue experiment as depicted in the flow scheme (Fig. 1). Cell-free coculture media were withdrawn at intervals and assayed for the release of transforming viruses by the focus-

Table 2

Transforming Virus Release in Cocultures of NRK_m^{153} Cells with Various Transformed Non-producer Cell Lines

RHHV/pBR Recombinant DNA Fragments Microinjected into the NRK^{153} Nucleus*

kb	8.8	5.8-6.0	4.0	2.65-2.85	2.5	3.5-3.7 Circ DNA
approximate no. DNA molecules**	205	260	500	700	400	100

FFU/ml of Cell-Free Media†

Coculture with						
Experiment No. 1						
NRK^{B101}	0	0				
HTC-H1	0	0				
KNRK	43(36-52)	18(10-18)				
Experiment No. 2						
HTC-H1	0	0	0	0	0	0
HT-1	42(28-56)	38(20-56)	0	0	0	0
KNRK	6(1-11)	3(0-7)	0	0	0	8(0-19)

formation assay. Results from two independent microinjection and subsequent genome rescue experiments were summarized in Table 2. Cocultivation between $NRK^{153}_{m8.8}$, or $NRK^{153}_{m6.0}$, or $NRK^{153}_{m3.5/cir}$ cells with K-NRK cells or with HT-1 cells resulted in the successful replication of infectious transforming viruses 3-4 days post cocultivation. Figure 7A, B and C shows the morphology of the normal NRK^{153} cells and of transformed foci of NRK^{153} cells infected and transformed by the viruses released in the cocultures

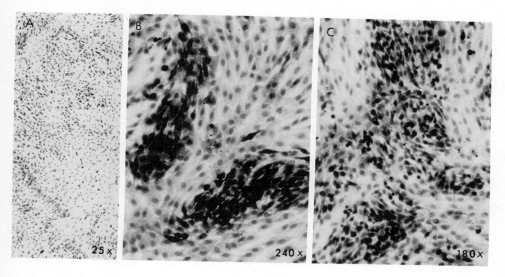

Figure 7

Morphology of (1) normal NRK^{153} cells and transformed NRK^{153} foci post infection by cell-free culture media prepared from cocultures of K-NRK cells with either (2) $NRK^{153}_{m8.8}$ cells or (3) $NRK^{153}_{m6.0}$ cells.

carrying $NRK^{153}_{m8.8}$ or $NRK^{153}_{m6.0}$ cells with K-NRK cells. The transformed foci resembled that of Kirsten murine sarcoma virus induced focus. This suggested that recombination between the rat helper viral DNA sequence of the microinjected cells with the Src sequence carried in the K-NRK cells occurred during the very first or second cell division, considering that the cell cycle of these co-cultures was approximately 20 hrs. The synthesis and replication of infectious transforming viruses probably occurred at the 2-3 generation of the coculture since the coculture media were found positive with transforming viruses on day 3-4. This suggested that the recombination and the replication events must have occurred efficiently and expediently.

As an alternative explanation, but lacking supporting evidence, it is possible that the synthesis of the helper virus protein independently could have provided the viral polypeptide

Table 3

Continuation of Progeny Virus Replication in Culture Media of Cloned Transformed NRK^{153} Cells

Virus Source	FFU/ml of culture medium*
	Primary transformation in NRK^{153} cells**
$NRK^{153}_{m8.8}$/K-NRK† coculture	43^a (36-52)
$NRK^{153}_{m6.0}$/K-NRK† coculture	18^b (10-26)
$NRK^{153}_{m3.5/cir}$ superinfected K-NRK culture	9^c (1-17)
	Secondary transformation in NRK^{153} cells
Transformed NRK clone a	236 (95-377)
Transformed NRK clone b	24 (7-41)
Transformed NRK clone c	13 (2-24)

* All samples were done in triplicate. The number of FFU/ml represented averaged values. The number in the bracket represented the range of FFU/ml among the triplicate samples.
** Transformed NRK^{153} cells were designated clone a, b or c and were isolated by either single-cell cloning or penny-cylinder cloning.
† Coculture was designated by the cells used: e.g. $NRK^{153}_{m8.8}$/K-NRK represented NRK^{153} cell microinjected with the 8.8 kb DNA cocultured with K-NRK cells.

essential for the encapsidation of the sarcoma virus genome.

Transformed cells from certain foci (fig. 7B,C) were isolated by pennycylinder or single-celled cloning and replication of transforming viruses into the culture media was further monitored by focus-formation assay. As summarized in Table 3, the daughter clones, designated a, b, and c continued to produced progeny infections transforming viruses at even higher titer than the parental cultures when culture media were assayed for focus-formation units. This increase in virus titer was probably due to viral gene amplification by multiple integrations during cell division.

NRK^{153} cells microinjected with all the other subgenomic DNA fragments such as the 4.0 kb pBR DNA, and the RHHV subgenomic DNA fragments, 2.5 kb at the center and 2.65-2.8 kb at the 3' terminus of the restriction map (fig.4), all failed to replicate transforming viruses upon cocultivation with K-NRK cells. Such an observation is consistent with the thesis that the complete helper viral sequence including gag, pol, and env genes and the LTR at the 5' terminus must be present in order to achieve successful rescue of the Src sequence resulting in the envolvement of infectious transforming and replication-competent pseudotype virus.

DISCUSSION

We have successfully cloned the proviral DNA of the rat hepatoma helper virus, RHHV, en toto and of subgenomic fragments. Furthermore, we were able to transfer by intranuclear microinjection the rat helper viral total or subgenomic DNA sequences into NRK^{153} cells. Expression of the helper viral function was positively identified in some of these microinjected cells with RHHV en toto or subgenomic DNAs (5,6). The DNA sequence encoding for complete helper viral biological activity that proved vitally essential in recombination with the Src DNA sequence for the evolvement of an infectious transforming virus, resided within the 6.0 kb proximal to the 5' end of the rat helper viral DNA. DNA fragments consisting of primarily LTR sequences such as the 2.65 kb proximal to the 3' terminus, or of incomplete viral information such as the 2.5 kb at the center of the RHHV DNA, failed to elicit complete helper biological activity when tested in microinjection and genome rescue experiments.

Moreover the observation that the continuation of the replication of transforming virus, along with preliminary observations of restriction enzyme analysis and Southern-blot filter hybridizations, suggested that the microinjected DNA sequence became integrated into the host cell genomic DNA. A recent paper (20) on microinjection studies of cloned viral DNA into RSV(-) transformed cells, a far more sensitive system, had documented successful release of infectious transforming virus at 3 hr post microinjection. In view of the rapid response in viral particle synthesis the authors proposed that integration of the viral DNA sequence was probably not necessary for the expression of viruses. Relatively little difference was observed with respect to viral replications when the viral DNA was intracytoplasmically microinjected. The efficiency of intranuclear as compared with intracytoplasmic microinjections was approximately 5-10 fold greater in the avian system (20). Intracytoplasmic microinjection was not attempted in this investigation. Albeit, numerous transfection attempts with cloned RHHV DNA by the calcium phosphate technique (28) had failed to establish permanent integration of the viral DNA within NRK^{153} nuclear DNA when examined by Southern blot-hybridization assay using RHHV ^{32}P-cDNA probe. Incorporation of the RHHV ^{32}P-DNA into the cytoplasma of NRK^{153} cells was readily observed. The viral DNA was presumably destroyed before it was successfully integrated into the nuclear DNA. In these experiments superinfection and genome rescue attempts were all unsuccessful. Problems inherent to research on rat type C tumor viruses are: (1) poor viral replication; and (2) a lack of sensitive indicator cell(s) for bioassays; and (3) relatively stringent host requirement. The failure of viral DNA transfection in the rat cell experiment may well be due to an insensitive host cell. Alternatively, the intrinsic difference between the cytoplasmic nuclease concentration of the rat cell versus the RSV(-) transformed cell could have accounted for the relative success in intracytoplasmic microinjection or transfection experiments.

In this era of rapidly advancing high technology, the combination of recombinant DNA technology and microinjection has allowed us to dissect out the vital genetic element within the genomic DNA of a rat helper leukemia virus that proved critical for the conservation of the helper biological activity. While alternative explanations such as discussed earlier are possible, this observation was consistent with the thesis that successful recombination with the Src sequence resulting in the evolvement of an infectious transforming and replication-competent virus, requires DNA sequence encoding for the 5' terminus sequence containing 1 copy of LTR, the gag, the pol, and the env genes, and which resided within the 6.0 kb proximal to the 5' terminus of the RHHV DNA.

ACKNOWLEDGMENTS

The authors gratefully thank V. Armstrong and G. Kington for typing the manuscript, and Janet Taub for technical assistance in this investigation.

REFERENCES

1. S.S. Yang, L-S.L. Yeh, J. Taub, N. Miller and M.B. Gardner, Integrated and proviral DNA sequences for wild rat tumor virus and for rat hepatoma helper virus in various laboratory and wild rat tumor cells: A restriction enzyme analysis. In Essex, M. Todaro, G., and Zue Hausen, H., ed. <u>Viruses in Naturally Occurring Cancers</u>. Cold Spring Harbor Conferences on Cell Proliferation, Vol. 7: pp. 1083-1092, 1980.

2. D. Stehelin, H.E. Varmus, J.M. Bishop and P.K. Vogt, DNA related to the transforming gene(s) of avian sarcoma viruses is present in normal avian DNA. Virology 108:177-188, 1976.
3. D.H. Spector, K. Smith, T. Padgett, P. McCombe, D. Roulland-Dussoix, C. Moscovici, H.E. Varmus and J.M. Bishop, Uninfected avian cells contain RNA related to the transforming gene of avian sarcoma viruses. Cell 13:371-380, 1978.
4. P.K. Vogt, Genetics of RNA tumor viruses. In Comprehensive Virology Kraenkel-Conrat, H., and Wagner, R. eds. 9:341, Plenum Press, New York, 1977.
5. S.S. Yang, J. Taub, R. Modali and M. Gardner, Molecular cloning of the rat leukemia-helper virus DNA sequences and functional analysis of some restricted DNA fragments. In Advances in Comparative Leukemia Research, Blakeslee, J.R. ed. Elsevier/North-Holland Biomedical Press, 1981.
6. S.S. Yang, R. Modali, R. We and M. Gardner, Molecular cloning of the endogenous rat C-type helper virus DNA sequence: Structural organization and functional analysis of some restricted DNA fragments. Virology, submitted, 1981.
7. S.S. Yang, C. Chan and E.B. Thompson, Rescue of a rat tropic hepatoma virus pseudotype Kirsten sarcoma virus by co-cultivation of hepatoma tissue culture cells with K-NRK cells. J. Gen. Virol. 32:239-243, 1976.
8. S.S. Yang and R.C. Ting, A comparative study of two rat viral complexes: MSV(RaLV) and KiMSV(RHHV). In Advances in comparative Leukemia Research, 1977 (Bentvelzen, P. et al. eds.), p. 264-267, Elsevier/North Holland, Amsterdam, 1978.
9. S.S. Yang, H.L. Malech, R.S. Wu and D.I. Woronow, Properties of a transforming virus, KiMSV(RHHV), isolated from a coculture of rat HTC-H1 cells with K-NRK cells. J. Gen. Virol. 38:209-221, 1978.

10. H. Duc-Nguyen, E.N. Rosenblum and R.F. Zeigel, Persistent infection of a rat kidney cell line with Rauscher murine leukemia virus. J. Bact. 92:1133-1140, 1966.

11. B. Hirt, Selective extraction of polyoma DNA from infected mouse cell cultures. J. Mol. Biol. 26:365-369, 1967.

12. S.S. Yang and N.A. Wivel, Physicochemical analysis of the deoxyribonucleic acid product of murine intracisternal A particle RNA-dependent DNA polymerase. Biochimica et Biophysica Acta 447:167-174, 1976.

13. R. Curtiss, M. Inoue, D. Pereira, J.C. Hsu, L. Alexander and L. Rock, Construction and use of a safer bacterial host strains for recombinant DNA research. Miami Winter Symposium 13:99-114, 1977.

14. M. Grunstein and D.S. Hogness, Colony hybridization: a method for the isolation of cloned DNA's that contain a specific gene. Proc. Nat. Acad. Sci. U.S.S. 72:3961-3965, 1975.

15. D. Steffin and R.A. Weinberg, The integrated genome of murine leukemia virus. Cell 15:1003-1010, 1979.

16. S.S. Yang and N.A. Wivel, Intracisternal A particle-specific DNA sequence in mammary tumor cells, hybrids, and cybrids derived from laboratory mice and from feral mice of Mus musculus and Mus cervicolor. Virology 96:167-176, 1979.

17. A.M. Maxam and W. Gilbert, Sequencing end-labeled DNA with base-sepcific chemical cleavages. Methods Enzymology 65:499-560, 1980.

18. P.W.J. Rigby, M. Dieckmann, C. Rhodes and P. Berg, Labeling deoxynuleic acid to high specific activity in vitro by nick trnaslation with DNA polymerase I. J. Mol. Biol. 113:237-251, 1977.

19. W.S. Stacey, Microinjectino of mRNA and other macromolecules into living cells. In Methods in Enzymology (in press), 1981.

20. J.J. Kopchik, G. Ju, A.M. Skalka and D.W. Stacey, Proc. Nat. Acad. Sci. U.S.A., 78:4383-4387, 1981.
21. S.S. Yang, C. Chen, J. Taub and R.C. Ting, Viral specific DNA sequence and appearance of aneuploidy and marker chromosomes in Fisher rat tumors and embryonic cell transformation by KSV(RHHV). In *Advances in Comparative Leukemia Research 1981*. Ed. by Yohn, D.S. and Blakesles, J.R., Elsevier/North-Holland Biomedical Press, 1981.
22. J. George, R.W. Blakesley and J.G. Chirikjian, Sequence-specific endonuclease Bam Hl. J. Biol. Chem. 255:6521-6524, 1980.
23. B. Polisky, P. Greene, D.E. Garfin, B.J. McCarthy, H. Goodman and H.W. Boyer, Specificity of substrate recognition by the Eco RI restriction endonuclease. Proc. Nat. Acad. Sci. 72:3310-3314, 1975.
24. E. Gilboa, S.W. Mitra, S. Goff and D. Baltimore, *In vitro* synthesis of a 9 kbp terminally redundant DNA carrying the infectivity of Moloney murine leukemia virus. Cell 16:863-874, 1979.
25. F. Hishinuma, P.J. DeBona, S. Astrin and A.M. Skalka, Nucleotide sequences of acceptor site and termini of integrated avian endogenous provirus ev 1: integration creates a 6 bp repeat of host DNA. Cell 23:155-164, 1981.
26. T.W. Hsu, J.C. Sabran, G.E. Mark, R.V. Guntaka and J.M. Taylor, Analysis of unintegrated avian RNA tumor virus double-stranded DNA intermediates. J. Virol. 28:810-818, 1978.
27. J.G. Sutcliffe, T.M. Shinnick, I.M. Verma and R.A. Lerner, Nucleotide sequence of Moloney leukemia virus: 3' end reveals details of replication, analogy to bacterial transposons, and unexpected gene. Proc. Nat. Acad. Sci. U.S.A. 77:3302-3306, 1980.
28. N.D. Stowe and N.M. Wilkie, J. Gen. Virol. 33:447-458, 1976.

ANALYSIS OF CAD GENE AMPLIFICATION USING A COMBINED APPROACH
OF MOLECULAR GENETICS AND CYTOGENETICS[1]

Geoffrey M. Wahl, Virginia Allen, Suzanne Delbruck,
Walter Eckhart, Judy Meinkoth, Bruno Robert de Saint
Vincent and Louise Vitto

Tumor Virology Laboratory
The Salk Institution for Biological Studies
La Jolla, California 92037

ABSTRACT

CAD[2] is a multifunctional protein which catalyzes the first three steps of de novo uridine biosynthesis. Rodent cells resistant to PALA,[3] a specific inhibitor of the ATCase activity of CAD, overproduce the CAD protein and CAD mRNA as a direct result of the amplification of the CAD gene. In order to study the mechanism of CAD gene amplification, a functional Syrian hamster CAD gene was inserted into a cosmid vector using molecular cloning techniques. The cloned genes were assayed for biological function by fusing CAD-deficient Chinese hamster ovary (CHO) cell mutants

---

[1] This work was supported by research grant numbers GH27754, CA 13884 and CA14195 from the NIH D.H.H.S. B.R.S.V. was supported by a NATO fellowship and by the French CNRS. J.M. was supported by NIH predoctoral training grant #PHS CA 09345.

[2] CAD is an acronym indicating the covalent association of carbamyl phosphate synthetase (CPSase), aspartate transcarbamylase (ATCase) and dihydro orotase (DHOase) in a single polypeptide chain.

[3] PALA is N-(phosphonacetyl)-L-aspartate, a transition state analog inhibitor of ATCase.

with protoplasts of E. coli containing the CAD cosmids. Two clones with functional CAD genes were isolated and shown to contain inserts 40 and 45 kb long. The cloned genes could also be introduced into wild type CHO cells by selecting for cells which became resistant to high PALA concentrations in a single step. Transformations of mutant and wild type CHO cells contained multiple active copies of the donated Syrian hamster CAD genes in addition to their endogenous CHO CAD genes. The cloned genes in all transformants analyzed are integrated into host cell chromosomes at single locations defined by in situ hybridization. Independently isolated transformants contain the donated genes in different chromosomes. Co-transformation of CHO cells with two different genes by protoplast fusion is also shown to be possible.

INTRODUCTION

Gene amplification has been shown to be a common mechanism of resistance to a variety of antiproliferative agents in both prokaryotes (1) and eukaryotes (2). It is conceivable that the overproduction of specific enzymes which mediate the resistance of eukaryotic tissue culture cells to a number of other agents may in many cases be due to gene amplification (2). Thus, it appears that the capacity to amplify genes is one that has been retained throughout evolution and is most likely not limited to a few loci. Recent observations of amplified sequences in tumor cells (3-6) have prompted speculations on the potential importance of gene amplification as a mechanism for elevating the level of expression of endogenous transforming genes (7,8). Gene amplification thus poses potentially significant clinical problems by engendering resistance to chemotherapeutic agents and by providing a possible

---

[4] MTX is methotrexate (amethopterin), a specific inhibitor of dihydrofolate reductase.

ANALYSIS OF CAD GENE AMPLIFICATION

mechanism for conversion to neoplasia. On the other hand, the ability to isolate mutant tissue culture cells with hundreds of copies of the amplified sequences facilitates the dissection of the molecular biology of gene amplification.

Extensive information is available concerning the amplification of dihydrofolate reductase genes in MTX[4] resistant cells and CAD genes in PALA resistant cells. Using multiple step selections with increasing drug concentrations, mutants were isolated which synthesize DHFR and CAD proteins at levels exceeding 10% of the total soluble protein (9,10). These mutants contain 100 or more DHFR (11) or CAD genes (12). Whereas all PALA resistant mutants isolated to date are stable (10), MTX resistant mutants can be either stable (13,14) or unstable (15).

Highly specific molecular probes have been used to localize the amplified genes in metaphase spreads of chromosomes from MTX and PALA resistant lines. Amplified DHFR genes in stable MTX-resistant cell lines (13,14) and amplified CAD genes in all lines studied thus far are contained in expanded chromosomal regions (16). Analysis of the trypsin-Giemsa banding patterns in these lines showed that the amplified DHFR genes are contained in a region which stains homogeneously (13,14,17) while amplified CAD genes are in a region which bands in a ladder-like pattern (16). Occasionally, amplified CAD genes are found on more than one chromosome, probably as the result of translocation from the major site of amplification. The amplified CAD genes at the major site originate from the site of the wild type CAD gene (16). Importantly, ribosomal RNA genes, which are located near the wild type CAD gene, are co-amplified with the CAD genes in the PALA resistant mutants (Wahl, Vitto and Runitz, in preparation). In both DHFR and CAD gene amplification, the average size of a single unit of gene amplification is approximately 500 kb, 10-20 times the size of a functional DHFR or CAD gene. In contrast, amplified

DHFR genes in mutants which are not stably resistant to MTX are contained in small paired acentromeric material (double minute chromosomes, DM's; 15).

A major emphasis in our studies is to elucidate the contribution of the sequences flanking the CAD gene in determining the size, structure, stability and location (i.e., intra- or extrachromosomal) in which it is amplified. Molecular cloning techniques have enabled us to isolate a functional CAD gene from the genome of PALA resistant Syrian hamster cells. Reintroduction of the purified gene into new chromosomal locations and amplification of the relocated genes should allow us to study how chromosomal location affects the amplification of this gene. Theoretically, this approach should be equivalent to analyzing the amplification of many different genes but it has the significant advantages of requiring a single set of molecular probes which are already available and avoids problems of comparing systems which may have subtle but profound differences in intracellular biochemical side effects.

METHODS

<u>Cells and cell culture</u>. Wild type Chinese hamster ovary cells (CHO-K1) and a uridine auxotroph subline with a lesion in the CAD gene (Urd^-A; 18) were kindly supplied by Dr. David Patterson. Both were propagated in Ham's F-12 medium (Gibco) supplemented with 8% fetal calf serum, and in addition, 30 μM uridine was included for Urd^-A cells. 165-28 cells, a PALA resistant mutant of Simian virus 40 (SV40) transformed baby Syrian hamster kidney cells (BHK21; 19) containing approximately 200 copies of the CAD gene (10,12) were grown in Dulbecco's modified Eagle's medium (DME) supplemented with 10% calf serum.

Construction of a cosmid library from EcoR1 partials of DNA from 165-28 cells, a Syrian hamster PALA-resistant mutant with 200 CAD genes. Detailed methods for the construction of cosmid libraries have been presented elsewhere (20,21). A brief summary of the procedure follows. DNA was isolated from 165-28 cells and cleaved partially with endonuclease EcoR1. The cleaved DNA was fractionated on a 10-40% sucrose gradient and 400 µl fractions were collected. Aliquots (20µl) from each fraction were analyzed by gel electrophoresis (0.4% agarose Tris acetate gel; 22) and those fractions containing DNA 30-50 kb long were pooled. The cell DNA was then ligated to the cosmid vector MUA3 (20) which had been cleaved with EcoR1 and treated with alkaline phosphatase to prevent oligomerization during the ligation. Treatment with alkaline phosphatase was found to be crucial for constructing large cosmids which did not delete at a high frequency during propagation in E. coli (21). Ligated molecules were then packaged into bacteriophage λ heads in vitro (21,23,24). Packaged cosmids were adsorbed to a recA$^-$ HB101, diluted into growth medium to allow expression of tetracycline (tet) resistance, and spread onto agar plates containing Luria broth and 10 µg/ml tet. Colonies developed after incubation at 37°C for 1 to 2 days; approximately 2.5×10^4 cosmids were obtained per 1 µg of hamster DNA under conditions yielding $0.5-1.0 \times 10^8$ λ plaques per 1 µg uncleaved λ DNA.

Cosmid screening. Each of 14 plates containing 500-2000 recombinants were harvested into separate sterile tubes. Overnight cultures were prepared by inoculating 20 µl from each tube into 5 ml of L-broth containing 10 µg/ml tetracycline. Cosmid DNA isolated by the cleared lysate procedure (12,25) was then fractionated on a 0.7% agarose gel (22) and transferred to diazotized paper (26). The DNA blot was then probed in two sequential steps using DNA fragments specific for the 3'- and 5'-ends of CAD mRNA (isolated from the recombinants p102 and p204, respectively; 27). Two pools contained clones hybridizing with both sequences, and the CAD clones within these pools were then located and purified by standard techniques (28). These clones are designated cCAD1 and cCAD6.

Protoplast fusion. The method of Shaffner (29) as modified by Sandri-Goldin et at. (30) was followed without modification.

Preparation of clones containing both cCAD1 and GPT genes. Recombinants containing both CAD and GPT genes were prepared by introducing purified PSV2 GPT DNA (31,32) into E. coli containing cCAD1. Cells resisting both 50 µg/ml ampicillin and 10 µg/ml tetracycline were selected. Plasmid and cosmid DNA was isolated (12,25) and analyzed on agarose gels to determine which clones contained both recombinants.

Restriction analysis of CAD sequences in genomic DNA from CHO-K1, Urd⁻A and cells transformed with cloned CAD genes. Genomic DNA was isolated (12) and digested to completion with the indicated enzymes according to the manufacturer's specifications. DNA fractionated in agarose gels was transferred to nitrocellulose or diazotized paper was described (26) except that the HCl treatment for DNA depurination was for 10 min with 0.25 M HCl.

Metaphase spreads and in situ hybridization. Metaphase spreads were prepared and in situ hybridizations were done in the presence of 10% dextran sulfate as described by Wahl et al. (16). The probe was made by nick translating the plasmid p102 (27) with [^{125}I[-dCTP to a specific activity of $0.5-1.0 \times 10^8$ cpm/µg. Approximately 2.5×10^5 cpm of probe in a volume of 25 µl was used per slide. Hybridization was overnight and exposures were for one week at 4°C.

ANALYSIS OF CAD GENE AMPLIFICATION

RESULTS

Cloning a functional CAD gene. CAD mRNA in Syrian hamsters is encoded by at least 25 kb of DNA which contains roughly 37 intervening sequences (27). Since it is not possible to clone DNA this large in vectors such as plasmids or phage, it is necessary to use a cosmid vector for this purpose (23,33). These vectors consist of the cohesive ends of bacteriophage λ (cos sites) attached to a bacterial plasmid which supplies replication sequences and drug resistance markers. The cos sites enable recombinant molecules with a total length of 37-53 kb to be packaged most efficiently. Since we used the cosmid MUA3, which is only 4.7 kb long (20), DNA inserts 33-48 kb long are needed to allow the recombinant molecules to be packaged (34).

Isolating recombinants with CAD inserts was simplified by using DNA from the PALA-resistant Syrian hamster mutant 165-28 which contains about 200 CAD genes (12). Approximately one in every 1,000 cosmids is expected to contain CAD sequences in a cosmid library prepared using 165-28 DNA with an average size of

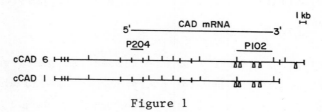

Figure 1

Cosmid clones containing CAD sequences. Two cosmid clones which exhibited hybridization to sequences specific for the 5'- and 3'-ends of CAD mRNA were isolated as described in Methods. The clones were digested with endonucleases EcoRl (———), BamHI (———) and Sst I (———, only the 3' Sst sites are shown) and the restriction maps obtained are shown. The portion of these clones encoding the CAD mRNA is indicated, as are the 3'- and 5'-specific clones used as the probes.

40 kb. Two clones in a library of approximately 10,000 clones were identified which contained sequences at both the 3' and 5'-ends of the CAD mRNA. Since these clones contained inserts 40 kb (cCAD1) and 45 kb (cCAD6) long, they were good candidates for containing a functional CAD transcriptional unit (see Fig. 1).

The cloned Syrian hamster CAD genes function in CHO cells. Several methods are available for transferring cloned genes into animal cells in culture to assess whether they are biologically active. Protoplast fusion (29) was chosen for these experiments since it does not involve DNA isolation procedures which might nick or degrade the 45-50 kb CAD cosmids and thus destroy their biological activity. The general features of the protoplast fusion technique are described in Fig. 2. E. coli containing a recombinant molecule are converted to protoplasts by treatment with lysozyme and ethelenediamine tetracetic acid in an isotonic buffer to prevent cell lysis. The protoplasts are then fused directly with a monolayer culture of animal cells in the presence of polyethylene glycol (PEG1000). Bacteria and PEG are removed and the animal cells are then placed under selective conditions after an appropriate length of time to allow expression of the donated genes.

The biological activity of the cloned CAD genes was first scored in mutant Chinese hamster ovary cells (CHO-K1) which lack a functional CAD gene (Urd$^-$A; 18). Due to this defect, these mutants fail to synthesize uridine <u>de novo</u> and, therefore, require an exogenous source of uridine for growth. If the cloned Syrian hamster CAD genes were functional, they should suppress the uridine auxotrophy of Urd$^-$A cells. Urd$^-$A cells grown in the presence of uridine were fused with protoplasts made from bacterial clones containing either cCAD1 or cCAD6 and then exposed to medium lacking uridine. No prototrophic colonies were obtained from 10^5 Urd$^-$A cells that were treated with the fusing agent

ANALYSIS OF CAD GENE AMPLIFICATION

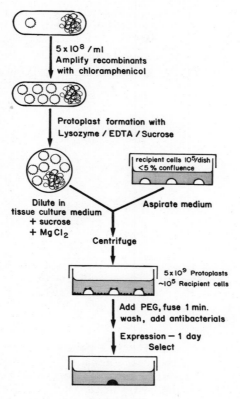

Figure 2

Important features of the protoplast fusion technique. The experimental procedure for fusing protoplasts of E. coli with monolayers of mammalian tissue culture cells is depicted here essentially as described by Shaffner (29) and Sandri-Goldin et al (30). The expression time required prior to application of the selective agent may vary with different genes.

(PEG1000) alone or when they were fused with protoplasts containing a selectable marker other than cCAD1 or cCAD6 (Table 1). In contrast, 100-500 colonies were obtained from 10^5 Urd$^-$A cells after fusion with protoplasts containing either of the CAD cosmids (Table 1). Under these conditions, 0.1-0.5% of the recipient cells were "transformed" to uridine-independence by the donated CAD genes.

Table 1

Introduction of two genes into a single cell
by protoplast fusion

	Frequency (colonies/10^5 recipient cells) of transformants containing indicated genes		
Marker in protoplast*	CAD	GPT	CAD + GPT
1. None	0	0	0
2. CAD (1)	250	0	0
3. CAD (1/10)	42	0	0
4. GPT (1)	0	175	0
5. GPT (1/10)	0	20	0
6. CAD (1) + GPT (1)	250	232	13,7
7. CAD (1) + GPT (1/10)	250	24	1,1
8. CAD (1/10) + GPT (1)	35	193	0,0
9. CAD (1/10) + GPT (1/10)	29	22	0,0
10. (CAD +GPT)	100	100	10

*Number in parentheses equals relative number of the indicated protoplasts used for fusion. Rows 6-9 indicate fusions with bacteria containing either gene in each cell, while row 10 indicates that the fusion was done with bacteria containing both genes in each cell.

Protoplasts containing the indicated genes were prepared and fused with Urd¯A cells as described in "Methods." The fused cells were then selected in uridine-deficient medium (CAD selection), medium containing 10 µg/ml mycophenolic acid (MPA), 250 µg/ml xanthine and 30 µM uridine (GPT selection) or medium lacking uridine and containing MPA and xanthine (CAD + GPT). Row 10 indicates that protoplasts were made from E. coli containing both markers in each bacterial cell. Plates were stained with Giemsa 10 days after the selections were started and the colonies were counted. Fusion of Urd¯A cells with protoplasts containing the GPT gene and selecting for CAD function allowed us to test for the reversion frequency of Urd¯A cells or the ability of bacteria to supply a function which could suppress the auxotrophy of the Urd¯A cells. Selection of parallel cultures for GPT function provided a positive control for fusion.

The number of Syrian hamster CAD genes donated by protoplast fusion was estimated in two ways. First, we determined the CAD protein specific activities in five independently isolated transformant clones. Whereas the parental Urd¯A line has less than 1% of the CAD activity (determined by measuring ATCase activity) of wild type CHP-K1 cells, the mutants have 2-15 times

ment occurred in this transformant. We looked for further rearrangements in these transformants by eluting the hybrids from the activity of the wild type cells. Due to their high CAD levels, the transformants are also resistant to significantly higher PALA concentrations than the wild type cells. A second method was to determine the number of Syrian hamster CAD sequences in the genomic DNA of the transformants using the DNA blotting technique of Southern (35). This was straightforward since CHO and Syrian hamster CAD genes exhibit different patterns after cleavage with a variety of restriction endonucleases (Wahl, unpublished observations and see below). Digestion with endonuclease Sst I gave a simple restriction pattern when hybridized with the 3'-specific probe p102 (see below and ref. 21). Comparison of hybridization band intensities from the transformants and an internal standard of cosmid DNA showed that all transformants have at least 10 copies of the donated genes (21).

We have begun to explore the structure of the donated genes by analyzing the restriction patterns of transformant DNA which has been digested with enzymes which cut frequently within the cosmid. Hybridization of Southern blots of digested transformant cell DNA with probes specific for different regions of the cosmid enables as to localize sequences used for rearrangements. An example of this type of analysis is shown in Fig. 3. Radioactive probes were prepared (36) from either p102 (Fig. 3a), the entire cCAD1 cosmid (Fig. 3b) or the vector MUA3 (Fig. 3c) and hybridized with Pvu II cleaved DNA isolated from three transformants. cCAD1 DNA (land 1 in all samples) was included as an internal control. Hybridization with the 3'-specific probe p102 reveals two bands from the endogenous CHO CAD genes (arrows) and 8 bands which co-migrate in two transformants and cCAD1. A third transformant contains one higher molecular weight band in addition to these 8 (compare lane 2 with lanes 1, 3, 4), indicating that a rearrange-

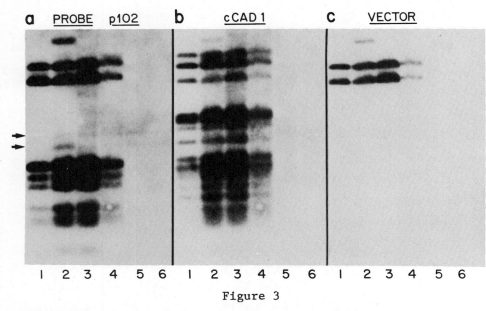

Figure 3

Restriction analysis of genomic DNA isolated from transformants of Urd⁻A cells. In each panel, lanes 2-6 contain 10 μg of DNA is isolated from transformants, Urd1-1, Urd1-2, Urd1-3, Urd⁻A and CHO-K1. Land 1 contains 1 ng (equivalent to 10 gene copies) of cCAD1 DNA. All DNA samples were digested to completion with a four-fold excess of endonuclease Pvu II, fractionated by electrophoresis through a 0.7% agarose gel, and then transferred to diazotized paper as described in Methods. After hybridization with [^{32}P]-p102 (prepared by nick-translation) the blots shown in panel (a) were autoradiographed and then washed with 0.5 M NaOH to elute the hybridized material. The blot was then sequentially hybridized with the [^{32}P]-labelled probes indicated in (b) and (c) with a NaOH wash included prior to rehybridization. This procedure enabled the same blot to be used for all experiments to allow direct comparisons. The arrows indicate the positions of the endogenous CAD sequences.

this blot and re-hybridizing with a probe prepared by nick-translating (36) the whole cCAD1 cosmid and removing the highly reiterative elements (Breson, Ardeshir and Stark, personal communication). Although many additional bands are revealed, only the single band described above is obviously different (Fig. 3b). We have localized the region contributing to this difference by removing the hybrids as described previously and rehybridizing the same blot with a probe prepared by nick-translating the vector alone. Fig. 3c shows that this probe, which does not contain any CAD sequences, hybridizes with the rearranged sequence, demonstrating that vector sequences must be involved in the rearrangement in this transformant. It is not clear if CAD gene sequences might also participate in such rearrangements.

The CAD gene is a dominant selectable marker. The experiments described above were facilitated by using recipient cells having a defective CAD gene. The usefulness of the cloned CAD gene would be expanded greatly if it could be reintroduced into any cell, independent of its endogenous CAD activity. Two observations suggested that the methods described above could be used to isolate clones of wild type CHO-K1 cells "transformed" with Syrian hamster CAD genes. First, multiple copies of the donated CAD gene were integrated into the genomes of all transformants of Urd$^-$A cells analyzed thus far, engendering high levels of PALA resistance. These lines exhibited relative plating efficiencies of 10-40% in the presence of 250 μM PALA compared to 0.001% for wild type CHO-K1 cells (data not shown). Second, there is a high efficiency of gene transfer by protoplast fusion (0.1-0.5% of the recipients are transformed). Thus, after fusion, transformants to CHO-K1 cells resistant to 250 μM PALA should be isolated at a frequency of about 10^{-4} (plating efficiency x transformation efficiency = ≥ 0.1 x ≥ 0.001 = $\geq 10^{-4}$), at least one order of magnitude higher than the frequency of spontaneous mutants resistant to this drug level. In one experiment, CHO-K1 cells were treated with PEG alone or were fused with protoplasts containing cCAD1 and selected for resistence to 250 μM PALA. Mock fused cultures yielded no colonies, while colonies were observed

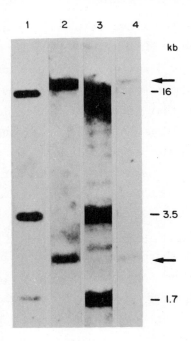

Figure 4

Hybridization patterns of CAD sequences in wild type CHO-K1, PALA-resistant CHO-K1, and PALA-resistant CHO-K1 isolated after fusion with cCAD1. Genomic DNA (10 μg) from the indicated cell lines or 1 ng of cCAD1 DNA was digested to completion with endonuclease Sst I and then analyzed by blot hybridization as described in the legend to Fig. 3 and Methods. The probe was [^{32}P]-p102. The arrows indicate the positions of the endogenous CAD sequences. Lane 1, cCAD1; lane 2, clone K1-2, a CHO-K1 clone selected for resistance to 250μM PALA after fusion with protoplasts containing cCAD1; lane 3, a CHO-K1 clone isolated directly for resistance to 250 μM PALA; lane 4, wild type CHO-K1.

when the cells were fused with protoplasts containing the cosmid. Colonies were picked and grown in selective medium (250 μM PALA) for 20-30 generations. In a parallel experiment, unfused CHO-K1 cells were subjected to the same selection (250 μM PALA). DNA extracted from these lines was digested with Sst I and analyzed as described above using nick-translated p102 as the probe. The hybridization patterns from one representative clone isolated after protoplast fusion or direct selection are shown in Fig. 4 (lanes 2,3) and the hybridization patterns of Sst I digested cCAD1 and CHO-K1 DNA (lanes 1 and 4, respectively) are shown for comparison. The hybridization patterns are identical in CHO-K1 DNA and DNA from CHO-K1 cells selected for PALA resistance, but the signal intensity is greater in the latter case. This demonstrates that amplification of the endogenous CHO CAD genes mediates PALA resistance in these cells. In contrast, DNA from the protoplast transformed clone exhibits strong hybridization to bands characteristic of the Syrian hamster CAD gene in addition to weak hybridization to bands characteristic of the CHO-K1 CAD genes.

Simultaneous introduction of two genes into a single recipient cell using protoplast fusion. Examination of many biochemical events at the molecular level is hampered by the scarcity of the products being assayed. Gene amplification is clearly one way to significantly increase the levels of such products. Since many genes cannot be selected for amplification directly, we have explored the possibility of transforming a single cell with the CAD gene in addition to a second gene to determine whether the latter could be co-amplified after selection of the transformants for resistance to PALA.

Two approaches were used for these experiments. First, bacteria containing either the cloned CAD gene or the GPT gene (for a description of this gene, see the legend to Table I and refs. 31 and 32) were converted to protoplasts and then fused alone or in combination to Urd$^-$A cells. The efficiency of transformation with the CAD or GPT genes was then measured by selecting for cells which grew in the absence of uridine or in the presence of mycophenolic acid + xanthine + uridine, respectively. Table I shows that transformants for either marker occurred at frequencies of approximately $200/10^5$. Reducing the number of bacterial cells used for fusion caused a corresponding decrease in the number of transformants obtained. When equal numbers of protoplasts containing each marker were mixed and then fused with the Urd⁻A cells, the frequency of obtaining double transformants was about $10/10^5$ while the frequency of obtaining single transformants for either marker measured in the same experiment was $200/10^5$. Decreasing the number of protoplasts containing either marker by 10 fold while keeping constant the number of protoplasts containing the other marker resulted in a parallel decrease in the number of double transformants.

A second approach for obtaining double transformants was to introduce both markers into a single bacterial cell. This was straightforward since the CAD cosmids confer tetracycline resistance and the GPT recombinants confer ampicillin resistance to the host bacterium. E. coli containing both markers were isolated, grown into mass culture, converted to protoplasts, fused with Urd⁻A cells and selected as described above. The frequency of obtaining single transformants was approximately the same as reported above. Interestingly, the frequency of obtaining double transformants was again approximately 10-fold lower than that for obtaining transfer of a single marker. The implications of these results will be considered in the Discussion.

⎯⎯⎯⎯⎯⎯⎯⎯⎯⎯⎯⎯⎯⎯⎯⎯⟶

Figure 5 In situ localization of the donate Syrian hamster genes in CHO chromosomes. a) Metaphase spread from cell line K1-4 hybridized in situ with 2.5×10^5 cpm of [^{125}I]-p102 prepared by nick- translation with [^{125}I]-dCTP. Dextran sulfate (10%) was included to accelerate hybridization rates as described in Methods. b) Trypsin-Giemsa banded chromosomes were prepared from the Urd⁻A recipient cells. Some chromosome heterogeneity exists in this line, as shown by the presence of chromosome Z-2 in a metaphase spread from another cell on the same slide (inset). The chromosome exhibiting significant and reproducible hybridization in the indicated lines is shown below its G-banded counterpart. The chromosomes which hybridize in transformants Urd 1-2 and Urd 1-3 cannot be distinguished between the X-chromosome Z-2 since unbanded chromosomes were used for hybridization.

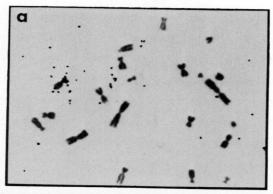
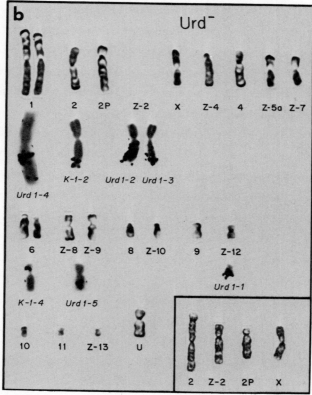

Genes donated by protoplast fusion ultimately reside in the chromosomes of the recipient cell. The DNA blotting experiments describing above cannot address the question of whether the donated genes are located within chromosomes or replicate extrachromosomally. More precise information concerning the cytogenic location of the donated genes was obtained by hybridizing metaphase spreads of chromosomes from the transformants with [^{125}I]-labeled probe prepared from p102. Fig. 5a shows a typical metaphase spread of chromosomes hybridized as described previously (16). It is clear that only a single chromosome exhibits specific hybridization. Fig. 5b presents a compilation of the results obtained by analyzing five Urd$^-$A and two CHO-K1 transformants. In each cell line examined, only a single chromosome exhibited specific hybridization. These chromosomes are shown below a banded karyotype of chromosomes from Urd$^-$A cells. It is clear that the donated genes are integrated into different chromosomes in each transformant. Furthermore, the data available indicates that the donated genes do not integrate at preferred chromosomal locations (i.e., ends of chromosomes in comparison with internal sites). Most of the transformants isolated thus far have approximately the same number of chromosomes as the parental cells although in one experiment pseudotetraploid transformants were isolated (data not shown and ref. 21). This may have occurred because the density of the mammalian cells was too high during fusion.

DISCUSSION

The functional CAD genes we have isolated are 40 and 45 kb long. At the time this work was initiated, it was unclear whether such large recombinant molecules could be reintroduced efficiently into CHO cells using calcium phosphate mediated gene transfer (37,38). We chose to use protoplast fusion for three reasons.

ANALYSIS OF CAD GENE AMPLIFICATION

First, up to 50% of the recipient cells take up material after fusion with protoplasts (30) and up to 6% of the recipient cells express donated DNA molecules (29). Second, high efficiency gene transfer occurs with a broad range of cell lines with protoplast fusion (29; 30; R. Sandri-Goldin, personal communication). Third, protoplast fusion does not involve manipulations which might destroy the biological activity of the CAD cosmids (e.g., nicking which occurs during DNA isolation or transfection). Our data shows that the high efficiency of gene transfer by protoplast fusion is independent of the size of the recombinant molecule over the range 8-50 kb. The high transformation efficiency of this method has made it possible to screen clone banks for functional eukaryotic genes. For example, we found one clone containing a functional CAD gene in a population of 1,000 clones.

By the criteria of CAD specific activity measurements, PALA-resistance, Southern blot analysis (35) and _in situ_ hybridization, we have shown that many functional copies of the CAD cosmid are transferred from the protoplasts to the recipient Urd⁻A or CHO-K1 cells. Most, if not all, of the donated genes are localized to a single chromosomal region in each transformant. Since the donated genes have been localized to six and possibly seven different chromosomes, our data suggests that CAD gene integration may occur randomly with respect to chromosome location. No information is available concerning the nucleotide sequences used for integration leaving open the question of whether sequence specific integration occurs in these cases. Since exogenous genes have also been introduced into chromosomes using calcium phosphate medated gene transfer (39) and mechanical microinjection (Wahl, Vitto and Capecchi, unpublished), it appears that the ability to integrate donated DNA sequences into chromosomes will not be limited by the technique used for gene transfer. These results indicate that it should be possible to use these techniques to isolate a collection of cell clones in which the CAD gene is introduced into every chromosome with each clone having the CAD genes integrated into one specific chromosome. A library of cell clones such as this would thus have a dominant amplifiable marker in each chromosome and would be useful for somatic cell hybridization studies where an easily selectable marker is not otherwise present on the chromosome of interest.

An intriguing question raised by the data described above is how the donated genes integrate into the recipient cell chromosomes. In our experiments, the integration of multiple copies of the donated gene was favored. This was not due to the selection used since only 25% of the wild type level of CAD is sufficient to allow cell multiplication under the conditions used (18). It is conceivable that many cosmid molecules could become linked prior to integration, perhaps through the formation of large random concatemers as happens in calcium phosphate mediated gene transfer (40). One observation that is consistent with this idea is that some of the minor bands in restriction analyses of DNA from some of the transformants comigrate with fragments from cosmids which had deleted sequences during passage in E. coli. However, as we have shown, if such concatemers do form, they do not cause significant rearrangements in the donated CAD genes (see Fig. 3). It is important to emphasize that we should have been able to detect such rearrangements since only one or a few of the many donated CAD genes are needed to retain function in order to allow cell survival. These data indicate that either such rearrangements could influence the functioning and/or expression of adjacent CAD genes and would be selected against, that integration of CAD cosmids occurs somewhat specifically with regard to the cosmid sequences employed, or that homologous recombination occurs between the donated sequences.

The integration of multiple functional copies of the CAD gene allowed us to transfer the cloned genes to wild type CHO cells by selecting transformants which resist high levels of PALA in a single step. These experiments show that the CAD gene can be used as a dominant marker. The integration of most, if not all, of the donated genes into single chromosomal locations in these transformants reemphasized the possibility that some of the donated genes became linked. In order to test this idea and expand the usefulness of the protoplast fusion technique, we fused Urd$^-$A cells with a mixture of protoplasts containing either the CAD gene or the GPT gene (another dominant marker) and determined whether two dissimilar, independently donated genes would also become linked. The frequency of obtaining double transformants ($5 \times 10^{-5} - 10^{-4}$) was approximately 5-10% of the frequency of obtaining single transformants (10^{-3}). The observed frequency of double transformants is thus 10-100 fold higher than expected for the two markers being donated and integrated on a totally random basis (i.e., $10^{-3} \times 10^{-3} = \underline{10^{-6}}$), but it is lower than expected if both markers become linked 100% of the time (i.e., 10^{-3}). We tested whether this result was due in part to donating the genes from two bacterial populations by inserting both genes into the same bacterium prior to fusion. The frequency of single or co-transformation was the same in this experiment as observed when two separate bacteria were used (see Table I). Our results suggest that either of these methods may be used to co-transform cells with sequences that are not linked at the time of transformation, but they also indicate that linkage between two genes donated by protoplast fusion will also depend upon factors other than whether both genes are introduced into the same mammalian cell. For example, GPT transformants of CHO cells grow very slowly (Robert de Saint Vincent, unpublished observation) whereas CAD transformants grow rapidly. It is conceivable that many GPT genes have to be donated in order to enable the cell to survive under the selective conditions used. If concatemer formation is necessary for transformation and/or integration, it is conceivable that concatemers of sufficient size to contain the requisite number of CAD and GPT genes to withstand the double selection arise at a lower frequency than those required for these genes donated independently. We are currently investigating whether the CAD and GPT genes are linked at the molecular level by DNA blotting experiments and by determining whether both are integrated into the same chromosomal location in the co-transformants. We are also assessing whether other gene combinations exhibit higher co-transformation frequencies. These results differ from those described for the co-transformation of sequences with the calcium phosphate technique since in the calcium phosphate method a vast excess of an unselected relative to a selected marker is used in order to insure that both will become associated (40).

It is unclear that the ability to clone functional mammalian genes (e.g., see 41), mutate purified DNA sequences at specific sites in vitro (e.g., see 42), and study the biological function of genes altered in vitro by reinserting them into animal cells and analyzing their expression in vivo (e.g., see 43) has opened up a new era in eukaryotic genetics. These technologies hold the promise for substantial leaps in our knowledge of the relationship between gene structure and gene expression. In a similar fashion, we anticipate that our studies concerning the role of chromosomal location in gene amplification may give insight into the molecular details of genome plasticity.

ACKNOWLEDGMENTS

We thank Dr. David Patterson for providing us with the wild type CHO-K1 and Urd$^-$A cells, Drs. M. Levine and R. Sandri-Goldin for their protocol for protoplast fusion and for valuable discussions about this technique, and Drs. R. Mulligan and P. Berg for supplying the recombinant plasmid containing the Eco GPT gene. We also thank Drs. O. Brison, J. Zieg, F. Ardeshir and G. Stark for providing us with their restriction analysis of cosmids cCAD1 and cCAD6 and for allowing us to present this data.

REFERENCES

1. R.P. Anderson and J.R. Roth, Tandem genetic duplications in phage and bacteria. Ann. Rev. Microbiol. 31:473-505, 1977.
2. R.A. Padgett, G.M. Wahl, O. Brison and G.R. Stark, Use of recombinant DNA techniques to examine amplified regions of DNA from drug-resistant mutant cells. In The Third Cleveland Symposium on Macromolecules: Recombinant DNA (Elsevier), in press.
3. P.E. Barker and T.C. Hsu, Double minutes in human carcinoma cell lines, with special references to breast tumors. J. Natl. Cancer Inst. 62:257-261, 1979.
4. A. Levan, G. Levan and F. Mittleman, Chromosomes and cancer. Hereditas 86:15-30, 1977.
5. G. Kovacs, Homogeneously staining regions on marker chromosomes in malignancy. Intl. J. Cancer 23:299-301, 1979.
6. D.L. George and V.E. Powers, Cloning of DNA from double minutes of 41 mouse aderenocortical tumor cells: Evidence for gene amplification. Cell 24:117-123, 1981.
7. M.L. Pall, Gene-amplification model of carcinogenesis. Proc. Natl. Acad. Sci. USA 78:2465-2468, 1981.
8. A. Varshavsky, Phorbal ester dramatically increases incidence of methotrexate-resistant mouse cells: Possible mechanisms and relevance to tumor promotion. Cell 25:561-572, 1981.
9. M.T. Hakala, S.F. Zakrgewski and C. Nichol, Relation of folic acid reductase to amethopterin resistance in cultured mammalian cells. J. Biol. Chem. 236:952-958, 1961.

10. T.D. Kempe, E.A. Swyryd, M. Bruist and G.R. Stark, Stable mutants of mammalian cells that overproduce the first three enzymes of pyrimidine nucleotide biosynthesis. Cell 9:541-550, 1976.

11. F. Alt, R. Kellems, J.R. Bertino and R.T. Schimke, Selective multiplication of dihydrofolate reductase genes in methotrexate-resistant variants of cultured murine cells. J. Biol. Chem. 253:1357-1370, 1978.

12. G.M. Wahl, R.A. Padgett and G.R. Stark, Gene amplification causes overproduction of the first three enzymes of UMP in N-(phosphonacetyl)-L-aspartate-resistant hamster cells. J. Biol. Chem. 254:8679-8689, 1979a.

13. J.H. Nunberg, R.J. Kaufman, R.T. Schimke, G. Urlaub and L.A. Chasin, Amplified dihydrofolate reductase genes are localized to a homogeneously staining region of a single chromosome in a methotrexate-resistant Chinese hamster ovary cell line. Proc. Natl. Acad. Sci. USA 75:5553-5556, 1978.

14. B.J. Dolnick, R.J. Berenson, J.R. Bertino, R.J. Kaufman, J.H. Nunberg and R.T. Schimke, Correlation of dihydrofolate reductase elevation with gene amplification in a homogeneously staining chromosomal region in L5178Y cells. J. Cell Biol. 83:394-402, 1979.

15. R.J. Kaufman, P.C. Brown and R.T. Schimke, Amplified dihydrofolate reductase genes in unstably methotrexate resistant cells are associated with double minute chromosomes. Proc. Natl. Acad. Sci. USA 76:5669-5673, 1979.

16. G.M. Wahl, L. Vitto, R.A. Padgett and G.R. Stark, Localization of single copy and amplified CAD genes in Syrian hamster chromosomes using a highly sensitive method for in situ hybridization. Mol. Cell. Biol. 2:308-317, 1982.

17. J.L. Biedler and B.A. Spengler, Metaphase chromosome anomaly: Association with drug resistance and cell-specific products. Science 191:185-187, 1976.

18. D. Patterson and D.V. Carnright, Biochemical genetic analysis of pyrimidine biosynthesis in mammalian cells: I. Isolation of a mutant defective in the early steps of de novo pyrimidine synthesis. Somat. Cell Genet. 3:483-495, 1977.
19. C.N. Wiblin and I.A. MacPherson, The transformation of BHK21 hamster cells by Simian virus 40. Int. J. Cancer 10:296-309, 1972.
20. E.M. Meyerowitz, G.M. Guild, L.S. Prestidge and D.S. Hogness, A new high capacity cosmid vector and its use. Gene 11:271-282, 1980.
21. B. Robert de Saint Vincent, S. Delbrück, W. Eckhart, J. Meinkoth, L. Vitto and G. Wahl, Cloning and reintroduction into animal cells of a functional CAD gene, a dominant amplifiable genetic marker. Cell, accepted for publication, 1981.
22. M.W. McDonnell, M.N. Simon and F.W. Studier, Analysis of restriction fragments of T7 DNA and determination of molecular weights by electrophoresis in neutral and alkaline gels. J. Mol. Biol. 110:119-146., 1977.
23. J. Collins and B. Hohn, Cosmids, a new type of plasmid gene-cloning vector which is packageable in vitro in lambda bacteriophage heads. Proc. Natl. Acad. Sci. USA 75:4242-4246., 1978.
24. F.G. Grosveld, H.-H.M. Dahl, E. deBoer and R.A. Flavell, Isolation of β globin-related genes from a human cosmid library. Gene 13:227-237, 1981.
25. D.B. Clewell and D.R. Helinski, Supercoiled circular DNA-protein complex in Excherichia coli: Purification and induced conversion to an open circular DNA form. Proc. Natl. Acad. Sci. USA 62:1159-1166, 1969.
26. G.M. Wahl, M. Stern and G.R. Stark, Efficient transfer of large DNA fragments from agarose gels to diazo-benzyloxy-methyl-paper and rapid hybridization by using dextran sulfate. Proc. Natl. Acad. Sci. USA 76:3683-3687, 1979.

27. R.A. Padgett, G.M. Wahl and G.R. Stark, Structure of the gene for CAD, the multifunctional protein that initiates UMP biosynthesis in Syrian hamster cells. Mol. Cell. Biol. 2:293-301, 1982.
28. M. Grunstein and D.S. Hogness, Colony hybridization: A method for the isolation of cloned DNAs that contain a specific gene. Proc. Natl. Acad. Sci. USA 72:3961-3965, 1975.
29. W. Shaffner, Direct transfer of cloned genes from bacteria to mammalian cells. Proc. Natl. Acad. Sci. USA 77:2163-2167, 1980.
30. R.M. Sandri-Goldin, A. Goldin, M. Levine and J.C. Glorisso, High-frequency transfer of cloned Herpes simplex virus type 1 sequences to mammalian cells by protoplast fusion. Mol. Cell. Biol. 1:743-752, 1981.
31. R. Mulligan and P. Berg, Expression of a bacterial gene in mammalian cells. Science 209:1422-1427, 1980.
32. R. Mulligan and P. Berg, Selection for animal cells that express the Escherichia coli gene coding for xanthine-guanine phosphoribosyl transferase. Proc. Natl. Acad. Sci. USA 78:2072-2076, 1981.
33. J. Collins and H.J. Bruning, Plasmids useable as gene-cloning vectors in an in vitro packaging by coliphage λ: "cosmids." Gene 4:85-107, 1978.
34. M. Feiss, R.A. Fisher, M.A. Crayton and C. Egner, Packaging of the bacteriophage chromosome: Effect of chromosome length. Virology 77:281-293, 1977.
35. E.M. Southern, Detection of specific sequences among DNA fragments separated by gel electrophoresis. J. Mol. Biol. 98:503-517, 1975.
36. P.W. Rigby, M. Dieckmann, C. Rhodes and P. Berg, Labeling deoxyribonucleic acid to high specific activity in vitro by nick translation with DNA polymerase I. J. Mol. Biol. 113:237-251, 1977.

37. F. Graham and A.J. van der Eb, A new technique for the assay of infectivity of human adenovirus 5 DNA. Virology 52:456-467, 1973.
38. M. Wigler, S. Silverstein, L.-S. Lee, A. Pellicer, Y.-C. Cheng and R. Axel, Transfer of purified Herpes virus thymidine kinase gene to cultured mouse cells. Cell 11:223-232, 1977.
39. D.M. Robins, S. Ripley, A.S. Henderson and R. Axel, Transforming DNA integrates into the host chromosome. Cell 23:29-39, 1981.
40. M. Perucho, D. Hanahan and M. Wigler, Genetic and physical linkage of oxogenous sequences in transformed cells. Cell 22:309-317, 1980.
41. I. Lowy, A. Pellicer, J.F. Jackson, G.K. Sim, S. Silverstein and R. Axel, Isolation of transforming DNA: Cloning the hamster aprt gene. Cell 22:817-823, 1980.
42. D. Shortle and D. Nathans, Local mutagenesis: A method for generating viral mutants with base substitutions in preselected regions of the viral genome. Proc. Natl. Acad. Sci. USA 75:2170-2174, 1978.
43. S.L. McKnight, E.R. Gavis, R. Kingsbury and R. Axel, Analysis of transcriptional regulatory signals of the HSV thymidine kinase gene: Identification of an upstream control region. Cell 25:385-398, 1981.

ALLOREACTIVE T CELL CLONES[1]

Frank W. Fitch

Department of Pathology
Committee on Immunology
The University of Chicago
Chicago, Illinois

ABSTRACT

T cell clones are useful models for studying lymphocyte function both at the level of the individual cell and in interacting systems. Murine cytolytic and non-cytolyic T cell clones have been obtained with relative ease, and the particular procedure used to derive and maintain T cell clones may influence profoundly the characteristics of the resulting cells. The method of choice depends on the specific question to be asked.

Although some clones have characteristics that would have been expected on the basis of results observed with bulk cell

---

[1]This research was supported by USPHS Grants AI-04197, AI-16558, and CA-19226.

Abbreviations: BCSF - B cell stimulating factor; CM - conditioned medium; CSF - colony stimulating factor; CTL - cytolytic T lymphocytes; IL-2 - interleukin 2; IL-3 interleukin 3; MHC - major histocompatibility complex; MLC - mixed leukocyte culture; MoLV - Moloney leukemia virus.

populations, other clones have rather unexpected properties. Although most T cell clones appear to be either cytolytic or non-cytolytic, this distinction is not always absolute. A high proportion of both cytolytic and non-cytolytic T cell clones have dual reactivity. This is true for cells which by other criteria appear to be true clones. The frequency of such cells is high enough to suggest that most if not all T cells may have reactivity for more than one antigenic determinant or that antigenic determinants recognized by T cells are shared widely and unexpectedly. It is not clear whether one or two different antigen receptors account for such dual reactivity. The nature of the T cell receptor for antigen remains obscure. T cell clones, because of their homogeneous nature, should make it easier to answer these important immunological questions.

Although it remains to be determined how many distinct molecules account for the numerous biological activities found in the culture supernatants from antigen-stimulated T cell clones, it is clear that these factors influence several different types of cells that are involved directly and indirectly in immune responses. IL-2 stimulates both cytolytic and non-cytolytic T cells to proliferate. BCSF causes polyclonal activation of B cells, and there may be other factors which influence B cell responses to antigenic stimulation. IL-3 apparently stimulates maturation of immature T cells. CSF stimulates production of macrophages from precursor cells found in the bone marrow. Supernatants also stimulate expression of Ia antigens by macrophages, and antigen-presenting cells have been found to bear Ia antigens. Interferon augments natural killer cell activity. Thus, regardless of how many molecules are involved in these effects, activated non-cytolytic T cells appear to be involved in a variety of ways in the modulation of immune responses. T cell clones will be useful in further studies to determine the detailed mechanisms of these cellular interactions.

A particularly significant advance in immunology during the past decade was the demonstration that there are multiple subsets of lymphocytes, some of which carry out effector functions of immune responses, others having a regulatory role to augment or inhibit the responses of effector cells. Although cell functions and mechanisms of cellular interaction have been defined in general terms, many details remain obscure. Research has been hampered because lymphocyte populations are heterogenous even after the use of positive and negative selection methods to enrich for particular cell types. Recently, however, it has become possible to derive clones of T lymphocytes having defined effector or regulatory functions. These T lymphocyte clones are being used in a variety of studies, and they promise to be especially useful in efforts to further characterize the molecular basis of immune responses.

The major development that was essential for T cell cloning was the serendipitous rediscovery of T cell growth factor, now usually called interleukin 2 (IL-2), by Dr. Robert Gallo and his colleagues at the National Institutes of Health in 1975 (1). Acutally, the observations that conditioned medium from activated T lymphocytes supported the growth of other lymphocytes had been made at least a decade earlier (2,3), but the paradigm in existence at that time could not accomodate this finding. Gillis and Smith (4) obtained long-term lines of murine cytolytic T lymphocytes (CTL) a year after description of T cell growth factor, and reports of successful cloning of T cells from long-term cultures appeared within a few months (5).

CLONING METHODS AND THEIR EFFECTS ON THE PROPERTIES OF T CELL CLONES

Several quite different approaches have been used to develop functional murine T cell clones. The earliest method was to culture cells from secondary mixed leukocyte culture (MLC) in conditioned medium (CM) which contained IL-2 (6,7). Usually, it has been necessary to passage cells in bulk cultures for weeks to months before cloning is successful when cultures are maintained only with CM (6,7). The frequency of cells capable of growing in the presence of IL-2 alone usually is low, and "self-cloning" appears to occur during the prolonged culture period which preceded cloning (6). The level and pattern of cytolytic activity of cloned CTL derived and maintained in CM alone may show fluctuations (5), and karyotypic abnormalities are frequent if not universal (8). Such cloned cells are dependent upon IL-2 for growth and may not be responsive to antigenic stimulation. A major advantage of this approach is that only one type of cell is present in culture; stimulating alloantigen or antigen-presenting cells are not present as possible contaminants. This type of cloned T cell may be preferable for biochemical studies, although such cells may be less useful in research concerned with mechanisms which regulate cell functions.

A second approach for deriving and maintaining T cell clones uses a combination of CM, stimulating antigen, and "filler cells" (9,10). Although clones usually have been derived using cells obtained from primary or secondary MLC, it is possible to clone cells directly from normal lymphoid tissue. However, the frequency of reactive cells is much greater in populations of cells obtained after immunization _in vitro_ or _in vivo_. The frequency of reactive cells in normal lymphoid populations usually is no greater than 1 per several hundred while the frequency in cells obtained from secondary MLC often is at least 1 per 3 cells (11). Cloning efficiency often is nearly 100% with cells from MLC (10). For alloreactive T cells, irradiated lymphoid cells provide both a source of stimulating alloantigen and filler cells. It is not clear what function is carried out by the filler cells. With soluble antigens, they may serve as antigen-presenting cells. Whatever their function, high initial cloning efficiency and ease of maintaining cells in long term culture is dependent upon filler cells. The phenotypic characteristics of both cytolytic and non-cytolytic T cell clones derived and maintained using this approach have been stable for several years, and karyotype has remained normal (10). The apparently "normal" characteristics of such cloned cells make them particularly useful for studies concerned with regulation of cellular functions and with mechanisms of cellular interactions. However, the inevitable but often minor contamination by stimulating and/or filler cells may complicate some experiments.

A third approach for obtaining functional T cell clones has been T cell hybridization. Functional helper (12), cytolytic (13), and suppressor (14) T cell hybrids have been developed. With T cell hybrids, it is possible to obtain very large numbers of functional T cells with great ease. Functional T cell lymphomas (15,16) are another alternative when numbers of cells are important. However, interpretation of results obtained with T cell hybrids or lymphomas is complicated by a possible contribution of tumor cell products to the effects observed. Also, it may be difficult to be certain that a particular cell hybrid has been obtained from two T cell parents since hybridomas formed by fusion of some T and B cell lymphoma cell lines have been shown to express the Thy-1 determinants of both parent cells (17).

Regardless of the method used to derive T cell "clones", it is essential to be certain that these cells are actually clones. Mixed populations of cells may persist in culture for long periods of time (18,19,20). Three different cloning procedures have been used. The soft agar method with an underlying feeder layer (5,21,22,23) and limiting dilution (24,25) have been used frequently. A statistical estimate of the likelihood that a given colony or well represents a clone can be obtained using these two methods (26). However, if plating efficiency is low, it may be impossible to be certain that cells having unusual characteristics actually represent clones. Micromanipulation probably is the only technique that always provides a high level of certainty for obtaining clones (27).

PROPERTIES OF CLONED CYTOLYTIC T CELLS

Cloned CTL are dependent upon IL-2 for growth and usually die within 24 hours if cultured in the absence of IL-2. Growth of cloned CTL often appears to be independent of antigen, and most cloned CTL do no proliferate when cultured with antigen alone. Although maximal proliferation can be induced using high levels of IL-2, synergism may be observed between antigen and suboptimal amounts of IL-2 (28). Some CTL clones do respond when stimulated by antigen alone, apparently because such stimulation induces release of IL-2 which in turn causes proliferation (29,30). Such clones are uncommon, and care must be taken to be certain that cells having this characteristic actually are true clones.

The level of cytolytic activity expressed by most cloned CTL usually is quite high; 50% lysis of ^{51}Cr-labelled cells is often observed at an effector to target cell ratio of less than 0.5 to 1 in a conventional 3 hour assay. In contrast to results observed with bulk MLC populations, the level of lytic activity per cell for cloned CTL appears to remain constant throughout the culture period (10). Cloned CTL can divide rapidly, often having doubling time shorter than 24 hours (10).

The patterns of lytic activity for alloreactive cloned CTL may be those which would be expected based upon results observed with MLC cells; only target cells bearing stimulating alloantigens are lysed (5,31). However, unexpected patterns of reactivity often have been observed, especially when a sufficiently large panel of target cells is employed. For example, when 9 separate CTL clones derived from secondary C57BL/6 anti-DBA/2 MLC were tested on a panel of 11 different target cells, at least 7 different patterns of lysis were observed (24).

The most extensive analysis of the CTL repertoire has been made by Sherman (32). She employed a model system involving B10.D2 ($H-2^d$) spleen cells responding to $H-2K^b$ alloantigens. This particular combination was chosen becuase of the availability of a number of mutant strains which express different antigenic determinants on $H-2K^b$ antigen. Forty-three individual clones from 7 different mice were selected on the basis of lysis of target cells bearing $H-2K^b$ alloantiogens. When assayed on a panel of target cells prepared from 7 different $H-2K^b$ mutant mouse strains, 23 different reactivity patterns were evident (32). In addition to indicating that there are at least 23 different determinants on the $H-2K^b$ molecule, these results also show that the apparently high frequency of T cells reacting with a given haplotype reflects at least partially the summation of the activity of cells reacting with different antigenic determinants on molecules specified by genes of the major histocompatibility complex (MHC).

Viral antigens appear to be recognized by CTL concomitantly with self-antigen specified by the MHC of the stimulating cells (33). Although MHC restriction of antigen recognition is observed with many cloned CTL, the restriction is often not as severe as results obtained with bulk cell populations would indicate. For example, at least 3 reactivity patterns were evident with CTL clones derived from C57BL/6 lymphoid cells which had been stimulated with Moloney leukemia virus (MoLV)-infected syngeneic cells (25). Some clones lysed syngeneic MoLV-derived tumor cells only, some lysed both syngeneic and allogeneic MoLV-derived tumor cells, while some lysed syngeneic MoLV-derived tumor cells and normal allogeneic cells. A further heterogeneity of antigen-recognition by these clones was evident in studies performed using monoclonal antibodies to block cytolysis. The lytic activity of many cloned CTL for MoLV-infected syngeneic cells was inhibited by anti-H-2D^b but not by anti-H-2K^b monoclonal antibodies (34). However, other CTL clones appeared to react with MoLV virus-associated antigens restricted by other regions of the MHC since their lytic activity was inhibited by anti-H-2^b antiserum but not by monoclonal anti-H-2D^b antibodies (34). Comparable ranges of lytic activity were observed for cloned CTL reactive with Friend leukemia virus-induced leukemic target cells (18), influenza virus-infected target cells (35,36), and syngeneic cells to which a variety of haptens had been coupled (6,7). Collectively, these results indicate a greater heterogeneity of CTL responses than is observed usually with bulk populations of cells after sensitzation either in vivo or in vitro.

PROPERTIES OF CLONED NON-CYTOLYTIC T CELLS

Non-cytolytic alloreactive cloned T cells and cloned T cells reactive with soluble antigens have been useful for studying both the requirements for stimulation of T cells and for evaluation of the products produced by stimulated T cells. In many instances, only proliferative responses of cloned T cells after antigen stimulation has been measured. Cloned non-cytolytic T cells have provided unambiguous answers to several important immunological problems. The existence of unique F_1-stimulating determinants was proven by clones of T cells which could be stimulated to proliferate only by F_1 alloantigens (21). The antigenic determinant recognized uniquely on the F_1 stimulating cells was shown to be related to the I-A region of the MHC (37,38). The stimulating determinant for these clones apparently results from the trans-complementation between the I-A region gene products of the two parental strains. Stimulating determinants for other clones which were reactive with both parental and F_1-stimulating cells was derived from cis-complementation of I-A-region gene products.

Products of I-A region genes also appear to be of major important in determining the MHC restriction of murine T cell clones which react with soluble antigens. It is necessary for cloned T cells and antigen-presenting macrophages to be histocompatible at the I-A subregion of the MHC in order for cloned T cells to be stimulated to proliferate after exposure to specific antigen. Similar to results observed with alloantigenic stimulation, some cloned T cells respond to soluble antigens only in the presence of antigen-presenting cells from F_1 mice. Clones of T cells from F_1 mice responding to soluble antigen can be divided into three categories: one type responds to antigen in the presence of macrophages from one parent or from F_1 mice, a second type responds to antigen in the presence of macrophages from the other parent or F_1 cells while the third type responds only to antigen in the presence of macrophages from F_1 hybrid mice (19,37). Genetic studies indicate that the required histocompatibility is at the I-A region of the MHC. Those clones reactive with antigen only in the presence of F_1 macrophages apparently recognize soluble antigens in association only with alloantigenic determinants resulting from trans-complementation of I-A region gene products of the two parents; the α_A chain is encoded by genes from one parent and the β_A chain is encoded by genes provided by the other parent (39). However, the complexity of MHC-restricted antigen recognition is greater than these results would indicate. T cell clones reacting to antigen only in the presence of β_{AE}-α_E Ia molecules also have been identified (40).

At least some antigen-reactive, non-cytolytic cloned T cells may also react with alloantigen. Some T cell clones which proliferate when cultured with soluble antigen in the presence of antigen-presenting cells which are histocompatible at the I-A region of the MHC also proliferate when cultured with allogeneic stimulating cells of a particular haplotype (20,41). Although different antigen-reactive clones are stimulated to proliferate by allogeneic cells of different haplotyes, only one haplotype is stimulatory for a given clone. Interestingly, the relevant stimulating alloantigens appear to by specific by genes of the I-A region of the MHC. Dual specificity also has been observed with some cloned T cells reactive with MHC and strong Mla-locus determinants (42).

BIOLOGICALLY ACTIVE FACTORS PRODUCED BY NON-CYTOLYTIC T CELL CLONES

Antigen-stimulated non-cytolytic T cell clones may release factors which influence the activities of other cells. This was shown first with a non-cytolytic clone reactive with MLs-locus determinants. Co-culture of cloned cytolytic and non-cytolytic T cell with stimulating alloantigen toward which the non-cytolytic cell reacted generated high levels of cytolytic activity, while culture of either type of cell alone with alloantigen did not (31). In this case, the non-cytolytic cloned T cells are stimulated by alloantigen to release IL-2 which in turn causes the cloned CTL to proliferate (10).

A wide variety of biological activities have been found to be present in culture supernantants obtained after antigen stimulation of T cell clones. These activities include IL-2 (43,44,45), IL-3 (46), B cell stimulating factor (BCSF) (43,47), colony stimulating factor (CSF) (43,44,45), erythroid and mixed myeloid-erythroid burst-promoting activities (43), interferon (46,48), recruitment of IA^+-macrophages into the peritoneal cavity (46), induction of Ia antigen expression by macrophages in vitro (46), induction of complement components by guinea pig macrophages (46). Antigen-specific suppressor factors also have been obtained from cloned T cells (49).

The time course for production of CSF and BCSF by cloned T cells differs from that of IL-2 (43,44) and IL-2 has been separated physically from CSF and BCSF (45). In addition, a variant T cell clone has been isolated which produces CSF and BCSF but does not release detectable IL-2 (44). Other clones have been described which produce different arrays of factor activities (50). Thus, it seems clear that several molecules account for these activities.

ACKNOWLEDGMENTS

The secretarial assistance of Mrs. Frances Mills is gratefully acknowledged.

REFERENCES

1. Morgan, D.A., Ruscetti, F.W. and Gallo, R. 1976. Selective in vitro growth of T lymphocytes from normal human bone marrows. Science 193:1007.
2. Gordon, J. and MacLean, L.D. 1965. A lymphocyte-stimulating factor produced in vitro. Nature 208:795.
3. Kasakura, S. and Lowenstein, L. 1965. A factor stimulating DNA synthesis derived from the medium of leucocyte cultures. Nature 208:794.
4. Gillis, S. and Smith, K.A. 1977. Long term culture of tumor-specific cytotoxic T cells. Nature 268:154.
5. Nabholz, M., Engers, H.D., Collavo, D. and North, M. 1978. Cloned T-cell lines with specific cytolytic activity. Curr. Top. Microbiol. Immunol. 81:176.
6. Haas, W., Mathur-Rochat, J., Pohlit, H., Nabholz, M. and von Boehmer, H. 1980. Cytotoxic T cell responses to haptenated cells. III. Isolation and specificity analysis of continuously growing clones. Eur. J. Immunol. 10:828.
7. von Boehmer, H. and Haas, W. 1981. H-2 restricted cytolytic and noncytolytic T cell clones: Isolation, specificity and functional analysis. Immunol. Rev. 54:27.
8. Engers, H.D., Collavo, D. North, M., von Boehmer, H., Haas, W., Hengartner, H. and Nabholz, M. 1980. Characterization of cloned murine cytolytic T cells lines. J. Immunol. 125:1481.

9. Hengartner, H. and Fathman, C.G. 1980. Clones of alloreactive T cells. I. A unique homozygous MLR-stimulating determinant present on B6 stimulators. Immunogenetics 10:175.
10. Glasebrook, A.L., Sarmiento, M., Loken, M.R., Dialynas, D.P., Quintans, J., Eisenberg, L., Lutz, C.T., Wilde, D. and Fitch, F.W. 1981. Murine T lymphocyte clones with distinct immunological functions. Immunol. Rev. 54:225.
11. MacDonald, H.R., Cerottini, J.-C., Ryser, J.-E., Maryanski, J.L., Taswell, C., Widmer, M.B. and Brunner, K.T. 1980. Quantitation and cloning of cytolytic T lymphocytes and their precursors. Immunol. Rev. 51:93.
12. Harwell, L., Skidmore, B., Marrack, P. and Kappler J. 1980. Concanavalin A-inducible, interleukin-2-producing T cell hybridoma. J. Exp. Med. 152:893.
13. Nabholz, M., Cianfriglia, M., Acuto, O. Conzelman, A., Weiss, A., Haas, W. and von Boehmer, H. 1980. Cytolytically active murine T-cell hybrids. Nature 287:437.
14. Taniguchi, M., Saito, T. and Tada, T. 1979. Antigen-specific suppressive factor produced by a transplantable I-J bearing T-cell hybridoma. Nature 278:555.
15. Farrar, J.J., Fuller-Farrar, J., Simon, P.L., Hilfiker, M.L., Stadler, B.M. and Farrar, W.L. 1980. Thymoma production of T cell growth factor (Interleukin 2). J. Immunol. 125:2555.
16. Shimizu, S., Kanaka, Y. and Smith, R. 1980. Mitogen-induced synthesis and secretion of T cell growth factors by a T lymphoma line. J. Exp. Med. 152:1436.
17. Taussig, M., Holliman, A. and Wright, L.J. 1980. Hybridization between T and B lymphoma cell lines. Immunology 39:57.

18. Baker, P.E., Gillis, S. and Smith, K.A. 1979. Monoclonal cytolytic T-cell lines. J. Exp. Med. 149:273.
19. Kimoto, M. and Fathman, C.G. 1980. Antigen-reactive T cell clones. I. Transcomplementing hybrid I-A-region gene products function effectively in antigen presentation. J. Exp. Med. 152:759.
20. Sredni, B. and Schwartz, R.H. 1980. Alloreactivity of an antigen-specific T-cell clone. Nature 287:855.
21. Fathman, C.G. and Hengartner, H. 1978. Clones of alloreactive T cells. Nature 272:617.
22. Sredni, B., Tse, H.Y. and Schwartz, R.H. 1980. Direct cloning and extended cultures of antigen-specific MHC-restricted, proliferating T lymphocytes. Nature 283:581.
23. von Boehmer, H., Hengartner, H., Nabholz, M., Lernhardt, W., Schreier, M. and Haas, W. 1979. Fine specificity of a continuously growing killer cell clone specific for H-Y antigen. Eur. J. Immunol. 9:592.
24. Glasebrook, A.L. and Fitch, F.W. 1980. Alloreactive cloned T cell lines. I. Interactions between cloned amplifier and cytolytic T cell lines. J. Exp. Med. 151:876.
25. Weiss, A., Brunner, K.T., MacDonald, H.R. and Cerottini, J.-C. 1980. Antigenic specificity of the cytolytic T lymphocyte response to murine sarcoma virus-induced tumors. III. Characterization of cytolytic T lymphocyte clones specific for Moloney leukemia virus-associated cell surface antigens. J. Exp. Med. 152:1210.
26. Taswell, C. 1981. Limiting dilution assays for the determination of immunocompetent cell frequencies. I. Data analysis. J. Immunol. 126:1614.

27. MacDonald, H.R., Maryanski, J.L. and Cerottini, J.-C. 1980. Cloning of cytolytic T lymphocytes: requirement for Interleukin 2 and irradiated spleen cells. Behring Inst. Mitt. 67:182.

28. Widmer, M.B. and Bach, F.H. 1981. Antigen driven helper cell-independent cloned cytolytic T lymphocytes. Nature 294:750.

30. Glasebrook, A.L. 1982. Cytolytic T cell clones which proliferate autonomously to specific alloantigenic stimulation. Frequency Interleukin-2 production and Lyt phenotype. (Submitted for publication.

31. Glasebrook, A.L. and Fitch, F.W. 1979. T-cell lines which cooperate in generation of specific cytolytic activity. Nature 278:171.

32. Sherman, L. 1980. Dissection of the B10.D2 anti-H-2K^b cytolytic T lymphocyte receptor repertoire. J. Exp. Med. 151:1386.

33. Zinkernagel, R.M. and Doherty, P.C. 1979. MHC-cytotoxic T cells: Studies on the biological role of polymorphic major transplantation antigens determining T-cell restriction specificity, function and responsiveness. Adv. Immunol. 27:51.

34. Weiss, A., MacDonald, H.R., Cerottinin, J.-C. and Brunner, K.T. 1981. Inhibition of cytolytic T lymphocyte clones reactive with Moloney leukemia virus-associated antigens by monoclonal antibodies: a direct approach to the study of H-2 restriction. J. Immunol. 126:482.

35. Braciale, T.J., Andrew, M.E. and Braciale, V.L. 1981. Simultaneous expression of H-2 restricted and alloreactive recognition by a cloned line of influenza virus-specific cytotoxic T lymphocytes. J. Exp. Med. 153:1371.

36. Braciale, T.J., Andres, M.E. and Braciale, V.L. 1981. Heterogeneity and specificity of cloned lines of influenza-virus specific cytotoxic T lymphocytes. J. Exp. Med. 153:910.
37. Fathman, C.G. and Kimoto, M. 1980. Studies utilizing murine T cell clones: Ir genes, Ia antigens and MLR stimulating determinants. Immunol. Rev. 54:57.
38. Fathman, C.G. and Hengartner, H. 1979. Crossreactive mixed lymphocyte reaction determinants recognized by cloned alloreactive T cells. Proc. Natl. Acad. Sci. USA 76:5863.
39. Fathman, C.G., Kimoto, M., Melvold, R. and David, C.S. 1981. Reconstruction of Ir genes, Ia antigens, and mixed leukocyte reaction determinants by gene complementation. Proc. Natl. Acad. Sci. USA 78:1853.
40. Sredni, B., Matis, L.A., Lerner, E.A., Paul, W.E. and Schwartz, R.H. 1981. Antigen-specific T cell clones restricted to unique F_1 major histocompatibility complex determinants. Inhibition of proliferation with a monoclonal anti-Ia antibody. J. Exp. Med. 153:677.
41. Sredni, B. and Schwartz, R.H. 1981. Antigen-specific, proliferating T lymphocyte clones. Methodology, specificity, MHC restriction, and alloreactivity. Immunol. Rev. 54:186.
42. Webb, S.R., Molnar-Kimber, K.L., Bruce, J., Sprent, J. and Wilson, D.B. 1982. T cell clones with dual specificity for Mls and various major histocompatible complex determinants. J. Exp. Med. 154:1970.
43. Schreier, M.H., Iscove,r N.N., Teese, R., Aarden, L. and von Boehmer, H. 1980. Clones of killer and helper T cells: growth requirements, specificity and retention of function in long-term culture. Immunol. Rev. 51:315.

44. Ely, J.M., Prystowsky, M.B., Eisenberg, L., Quintans, J., Goldwasser, E. and Fitch, F.W. 1981. Alloreactive cloned T cell lines. IV. Differential kinetics of IL-2, CSF, and BCSF release by a cloned T amplifier cell and its variant. J. Immunol. 127:2345.

45. Nabel, G., Greenberger, J.S., Sakakeeny, M.A. and Cantor, H. 1981. Multiple biologic activities of a cloned inducer T-cell population. Proc. Natl. Acad. Sci. USA 78:1157.

46. Prystowsky, M.D., Ely, J.M., Beller, D.I., Eisenberg, L., Goldman, M., Goldman, J., Ihle, J., Quintans, J., Remold, H., Vogel, S.N., Goldwasser, E. and Fitch F.W. 1981. Alloreactive cloned T cell lines. VI. Multiple lymphokine activities secreted by helper and cytolytic cloned T lymphocytes. J. Immunol. 129:2337.

47. Glasebrook, A.L., Quintans, J., Eisenberg, L. and Fitch, F.W. 1981. Alloreactive cloned T cell lines. II. Polyclonal stimulation of B cells by a cloned helper T cell line. J. Immunol. 126:240.

48. Marcucci, F., Waller, M., Kirchner, H. and Krammer, P. 1981. Production of immune interferon by murine T-cell clones from long-term cultures. Nature 291:79.

49. Fresno, M., Nabel, G., McVay-Bourdreau, L., Furthmayer, H. and Cantor, H. 1981. Antigen-specific T lymphocyte clones. I. Characterization of a T lymphocyte clone expressing antigen-specific suppressive activity. J. Exp. Med. 153:1246.

50. Glasebrook, A.L., Kelso, A., Zubler, R.H., Ely, J.M., Prystowsky, M.B. and Fitch, F.W. 1982. Lymphokine production by cytolytic and noncytolytic alloreactive T cell clones. In: Isolation, Characterization and Utilization of T Lymphocyte Clones. (Fathman, C.G. and Fitch, F.W., eds.) Academic Press, Inc. New York, p. 342.

PRODUCTIVE MURINE LEUKEMIA VIRUS (MuLV) INFECTION OF EL4
T-LYMPHOBLASTOID CELLS: SELECTIVE ELEVATION OF H-2 SURFACE
EXPRESSSION AND POSSIBLE ASSOCIATION OF THY-1 ANTIGEN WITH VIRUSES

Susanne L. Henley[1,3], Kim S. Wise[1,3] and
Ronald T. Acton[1,2,3]

Department of Microbiology[1], Public Health[2] and
Diabetes Research and Training Center[3]
University of Alabama in Birmingham
Diabetes Research and Education Bldg., Room 817
Birmingham, Alabama 35294

ABSTRACT

This study evaluated differences in the expression of three surface antigens ($H-2K^b$, $H-2D^b$, and Thy 1.2) of two EL4 T-lymphoblastoid cell lines. The two cell lines, EL4(G-) which is not virus infected and the MuLV-producing subline EL4G+, had the same cytographic size distribution and a doubling time of twenty-four hours. Reflective of the morphological similarities, the expression of the T-cell specific alloantigen Thy 1.2 for the two lines did not differ. In contrast to the Thy 1.2 expression, antibody-detectable $H-2K^b$ and $H-2D^b$ activity was much greater on the MuLV-producing EL4G+ subline as indicated by fewer cells needed to remove 50% of the cytotoxic activity of the antibody (AD50). Comparison of the EL4(G-)AD_{50}/EL4G+AD_{50} ratio indicated

---

[1] Supported by U.S. Public Health Service Grants CA18609, CA15338 as well as GB43575X from the Human Cell Biology section of the NSF and the Diabetes Trust Fund.

an increase in $H-2^b$ antigen expression on EL4G+ of 15.6 fold for $H-2K^b$ and 7.81 fold for $H-2D^b$. To investigate the possibility of a selective association between MuLV and EL4G+ H-2 antigens, the specific activity for $H-2K^b$, $H-2D^b$ and Thy 1.2 of EL4G+ membrane and MuLV preparations obtained from EL4G+ cultures was compared. The relative proportion of (Thy 1: $H-2K^b$: $H-2D^b$) in membrane preparation was (1.14:2.45:1.95) compared to the proportion (0.31:3.15:0.95) in virus preparations. Although the amount of Thy-1 activity was markedly elevated in the virus preparation, $H-2^b$ antigenic activities were not significantly different in the two preparations. These data are consistent with an active association of lymphoid surface glycoproteins and MuLV, and suggest an association between productive MuLV infection and an increase in the expression of $H-2$ antigens in the EL4G+ cell cultures.

INTRODUCTION

The requirement of histocompatibility ($H-2$) antigen expression for recognition and elimination of virus-infected target cells may be of some significance in regulating effective immunological control of retrovirus-induced leukemias (1,2,3). The mechanisms ultimately determining host susceptibility to leukemia remain unclear (4,5,6,7,8,9), although co-capping studies have indicated that antibody-induced redistribution of either viral or histocompatibility antigens leads to concommitant movement of putatively associated surface consituents (1,2,3,9). While the association between MHC products and MuLV constituents at the cell surface remains to be clearly defined, retrovirus infections alter the spatial arrangement of cell surface antigens by inducing the expression of virus-associated antigens. Such topographical perturbations may alter the immunogenicity of the cell (10,11).

Meruelo et al. (12) and Meruelo (13) have reported features associated with infection of mice with the leukeogenic radiation leukemia virus (RadLV), suggesting additional immunological consequences of virus infections. Their observation of elevated levels of H-2 antigen expressed on thymocytes of mice during the early stages of RadLV infection underscores the possibility that virus-induced alteration of lymphoid cell surface antigen expression may determine properties of immunocompetent cells affecting the host response to virus-induced leukemias.

The elucidation of the mechanisms by which murine retroviruses can alter T-lymphoid cell surface antigen expression, and identification of cell surface constituents associating with these viruses, are relevant to the understanding of immunologic control of virus-induced lymphoproliferative disorders. This report describes properties of two variants of a T-lymphoblastoid cell line, differing in the presence (or absence) of productive infection with murine leukemia virus. Selective elevation of H-2 antigen expression and possible preferential association of the Thy-1 T-cell differentiation alloantigen with MuLV virions in the productively infected variant line suggest both the alteration of quantitative expression and structural rearrangement of cell surface antigen associated with retrovirus replication in T-lymphoid cells. This in vitro system should enable more detailed evaluation of mechanisms by which murine retroviruses alter T-lymphoid cell populations.

MATERIALS AND METHODS

Cell lines. Lymphoblastoid cell lines EL4 (referred to here as EL4G- and EL4G+) were obtained, respectively, from the Cell Distribution Center, Salk Institute, San Diego, CA, and from Dr.

Bruce Chesebro, NIH Rocky Mountain Lab, Hamilton, MT. Cells were grown in suspension cultures with modified Dulbecco's medium (GIBCO, Grand Island, NY) containing 10% heat inactivated fetal calf serum, and were taken for analysis during exponential phase of growth. These cultures were free of contamination with mycoplasmas according to criteria previously described (14). Both cell lines express the MHC antigens $H-2K^b$, $H-2D^b$, and the T-cell differentiation antigen Thy-1.2 (15). The EL4G+ line, in contrast to the EL4G- line, produces infectious MuLV (15,16) and expresses elevated levels (17) of both the MuLV surface glycoprotein gp70 and the Gross MuLV-related Gross Cell Surface Antigen (GCSA) (18,19), a glycosylated surface constituent coded by the MuLV _gag_ gene (20,21,22).

Virus and Membrane Preparations. A slight modification of a previously described procedure (23) was used to isolate viruses from the supernatant fluid of EL4G+ cultures. Briefly, virus-containing medium was obtained by centrifugation of EL4G+ suspension cultures at 650 x g for 20 minutes. This supernatant was further centrifuged at 6,000 x g for 30 minutes and viruses were pelleted from the resulting clarified supernatant by subsequent centrifugation at 100,000 x g for 20 minutes. The virus-containing pellet was resuspended in STE buffer [0.15 M NaCl, 1 mM ethylenediaminetetraacetic acid (EDTA), 0.01 M tris (hydroxymethyl) aminomethane, pH 7.4) and subjected to isopycnic centrifugation for 90 minutes at 100,000 x g on a 5-40% (w/w) linear potassium tartrate gradient. Viruses banding at $\rho = 1.15 - 1.17$ g/cm^3 were collected, diluted with STE buffer and repelleted by centrifuging at 100,000 x g for 60 minutes. The final MuLV pellet was resuspended in phosphate buffered saline (PBS) and stored at -70°C.

Membranes were obtained from EL4G+ cultures using modifications of previously described procedures (23). Cells were pelleted by centrifugation, washed twice with PBS, and resuspended in ten volumes of 0.32 M sucrose containing 0.5 mM phenylmethylsulfonylfluoride (PMSF), pH 7.5. The cells were disrupted in a Dounce homogenizer with a tight-fitting Teflon piston until 50% of the cells were ruptured. A membrane preparation was obtained by sequential differential centrifugation as previously reported (23). Membranes were suspended in PBS containing both $MgCl_2$ and $CaCl_2$ at 0.001 M concentration, frozen at -70°C and thawed only once before use. When required, EL4G+ membrane and virus preparations were disrupted by sonic treatment at 4°C in a Biosonik IV (Bronwill, Danbury, NY) for 5 minutes using the "high" setting at 50% output.

Animals. AKR/J and C57BL/6J mice were obtained from the Jackson Laboratory, Bar Harbor, Maine. AKR/J thymocytes and C57BL/6J splenocytes were prepared as previously described (24,25) and used as target cells in assays for Thy-1 and H-2 antigens, respectively.

Antisera. Goat antiserum against purified Thy-1.1 (obtained from Dr. Robert Zwerner) was prepared as previously described (26). The titer of this antiserum at the 50% cytotoxicity endpoint using AKR/Jax (Thy-1.1) or AKR/Cum (Thy1.2) thymocytes was the same. Anti-$H-2K^b$ antiserum D33 [(B10.D2 x A) F_1 anti-B10.A(5R)] was kindly provided by Dr. John Ray of the Research Resources Branch, National Institutes of Health, Bethesda, MD. Anti-$H-2D^b$ antiserum [(BALB/c x HTI) F_1 anti-EL4] (9) was a generous gift from Dr. Frank Lilly, Albert Einstein College, Bronx, NY. Antiserum dilutions used for absorption assays (anti-$H-2K^b$ 1:75; anti-$H-2D^b$, 1:8; and anti-Thy-1, 1:60) contained approximately four times the concentration of serum necessary to lyse 50% of the labeled target cells with added complement in a direct cytotoxicity assay.

Absorption studies. The relative expression of $\underline{H\text{-}2D^b}$, $\underline{H\text{-}2K^b}$, and Thy-1 antigens on the EL4 cell lines was determined quantitatively by measuring the ability of these cells to absorb the cytotoxic activity of the appropriate antibody (26,27,28). Briefly, absorptions were performed at 4°C for two hours after which cells were pelleted and the supernatant containing residual antibody was added to splenocytes or thymocytes in the presence of guinea pig complement (25,28). The percent absorption was calculated as previously described (26,27). Quantitation of antigenic expression was based on the number of cells necessary to reduce the cytotoxic activity of the antiserum by 50% (absorption dose $_{50}$, AD_{50}). This allowed a direct comparison of the relative amount of antigen on each cell line. The AD_{50} for purified membrane and virus preparations, was the amount of protein (μg) which removed 50% of the cytotoxic activity of the antiserum.

Statistical analysis. For comparison of the relative expression of each antigen on the two cell lines, an AD_{50} ratio was obtained by dividing the AD_{50} value of the EL4G− subline by the AD_{50} value of the EL4G+ subline. When required, the AD_{50} ratio of each antigen was compared to 1 using a specialized Student's t-test with N-1 degrees of freedom (29).

RESULTS

Expression of cell surface antigens. EL4G+ and EL4G− cells exhibited similar growth, size and morphological characteristics. Growth curves of the two cell lines revealed a concurrent increase in cell density with a similar doubling time of approximately 24 hours (Figure 1). Scatter/absorption histograms of these lines have shown indistinguishable cell size characteristics, indicating a comparable surface area for EL4G− and EL4G+ cells (30). As illustrated by the quantitative absorption assay depicted in

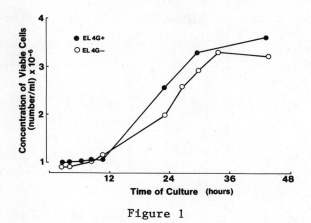

Figure 1

Growth curve of EL4 Cultures

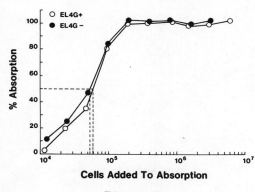

Figure 2

Absorption of Cytotoxic Anti-Thy-1 Antiserum by EL4 Cells

Figure 2, the amount of Thy-1 expressed per cell was virtually identical (EL4G+ AD_{50} = 5.9 x 10^4 cells: EL4G- AD_{50} = 5.4 x 10^4 cells). In a series of experiments (Table 1), the AD_{50} ratio calculated for the Thy-1 antigen did not differ significantly from 1.

In contrast to the similarity in expression of Thy-1 between the two cell lines, the expression of $H-2K^b$ histocompatibility (Figure 3) was markedly elevated on the productivity infected

Table 1
COMPARISON OF THY 1.2 EXPRESSION ON EL4G− and EL4G+ CELLS

Exp. No.	AD_{50} EL4G+	AD_{50} EL4G−	AD_{50} Ratio ± S.E.
(1)	1.60×10^4	1.10×10^4	.69
(2)	10.0×10^4	6.20×10^4	.62
(3)	5.00×10^4	5.20×10^4	1.04
(4)	3.00×10^4	3.55×10^4	1.18
(5)	1.60×10^4	1.95×10^4	1.22
			[a] $0.95 \pm .21$

[a] $p > 0.7$: The probability that the difference between the average AD_{50} ratio and 1 is due to random chance.

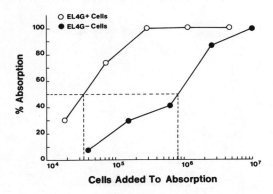

Figure 3

Absorption of Anti-H-2K[b] Cytotoxicity by EL4 Cells.

EL4G+ cells ($AD_{50} = 3.4 \times 10^4$ cells) compared to EL4G− cells ($AD_{50} = 81 \times 10^4$ cells). For the four experiments illustrated by Table 2, the mean AD_{50} ratio obtained from absorptions of anti-H-2K[b] antiserum by the EL4 variants was 15.6 indicating an average 15-fold greater expression of the H-K[b] antigen on the virus-producing EL4G+ cells.

Similarly, absorption studies using anti-H-2D[b] antiserum (Figure 4) indicated that this antigen was also expressed in greater quantities on EL4G+ cells ($AD_{50} = 6.5 \times 10^4$ cells)

Table 2
COMPARISON OF H-2K^b EXPRESSION ON EL4G- and EL4G+ CELLS

Exp. No.	E.4G+	AD$_{50}$ EL4G-	AD$_{50}$ Ratio ± S.E.
(1)	8.33 x 10^4	105. x 10^4	12.6
(2)	5.50 x 10^4	33.0 x 10^4	6.00
(3)	3.32 x 10^4	80.0 x 10^4	24.1
(4)	4.09 x 10^4	80.0 x 10^4	19.6
			[a]15.6±3.97

[a]$p<0.05$

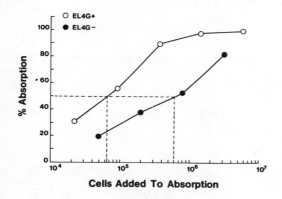

Figure 4
Absorption of Anti-H-2D^b Cytotoxicity by EL4 Cells.

compared to EL4G- cells (AD$_{50}$ = 60 x 10^4 cells). Expression of the H-2D^b antigen on EL4G+ cells was 7-fold higher than that of the EL4G cell line (Table 3). Thus, while Thy-1 antigen expression, size and growth parameters were nearly identical for both cell lines, an elevated expression of H-2K and H-2D region products was observed with the MuLV-producing EL4G+ cells.

Table 3

COMPARISON OF $\underline{H-2D^b}$ EXPRESSION ON EL4G- and EL4G+ CELLS

Exp. No.	EL4G+ AD_{50}	EL4G- AD_{50}	Ratio + S.E.
(1)	6.80×10^4	67.0×10^4	9.85
(2)	1.00×10^4	6.81×10^4	6.81
(3)	1.70×10^4	26.8×10^4	15.8
(4)	3.35×10^4	15.6×10^4	4.66
(5)	13.5×10^4	26.0×10^4	1.93
			[a]7.81 ± 2.38

[a]$p=0.05$

Table 4

ANTIGEN ACTIVITY OF THY-1.2, $\underline{H-2K^b}$ AND $\underline{H-2D^b}$ in EL4G+ MEMBRANES and MuLV

	Exp. No.	$AD_{50} \times 10^{-1}$ (micrograms protein) $\pm$ S.E.		
		Thy-1	$H-2K^b$	$H-2D^b$
Sonically treated membrane	(1)	1.18	2.75	1.90
	(2)	1.10	4.15	2.00
		$1.14 \pm .05$	$3.45 \pm .70$	$1.95 \pm .05$
Intact Virus	(1)	.28	1.75	.90
	(2)	.34	4.55	1.00
		$.31 \pm 04$	3.15 ± 1.40	$.95 \pm .05$
Sonically treated virus	(1)	.37	3.55	1.08
	(2)	.39	3.80	1.48
		$.38 \pm .01$	$3.68 \pm .13$	$1.28 \pm .20$

<u>Antigen Expression in MuLV Recovered from EL4G+ Cultures</u>

Since antigen expression in a purified membrane preparation can be used as an index of the antigen expression of the cells from which the preparation was obtained (26), comparison of specific activities (AD_{50}) of multiple antigens in virus and EL4G+ membrane preparations might reveal preferential associations occurring between the virus and selected cell surface constituents. These would be manifest as differences in relative proportions of antigen activities between the two preparations.

Comparisons of $H-2K^b$ and $H-2D^b$ and Thy-1 expression of EL4G+ membrane and virus preparations is depicted in Table 4. The relative proportion of (Thy-1:$H-2K^b$:$H-2D^b$) in membrane preparations is (1.14:3.45:1.95) compared to the proportion (0.31:3.15:0.95) in virus preparations. Relative to either $H-2$ antigen specificity, the proportion of Thy-1 is markedly elevated (lower AD_{50}) in the MuLV preparations when compared to that found in membrane preparations. However, neither $H-2K^b$ nor $H-2D^b$ antigens are represented in virus preparations in proportions significantly different from their relative expression in membranes. This is apparently not due to differences in availability of these antigens at the virus surface, or "cryptic" presence of antigens in viruses, since the specific activities of all antigens in sonically treated virus were similar to those observed with intact preparations.

DISCUSSION

This investigation has provided evidence for a relationship between productive MuLV infection and a selective, elevated expression of H-2 antigens on T-lymphoblastoid cells in vitro. Increased $H-2$ antigen expression on T-lymphoid cells has been previously reported in association with productive murine retrovirus infections in vivo, generally in systems where the virus is leukanogenic and infects primarily cells of the thymus (12,31). Age-related changes in the relative quantities of cell surface antigens on preleukemic AKR thymocytes include an increase in expression of $H-2$ antigens and a decrease in the expression of

Thy-1 (31). Additionally, Meruelo, et al., have demonstrated a close temporal association between infection with RadLV and elevated levels of H-2 antigens on thymocytes, in strains of mice both resistant and susceptible to the oncogenic effects of this virus (12,13). Perhaps significantly, the degree of increase in H-2 antigen expression was greater on thymocytes of animals resistant to virus-induced leukemia.

Meruelo (13) has suggested that elevated H-2 antigen expression occurring concomitantly with viral antigen appearance at the cell surface may enhance the effectiveness of the host's immune response against virus-infected cells, by facilitating recognition by cytotoxic T-cells. The unusual propensity of H-2 antigens to "co-cap" with a variety of other surface antigen molecules (10,11) emphasizes the potential importance of altering surface antigen distribution by regulating H-2 antigen expression.

Although both $H-2^b$ and Thy-1 antigen specificities were demonstrated in the virus preparation isolated from EL4G+ cultures, only the specific activity of Thy-1 was unusually high when compared to its relative expression in EL4G+ membranes. The elevated expression of Thy-1 could be due to 1) a difference in the rate of degradation of $H-2K^b$, $H-2D^b$, and Thy-1 antigen (32), 2) the concentration of cellular membrane fragments in the virus preparation, 3) adhesion of shed Thy-1 to the released virions, or 4) selective incorporation of surface Thy-1 into the virion as it buds from the cell. Protease inhibitors were used throughout the isolation of EL4G+ membranes to retard proteolysis (23). If only degradation of $H-2^b$ antigens had occurred it would have to occur in only one preparation to increase the specific activity of Thy-1 to H-2 as seen in the virus preparation. The specific activity of the $H-2K^b$ and $H-2D^b$ in one preparation (MuLV) should be significantly lower than that of the other sample (EL4G+

membrane). The specific activity for these antigens are very similar in all preparations tested, so selective degradation of H-2 in the membrane preparation apparently does not occur. Cell membrane fragments in varying amounts are often co-isolated with the MuLV and probably contribute antigenic activity to the virus preparation. However, if membrane contaminants contributed significantly to the specific activity of the antigens isolated with MuLV, the $H-2K^b$:$H-2D^b$:Thy-1 ratio would be expected to resemble more closely the $H-2K^b$:$H-2D^b$:Thy-1 ratio of the EL4G+ membrane preparations. It is, therefore, difficult to explain the high Thy-1 activity in MuLV preparations by cellular membrane contamination or preferential H-2 antigen degradation. Assessment of the last two possibilities has not been addressed by our investigation. Of the three antigens monitored, only Thy-1 has been reported to be shed in a form which remains detectable by antibody (32,33). Neither adhesion of Thy-1 to intact virions nor the selective concentration of the antigen on the cell surface at the site of virion assembly can, at present, be excluded.

ACKNOWLEDGEMENT

The authors are grateful to: Dr. Catherine Kirk for suggestions of statistical treatment of the data; Mrs. Barbara Blackmon and Mr. Jim Bradac for technical expertise; and Mrs. Susan Snead, Mrs. Mary Estock, Ms. Eleanor Hart and Mrs. Edie Vetter for preparation of this manuscript.

REFERENCES

1. J.W. Schrader, B.A. Cunningham and G.M. Edelman, 1975, Functional Interactions of Viral and Histocompatibility Antigens at Tumor Cell Surfaces, Proc. Nat. Acad. Sci. 72:5066.

2. R. Henning, J.W. Schrader and G.M. Edelman, 1976, Antiviral antibodies inhibit the lysis of tumor cells by anti-H-2 sera, Nature 263:689.
3. D.A. Zarling, I. Keshet, A. Watson and J.H. Bach, 1978, Association of Mouse Major Histocompatibility and Rauscher Murine Leukemia Virus Envelope Glycoprotein Antigens on Leukemia Cells and their Recognition by Syngeneic Virus-Immune-Cytotoxic T-lymphocytes, Scand. J. Immunol. 8:497.
4. F. Lilly, 1969, III Murine Leukosis. The Role of Genetics in Gross Virus Leukemogenesis, Comp. Leut. R. Bibl. Haemat. 36:213.
5. B. Chesebro, K. Wehrly and J. Stimpfling, 1974, Host Genetic Control of Recovery From Friend Leukemia Virus-Induced Splenomegaly Mapping of a Gene Within the Major Histocompatibility Complex, J. Exp. Med. 140:1457.
6. J. Klein, 1975, Biology of the Mouse Histocompatibility-2 Complex Principles of Immunogenetics Applied to a Single System. Springer-Verlag Inc., New York, pp. 389-410.
7. H.C. Ertl and U.H. Koszinowski, 1976, Modification of H-2 Antigenic Sites by Enzymatic Treatment Influences Virus-Specific Target Cell Lysis, J. Immunol. 117:2112.
8. M.J. Bevan and R. Hyman, 1977, The Ability of $H-2^+$ and $H-2^-$ Cell Lines to Induce or be Lysed by Cytotoxic T Cells, Immunogenetics 4:7.
9. J.E. Bubbers and F. Lilly, 1978, Anti-Friend Virus Antibody-induced Resistance of Friend Virus-infected Spleen Cells to Lysis Mediated by Anti-H-2 Antisera, Cancer Research 38:2722.
10. L.Y.W. Bourguignon, R. Hyman, I. Trowbridge and S.J. Singer, 1978, Participation of Histocompatibility Antigens in Capping of Molecularly Independent Cell Surface Components by Their Specific Antibodies, Proc. Nat. Acad. Sci. 75:2406.

11. B. Geiger, K.L. Rosenthal, J. Klein, R.M. Zinkernagel and S.J. Singer, 1979, Selective and Unidirectional Membrane Redistribution of an H-2 Antigen with an Antibody-Clustered Viral Antigen: Relationship to Mechanisms of Cytotoxic T-Cell Interactions, Proc. Nat. Acad. Sci. 76:4603.

12. D. Meruelo, S.H. Nimelstein, P.P. Jones, M. Lieberman and H.O. McDevitt, 1978, Increased Synthesis and Expression of H-2 Antigens on Thymocytes as a Result of Radiation Leukemia Virus Infection: A Possible Mechanism for H-2 Linked Control of Virus-Induced Neoplasia, J. Exp. Med. 147:470.

13. D. Meruelo, 1979, A Role for Elevated H-2 Antigen Expression in Resistance to Neoplasia Caused by Radiation-Induced Leukemia Virus: Enhancement of Effective Tumor Surveillance by Killer Lymphocytes, J. Exp. Med. 149:898.

14. K.S. Wise, G.H. Cassell and R.T. Acton, 1978, Selective Association of Murine T-Lymphoblastoid Cell Surface Alloantigens with Mycoplasma hyorhinis, Proc. Nat. Acad. Sci. 75:4479.

15. D. Chesebro, K. Wehrly, K. Chesebro and J. Portis, 1976, Characterization of Ia8 Antigen, Thy 1.2 Antigen, Complement Receptors, and Virus Production in a Group of Murine-Induced Leukemia Cell Lines, J. Immunol. 117:1267.

16. T. Aoki, R.B. Herberman, J.W. Hartley, M. Liu, M.J. Walling and M. Nunn, 1977, Surface Antigens on Transplantable Tumor Cell Lines Producing Mouse Type C Viruses, J. Natl. Cancer Inst. 58:1069.

17. K. Wise and R.T. Acton, 1978, Endogenous Virus Antigens of Murine Lymphoblastoid Cell Lines. Protides of the Biological Fluids. Proceeding of the 25th Colloquium. Ed. by H. Peeters. Pp. 707-714.

18. L.J. Old, E.A. Boyse and E. Stockert, 1965, The G (Gross) Leukemia Antigen, Canc. Res. 25:813.

19. J. Ledbetter and R.C. Nowinski, 1977, Identification of the Gross Cell Surface Antigen Associated with Murine Leukemia Virus-Infected Cells, J. Virol. 23:315.
20. N.G. Famulari, D.L. Buchagen, H.D. Klenk and E. Fleissner, 1976, Presence of Murine Leukemia Virus Envelope Proteins gp 70 and p15(E) in a Common Polyprotein of Infected Cells, J. Virol. 20:501.
21. E. Fleissner, H. Ikeda, J-S. Tung, E.S. Vitetta, E. Tress, W. Hardy, E. Stockert, E.A. Boyse, T. Pincus and P. O'Donnell, 1976, Characterization of Murine Leukemia Virus-Specific Proteins. In Cold Spring Harbor Symp. on Quant. Biol. Vol. 39. Tumor Virus Pp. 1057-1066.
22. J-S. Tung, A. Pinter and E. Fleissner, 1977, Two Species of Type C Viral Core Polyprotein on AKR Mouse Leukemia Cells, J. Virol. 23:430.
23. R.K. Zwerner, K.S. Wise and R.T. Acton, 1979, Harvesting the Products of Cell Growth. In Methods in Enzymology. Vol. LVIII. Edited by William B. Jakoby and Ira E. Paston. Academic Press, New York, USA. P. 221.
24. E.A. Boyse, L.J. Old and I. Chouroulinkov, 1964, Cytotoxic Test for Demonstration of Mouse Antibody. Methods Med. Res. 10:39.
25. J.L. Williams, T.H. Stanton and R.M. Wolcott, 1975, Adaptation to Tissue Culture of the Murine Leukemia ASL1, a High Producer of TL (Thymus Leukemia) Antigen, Tissue Antigens 6:335.
26. R.K. Zwerner, P.A. Barstad and R.T. Acton, 1977, Isolation and Characterization of Murine Cell Surface Components. I. Purification of Milligram Quantities of Thy 1.1, J. Exp. Med. 147:986.
27. R.K. Zwerner and R.T. Acton, 1975, Growth Properties and Alloantigenic Expression of Murine Lymphoblastoid Cell Lines, J. Exp. Med. 142:378.

28. P.A. Barstad, S.L. Henley, R.M. Cox, J.D. Lynn and R.T. Acton, 1977, Production of Milligram Quantities of H-2K and Thy-1 Alloantigens by Large-Scale Mammalian Cell Culture, Proc. Soc. Exp. Bio. Med. 155:296.

29. R.S. Sokal and F.J. Rohlf, 1969, Biometry. W.H. Freeman and Co. San Francisco, CA. P. 168.

30. K.S. Wise, S.L. Henley and R.T. Acton, 1979, Association of Murine Lymphoid Cell Surface Antigen Expression with Virus Production in ICN-UCLA Symposia on Molecular and Cell Biology. T and B Lymphocytes: Recognition and Function. Edited by Fritz Bach, B. Bonavida, and F.C. Fox. Academic Press, New York, USA. Pp. 495-504.

31. K. Kawashima, H. Ikeda, E. Stockert, T. Takahashi and L.J. Old, 1976, Age-Related Changes in Cell Surface Antigens of Preleukemic AKR Thymocytes, J. Exp. Med. 144:193.

32. E.S. Vitetta, J.W. Uhr and E.A. Boyse, 1974, Metabolism of H-2 and Thy-1 alloantigens in Murine thymocytes, Eur. J. Immunol. 4:276.

33. W.W. Freimuth, W.J. Esselman and H.C. Miller, 1978, Release of Thy-1.2 and Thy-1.1 from Lymphoblastoid Cells: Partial Characterization and Antigenicity of Shed Material, J. Immunol. 120:1651.

CONSTRUCTION OF HUMAN T-CELL HYBRIDS WITH HELPER FUNCTION

Oscar Irigoyen, Philip V. Rizzolo, Yolene Thomas, Linda Rogozinski and Leonard Chess

Division of Rheumatology; Department of Medicine
Columbia University
College of Physicians and Surgeons
New York, New York 10032

ABSTRACT

Human T-cell hybrids with helper activity were obtained after fusion of phytohemagglutinin-activated normal human T cells with a 6-thioguanine-resistant, aminopterin-sensitive human T-cell line. This mutant line, designated CEM-T15, was derived from the human T-cell line CEM after mutagenesis with ethyl methanesulfonate. The polyethylene glycol induced fusion and the selection in hypoxanthine-aminopeterin-thymidine medium were performed by modification of standard somatic cell hybridization techniques. After fusion, the strategy for selecting hybrids consisted in screening growing cultures for the presence of cells expressing

---

Supported in part by: NIH Grants AI 14969 and 15524 and The Arthritis Foundation.
Abbreviations: BrdUrd, 5-bromo-2'-deoxyuridine; TG, 6-thioguanine; HAT, medium contain hypoxanthine, aminopterin and thymidine; PHA, phytohemagglutinin; EMS, ethyl methanesulfonate; HGPRT, hypoxanthineguanine phosphoribosyl transferase; RHPA, reverse hemolytic plaque assay; PFC, plaque forming cells; PWM, pokeweed mitogen.

the OKT3 cell surface differentiation antigen. OKT3 was chosen because it is present in 85-95% of normal human T cells but absent from CEM-T15 cells. Thus, $OKT3^+$ cells growing 5-7 weeks after fusion most likely represented hybrids between normal T cells ($OKT3^+$) and continuously growing CEM-T15 cells ($OKT3^-$). Several of the hybrids were tested for their capacity to promote pokeweed mitogen-induced antibody production by B cells. These experiments demonstrated that many of the hybrids had helper activity. Periodical testing of these uncloned hybrids for helper activity revealed functional instability, with most of the hybrids losing helper activity after 20 weeks of continuous culture. However, early and repeated cloning of the same hybrids resulted in a series of hybrid clones with helper activity still present more than 8 months after fusion. In more recent fusions, we have demonstrated that human helper hybrids producing helper factor(s) can also be obtained. These and similar hybrids with different functions will be of considerable importance in further studies of the immunobiology of human T lymphocytes.

INTRODUCTION

In the last decade, with the enormous advance in our understanding of the physiology of the immune system, the extreme heterogeneity, specialization and functional diversity of immunocompetent cells, particularly T lymphocytes, have become evident. For example, T cells not only have receptors specific for a particular antigen, but importantly, cells with identical antigen specificity can respond to antigenic triggering in different ways. Thus prior to antigen stimulation T lymphocytes are programmed to either release immunoregulatory molecules (helper or suppressor factors) or to become effector (cytotoxic) cells. The T lymphocytes performing these different functions can be identified by virtue of the expression of distinct genetically determined cell surface differentiation antigens (1,2).

In human systems, for instance, our laboratory and others have been successful in identifying and isolating functionally distinct T-lymphocyte subsets by the use of antibodies directed against unique cell surface antigens (3-7). For example, the T-cell subset with cytotoxic and suppressor functions is recognized by the monoclonal antibody OKT8, whereas the helper T-cell subset can be identified by the monoclonal antibody OKT4 (8-10). More detailed analysis of these subsets, however, has demonstrated that this preliminary subdivision does not take into account the real scope of T-cell diversity. For example, further studies in our laboratory have revealed that the $OKT4^+$ T-cell subset contains at least four distinct sets of cells: 1) radiosensitive helper cells; 2) radioresistant helper cells; 3) radiosensitive inducers of suppressor cells; and, after activation, 4) radiosensitive suppressor cells (11,12). Furthermore, the radioresistant $OKT4^+$ helper T cells can be distinguished from the radiosensitive $OKT4^+$ helper cells by the expression of still another cell surface antigen, OKT17, recognized by a monoclonal antibody (13). Thus, although the analysis of T cell heterogeneity is greatly enhanced by the detection and isolation of cells with respect to a particular cell surface antigen, the detailed analysis of the immune system will be extremely difficult unless homogeneous populations of T lymphocytes can be obtained.

In murine systems, one of the approaches to isolate monoclonal populations of T cells uses techniques similar to those defined by Kohler and Milstein for the production of monoclonal antibodies (14). Thus, by fusing normal T lymphocytes with continuously-growing aminopterin-sensitive lymphoma lines, a number of investigators have created continuously-growing murine T-cell hybrids with functions covering most of the functional repertoire of normal murine T lymphocytes (15-21).

It is obvious that extension of this hybridization technology to the analysis of human T cells will be important for precise and detailed studies of immunoregulatory cells and their products. In this regard, Grillot-Courvalin et al. (22) and ourselves (23) have been successful in constructing human T cell hybrids with functional activities. To that end, we first derived a series of 5-bromo-2'-deoxyuridine (BrdUrd) and 6-thioguanine (TG)-resistant, aminopterin-sensitive mutant lines from the human T-cell line CEM. One of the TG-resistant lines, CEM-T15, was fused with PHA-activated normal human T lymphocytes, and hybrids were selected by growth in medium containing hypoxanthine, aminopterin and thymidine (HAT). Many of these hybrids express OKT3 antigens, which are present on normal T cells but not on the CEM-T15 mutant line. Furthermore, functional analysis demonstrated that many of these human T-cell hybrids and their clones exhibit helper function <u>in vitro</u> (23). In addition, more recent hybridizations resulted in a series of T-cell hybrids secreting factors with helper activity.

MATERIALS AND METHODS

<u>Lymphocyte preparation and fractionation</u>: Fresh peripheral blood mononuclear cells were isolated from healthy human volunteers by Ficoll-diatrizoate density gradient centrifugation. Highly enriched populations of T and B lymphocytes were obtained by E rosetting and passage through a Sephadex G-200 rabbit anti-human (Fab')$_2$ column, as described (24). The B cells were treated with the monoclonal antibody OKT3 plus complement in order

to eliminate residual T cells. OKT4$^+$ cells were obtained by negative selection using complement-mediated lysis of the T cell population by the OKT8 monoclonal antibody as previously described (11).

T-lymphocyte activation: For activation, normal human peripheral blood T-lymphocytes were incubated at a concentration of 2 x 10^6 cells/ml with 10 µg/ml of phytohemagglutinin (PHA, Grand Island Biological Co., Grand Island, NY). The cultures were carried out for 4 days in a 95% air-5% CO_2 humidified incubator.

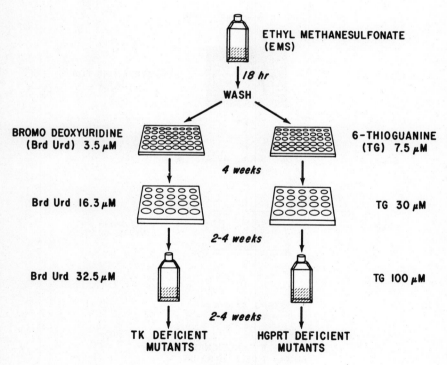

Figure 1

Derivation of BrdUrd and TG resistant mutants from the human T cell line CEM. EMS was used at concentrations of 5 and 10 µg/ml.

Generation of TG-resistant mutants: The procedures and culture conditions for the generation of TG-resistant mutants have been previously described (23). BrdUrd mutants were obtained in analogous manner (Fig.1). Briefly, ethyl methanesulfonate (EMS, Sigma Chemical Co., St. Louis, MO) treated CEM cells were grown in final medium containing increasing concentrations of TG (Fig.1). Final medium consisted of Iscove's modified Dulbecco's medium

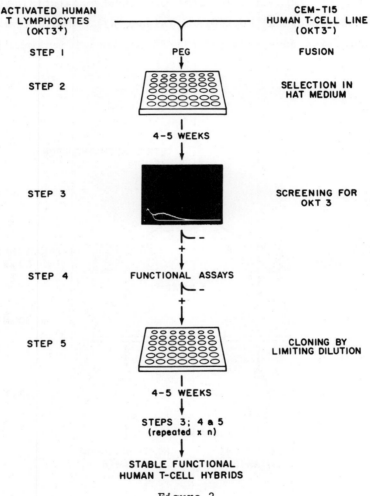

Figure 2

Construction of human T-cell hybrids.

(IMDM, Gibco) with 10-20% horse serum and 50 µg/ml of chlortetracycline (Gibco). After approximately 3 months in culture, CEM mutants growing in 100 µM TG were considered to be hypoxanthine-guanine phosphoribosyl transferase (HGPRT) deficient. One of the TG-resistant mutants with better growing characteristics, designated CEM-T15, was shown to be aminopterin sensitive when compared to the parent line CEM (23). CEM-T15 was therefore used in subsequent hybridization experiments. Testing of the CEM-T15 mutant line by the method of Chen (25) failed to detect mycoplasma contamination.

Hybridization procedure: The hybridization procedure for fusions 26 and 27 were previously described in detail (23). Briefly, PHA-activated normal human T cells were fused at a ratio of 4:1 to CEM-T15 by standard techniques (26) using polyethylene glycol 1,000 mol wt at 50% concentration. After fusion, the cells were washed and resuspended in hybridization medium containing 100 µM hypoxanthine, 30 µM thymidine and 33 nM aminopterin (HAT medium). Hybridization medium consisted of IMDM medium containing 10-20% horse serum, 50 µg/ml of chlortetracycline, 10% of NCTC 109 medium, 0.15 mg/ml of oxalacetic acid and 0.2 U/ml of bovine insulin (27). The cells were cultured in microtiter plates at a concentration of 4 to 10×10^4 mutant cells per well in 0.2 ml. The cultures were maintained at 37°C in a humidified 95% air, 5% CO_2 incubator. After 5 days in HAT medium, the cultures were switched to medium containing hypoxanthine and thymidine without aminopterin (HT medium). After 2 weeks in HT medium, the cultures were finally switched to final medium.

Surface antigens analysis: Phenotypic cell surface analysis was performed by flow cytometry using the monoclonal antibody OKT3 and a fluorescein-conjugated goat anti-mouse $F(ab')_2$ antibody (Cappel Laboratories, Cochranville, PA). The cells were analyzed on a Cytofluorograf (Model 30-H, Ortho Instruments, Westwood, MA) as previously described (12). A cell population was considered positive if greater than 20% of the cells expressed the OKT3 antigen. All experiments included a negative ascites control.

Chromosome analysis: Chromosome preparations were performed according to standard techniques (28). Briefly, 4×10^6 exponentially growing cells were cultured with 0.01 µg/ml of Colcemid (Gibco). After two or three hour incubation, the cells were centrifuged and treated with 0.075 M potassium chloride for 12 min and then fixed with methanol/glacial acetic acid, 3:1. Drops of the fixed cell suspension were allowed to dry on slides and were stained with Giemsa for microscopic examination.

Polyclonal induction of antibody-secreting cells: The culture conditions and the reverse hemolytic plaque assay (RHPA) for the assessment of induction and regulation of plaque-forming cell (PFC) generation was previously described in detail (11). In brief, 10^6 B cells were suspended in 1 ml of final medium containing 5 µg of pokeweed mitogen (PWM) and were cultured in 16 x 150 mm tissue culture tubes. To these tubes were added 10^5 of either normal T cells or the different cell hybrids to be tested. Control cultures consisted of B cells cultured alone or in the presence of PWM without added T cells. All cultures were incubated for 5-6 days in a humidified 5% CO_2 incubator at 37°C, and then assayed for antibody production using the RHPA as described (11). The results are expressed as PFC per 10^6 B cells at the beginning of culture.

RESULTS

Hybridization of PHA-activated human T cells with the CEM-T15 mutant line. Our major interest when we began these human T cell hybridization experiments was to examine its feasibility and to establish the best technical conditions for hybrid production. Thus, in our first series of hybridizations we chose to fuse CEM-T15 with polyclonally-activated normal peripheral blood T cells. For instance, in fusions, 26 and 27, from which helper hybrids were obtained, CEM-T15 was fused to PHA-activated normal T cells. After fusion, our strategy consisted in screening growing cultures for the presence of cells expressing the OKT3 cell surface antigen, which is present in 85-95% of normal human T cells but absent from the mutant CEM-T15 cells. We considered this additional step important because growth of cells in HAT medium following fusion does not necessarily imply successful hybridization. Survival of cells can be due, for example, to appearance of revertants of CEM-T15 that have regained HGPRT, or to metabolic cooperation between enzyme-bearing normal T cells and enzyme deficient CEM-T15 cells (29,30).

Fusions 26 and 27 resulted in a high yield of hybrids, with more than 80% of the wells containing growing cultures. Furthermore, screening for expression of OKT3, performed 5-7 weeks after fusion, demonstrated that 55-60% of the growing cultures contained OKT3 positive cells. These data strongly suggested that the growing cells represented T cell hybrids since CEM-T15 does not express OKT3 and the normal T cells would not be rapidly dividing after 5-7 weeks in culture.

In order to obtain further evidence for hybridization, samples from fusions 26 and 27 were also analyzed for chromosome number. When screened 10 weeks after fusion, 90% of these cultures contained polyploid cells, while the CEM and CEM-T15 lines contained diploid cells. These results, together with the phenotypic analysis for OKT3, provided further evidence of hybridization.

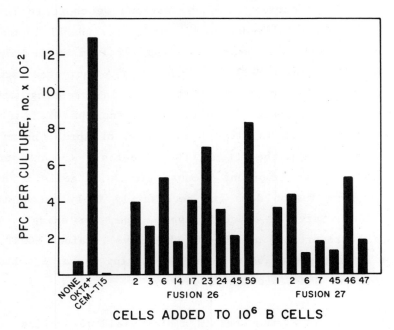

Figure 3

Helper activity of human T-cell hybrids. 10^6 B cells were cultured with 10µg/ml of PWM and 10^5 cells from either CEM-T15 or from different T-cell hybrid cultures. PFC were measured by an RHPA after 6 days in culture. Controls consisted of B cells cultured with PWM without T cells (negative) and plus OKT4$^+$ T cells (positive).

Functional analysis of human T-cell hybrids. To determine whether these growing T cell hybrids maintained functions known to be present in the normal mature parent T cells, cultures from fusions 26 and 27 were analyzed for polyclonal induction of antibody production by B cells (Fig. 3). In these experiments, growing hybrids were added to purified B cells in the presence of PWM, and PFC activity was measured six days later. Controls included normal helper T cells (OKT4$^+$) as well as CEM-T15 mutant cells. The experiments demonstrated that several different hybrid cultures from both fusions expressed helper activity. In contrast, CEM-T15 cells did not generate helper activity and in fact CEM-T15 cells and their supernatants inhibited the background level of PFCs generated in cultures containing B cells and PWM.

Following these results, we wanted to determine whether the helper T cell hybrids maintained suppressive activity derived from the parental mutant line CEM-T15. Thus, in a series of experiments designed to assess suppressor activity, we directly compared supernatants from CEM-T15 cultures with supernatants from several of the helper hybrids. These experiments consisted in culturing 10^6 B lymphocytes with 10^5 normal T cells and 10 µg/ml of PWM. To these cultures were added supernatants from CEM-T15 and from T-cell hybrids at a 1:4 dilution. PFC were measured by

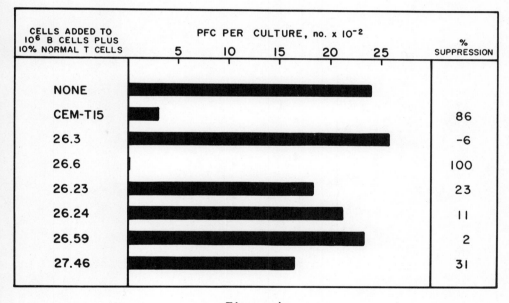

Figure 4

Most human helper T-cell hybrids lack suppressor activity. To 10^6 B cells cultured with 10^5 T cells and 10 µg/ml of PWM were added supernatants (25% concentration) from either CEM-T15 or from different helper T-cell hybrids. PFC were measured by an RHPA after 6 days in culture.

an RHPA after 6 days in culture in a 95% ail-5% CO_2 humidified incubator. The representative experiment shown in Figure 4 demonstrated that in contrast to CEM-T15, most of the human T cell hybrids with helper activity did not have suppressive activity.

Assessment of functional stability of human T-cell hybrids. One of the major concerns in somatic cell hybridization pertains to the functional stability of the created hybrids. This issue was addressed with respect to helper hybrids from fusions 26 and 27 by periodically testing the hybrid cultures for helper activity (Fig. 5). The experiments demonstrated that although the hybrids maintain helper activity for at least 13 weeks in culture, by 20

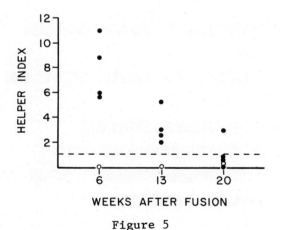

Figure 5

Instability of uncloned human helper T-cell hybrids. 10^6 B cells were cultured for 6 days with 10 µg/ml of PWM plus 10^5 cells from either CEM-T15 (o) or helper T-cell hybrid cultures (●). Helper index represents number of PFC generated by B cells in the presence of T cell hybrids divided by number of PFC generated by B cells without added T cells.

weeks most of them have lost their helper function. This indicates that uncloned hybrid cultures are highly unstable.

Rescue of functionally stable human T-cell hybrid clones by repeated cloning. In order to prevent loss of activity, we cloned a few of the functional hybrids as soon as evidence for function was obtained; that is, at 5 to 7 weeks after fusion. The cloning was performed by limiting dilution at a concentration of 1 cell/ml in 0.2 ml of final medium in microtiter plates. In most cases, after 3 weeks in culture at 37°C in a 95% air-5% CO_2 humidified incubator, 10 to 40% of the wells contained growing cells, which indicates that nearly 100% of the cells seeded resulted in growing

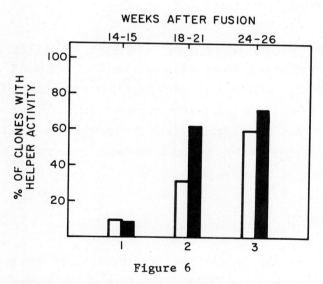

Figure 6

The number of T-cell hybrid clones with helper activity increased after repeated cloning. The helper T cell hybrids 26.23 (□) and 26.17 (■) were cloned (1) and helper clones were recloned (2;3). The number of cultures with helper activity was determined after each cloning step.

cultures. Our cloning strategy consisted in sequentially recloning new helper clones in order to select for those with functional stability (Fig.2). At every cloning step, functional clones were frozen and stored. The results of these experiments demonstrate that the frequency of helper hybrids, and therefore functional stability, increases as cloning proceeds (Fig.6).

Similarly, the amount of $OKT3^+$ cells in the cultures also increased from 30 to 40% of the cells in the initial cultures to 70 to 80% in the "4th generation" clones. Interestingly, when the chromosome number was reassessed at 20 weeks after fusion, many of these hybrids and their clones had a predominantly diploid karyotype. Taken together, these data indicate that despite significant loss of chromosomes during the first few months in culture, functional stability of these human T-cell hybrids can be preserved and consolidated by early and repeated cloning of the relevant hybrids.

<u>Construction of human T-cell hybrids secreting helper factor(s)</u>. Recently, human leukemic cells previously shown to secrete factor(s) with helper activity (31) were fused to CEM-T15. Preliminary functional analysis of these fusions (43 and 46) demonstrated that supernatants from several of the hybrid cultures contained factor(s) which induced antibody production by B cells in the absence of T cells (Fig. 7).

DISCUSSION

The first step toward the construction of human T-cell hybrids was to select a T-cell line from which enzyme deficient mutants could be obtained. We reasoned that if we selected a T-cell line not expressing a cell surface antigen normally present on functioning peripheral blood T lymphocytes, we could use that

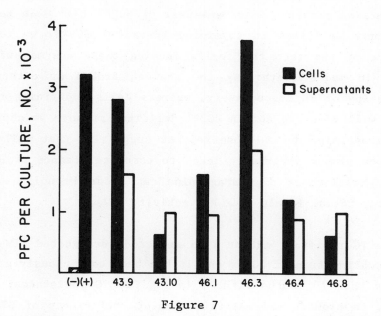

Figure 7

Human helper T-cell hybrids produce factor(s) with helper activity. 10^6 B cells were cultured with either supernatants (□) or 10 μg/ml of PWM plus 10^5 hybrid cells (■) from fusions 43 and 46. PFC were measured by an RHPA after 6 days in culture. Controls included B cells cultured alone (-) and with 10^5 normal T cells (+).

differentiation antigen as an additional selective marker after hybridization. Based on results of previous phenotypic screening of several human T-cell lines (7), we chose OKT3 as the selective marker because it is present on 85-95% of normal peripheral T lymphocytes but absent from the majority of human T-cell lines (7). Of all the OKT3⁻ T-cell lines, we selected CEM as the potential partner for fusion because it has excellent growing characteristics, can be grown in different media and types of sera and withstands controlled freezing and thawing with minimal cell damage.

Of the many hybrid selection systems described (33-36), we chose that described by Szybalski et al (32) which selects against cells lacking either thymidine kinase or HGPRT (33). To select for mutants deficient in thymidine kinase or HGPRT, we took advantage of the fact that cells lacking these enzymes will survive in media containing the analogs BrdUrd or either 8-azaguanine or TG respectively, whereas cells containing these enzymes will die. To obtain HGPRT deficient mutants we chose TG over 8-azaguanine due to some reports suggesting that while TG resistance almost invariably leads to complete absence of HGPRT activity, resistance to 8-azaguanine can occur in cells with apparently normal levels of HGPRT activity (37).

The TG resistant mutant line selected, designated CEM-T15, has a doubling time of 20 hr and it has been continuously growing for more than 10 months without losing its TG resistance. At present, to promote and maintain absolute deficiency of HGPRT, CEM-T15 is being continuously grown in medium containing 100 µM TG. We emphasize that mycoplasma contamination was not detected in CEM-T15. This is important because mycoplasma contamination has been associated with decreased hybrid production after fusion (38).

Our major interest when we began these experiments was to examine the feasibility of hybridization of human T cells and to establish the best technical conditions for hybrid production. In the first series of experiments we used as mature partners normal human peripheral blood T cells that had been polyclonally activated by lectins. In the case of fusions 26 and 27, for example, the lectin used was PHA. After fusion, four independent criteria suggested successful hybridization: 1) many cultures selectively grew in HAT medium; 2) the majority of the cultures contained cells expressing the OKT3 surface antigen; 3) most of the cultures contained polyploid cells; and 4) several of the cultures contained cells which provided helper activity in the PWM-induced antibody production by B cells.

The presence of $OKT3^+$ cells, although providing important evidence for hybridization, did not necessarily correlate with functional activity of the hybrids. For example, one of the hybrids with helper activity, 27.2 (Fig. 3), on initial testing contained $OKT3^+$ cells. However, by 3 months after fusion, 27.2 and its clones had lost expression of OKT3 but still maintained helper function, which has been preserved for more than 8 months. A similar phenomenon occurred regarding chromosome number. Thus, although the majority of the cultures contained polyploid cells when tested 13 weeks after fusion, seven weeks later most of them have lost all supernumerary chromosomes. Importantly, however, they did not lose their functional activity.

We consider that the key point for the successful rescue of functional hybrids despite heavy loss of chromosomes and, in some cases, of cell surface differentiation antigens, was early and repeated cloning of the relevant cultures. For example, we have shown that uncloned hybrids from fusions 26 and 27 begin to lose helper activity by 13 weeks of continuous culture. In contrast, early cloning of the same hybrids resulted in a progeny of T-cell hybrid clones in which helper activity is maintained for more than 8 months of continuous culture.

We have recently used the same approach to obtain human T cell hybrids which in preliminary experiments were shown to produce helper factors for B cell activation (Irigoyen et al., manuscript in preparation). These helper hybrids and their supernatants are currently being biologically and biochemically characterized in more detail. We also have preliminary evidence indicating that supernatants from several other cultures from fusions 43 to 46 contain Interleukin-2 activity. We are now in the process of constructing human T-cell hybrids with antigen specificity. These different hybrids and their products will be of enormous value for the isolation and characterization of the T cell receptor and of T cell products, and for detailed studies of cell interactions during immune responses.

REFERENCES

1. H. Cantor and E.A. Boyse, 1975, Functional subclasses of T lymphocytes bearing different Ly antigens. I. The generation of functionally distinct T-cell subclasses is a differentiative process independent of antigen, J. Exp. Med. 141:1376-1389.
2. J. Jadinski, H. Cantor, T. Tadakuma, D.L. Peavy and C.W. Pierce, 1976, Separation of helper T cells from suppressor T cells expressing different Ly components. I. Polyclonal activation: suppressor and helper activities are inherent properties of distinct T-cell subclasses, J. Exp. Med. 143:1382-1390.
3. J.M. Woody, A. Ahmed, R.C. Knudsen, D.M. Strong and K.W. Sell, 1975, Human T-cell heterogeneity as delineated with a specific human thymus lymphocyte antiserum. In vitro effects on mitogen response, mixed leukocyte culture, cell-mediated lymphocytotoxicity, and lymphokine production, J. Clin. Invest. 55:956-966.
4. R.L. Evans, J.M. Breard, H. Lazarus, S.F. Schlossman and L. Chess, 1977, Detection, isolation and functional characterization of human T-cell subclasses bearing unique differentiation antigens, J. Exp. Med. 145:221-233.

5. E.L. Reinherz, P.C. Kung, G. Goldstein and S.F. Schlossman, 1979, Separation of functional subsets of human T cells by a monoclonal antibody, Proc. Natl. Acad. Sci. USA 76:4061-4065.

6. S.M. Friedman, O. Irigoyen, Y. Thomas and L. Chess, 1980, Heteroantibodies to human T cell tumors: recognition of normal and malignant T cell subsets. In Regulatory T Lymphocytes, Edited by B. Pernis and H.J. Vogel, Academic Press, New York pp. 251-260.

7. O. Irigoyen, P.V. Rizzolo, Y. Thomas, M.E. Hemler, H.H. Shen, S.M. Friedman, J.L. Strominger and L. Chess, 1981, Dissection of distinct human immunoregulatory T-cell subsets by a monoclonal antibody recognizing a cell surface antigen with wide tissue distribution, Proc. Natl. Acad. Sci. USA 78:3160-3164.

8. E.L. Reinherz, P.C. Kung, G. Goldstein and S.F. Schlossman, 1979, Further characterization of the human inducer T cell subset defined by monoclonal antibody, J. Immunol. 123:2894-2896.

9. S.M. Friedman, S.B. Hunter, O. Irigoyen, P.C. Kung, G. Goldstein and L. Chess, 1981, Functional analysis of human T cell subsets defined by monoclonal antibodies. II. Collaborative T-T interactions in the generation of TNP altered-self reactive cytotoxic T lymphocytes, J. Immunol. 126:1702-1706.

10. Y. Thomas, J. Sosman, L. Rogozinski, O. Irigoyen, P.C. Kung, G. Goldstein and L. Chess, 1981, Functional Analysis of human T cell subsets defined by monoclonal antibodies. III. Regulation of helper factor production by T cell subsets, J. Immunol. 126:1948-1951.

11. Y. Thomas, J. Sosman, O. Irigoyen, S.M. Friedman, P.C. Kung, G. Goldstein and L. Chess, 1980, Functional analysis of human

T cell subsets defined by monoclonal antibodies. I. Collaborative T-T interactions in the immunoregulation of B cell differentiation, J. Immunol 125:2402-2408.

12. Y. Thomas, L. Rogozinski, O. Irigoyen, S.M. Friedman, P.C. Kung, G. Goldstein and L. Chess, 1981, Functional analysis of human T cell subsets defined by monoclonal antibodies. IV. Induction of suppressor cells within the $OKT4^+$ population, J. Exp. Med. 154:459-467.

13. Y. Thomas, L. Rogozinski, O. Irigoyen, H.H. Shen, M.A. Talle, G. Goldstein and L. Chess, 1981, Functional analysis of human T cell subsets defined by monoclonal antibodies. V. Suppressor cells within the activated $OKT4^+$ population belong to a distinct subset, J. Immunol. 128:1386-1390.

14. G. Kohler and C. Milstein, 1975, Continuous cultures of fused cells secreting antibody of predefined specificity, Nature(Lond.) 256:495-497.

15. R.A. Goldsby, B.A. Osborne, E. Simpson and L.A. Herzenberg, 1977, Hybrid cell lines with T-cell characteristics, Nature, 267:707-708.

16. S. Kontianinen, E. Simpson, E. Bohrer, P.C.L. Beverley, L.A. Herzenberg, W.C. Fitzpatrick, P. Vogt, A. Torano, I.F.C. McKenzie and M. Feldman, 1978, T-cell lines producing antigen-specific suppressor factor, Nature 274:477-480.

17. M. Taniguchi and J.F.A.P. Miller, 1978, Specific suppressor factors produced by hybridomas derived from the fusion of enriched suppressor T cells and T lymphoma cell line, J. Exp. Med. 148:373-381.

18. M.J. Taussig, J.R.F. Corvalan, R.M. Binns and A. Holliman, 1979, Production of an H-2-related suppressor factor by a hybrid T-cell line, Nature 277:305-308.

19. L. Harwell, B. Skidmore, P. Marrak and J. Kappler, 1980, Concanavalin-A-inducible Interleukin-2-producing T cell hybridoma, J. Exp. Med. 152:893-904.

20. Z. Eshhar, R.N. Apte, I. Lowy, Y. Ben-Neriah, D. Givol and E Mozes, 1980, T-cell hybridoma bearing heavy chain variable region determinants producing (T,G)-A--L-specific helper factor, Nature 286:270-272.

21. Y. Kaufman, G. Berke and Z. Eshhar, 1981, Cytotoxic T lymphocyte hybridomas that mediate specific tumor-cell lysis in vitro, Proc. Natl. Acad. Sci. USA 78:2502-2506.

22. C. Grillot-Courvalin, J-C. Brouet, R. Berger and Bernheim, 1981, Establishment of a human T-cell hybrid line with suppressive activity, Nature 292:844-845.

23. O. Irigoyen, P.V. Rizzolo, Y. Thomas, L. Rogozinski and L. Chess, 1981, Generation of functional human T cell hybrids, J. Exp. Med. 154:1827-1837.

24. L. Chess, R.P. MacDermott and S.F. Schlossman, 1974, Immunologic functions of isolated human lymphocyte subpopulations. I. Quantitative isolation of human T and B cells and response to mitogens, J. Immunol. 113:1113-1121.

25. T.R. Chen, 1977, In situ detection of mycoplasma contamination in cell cultures by fluorescent Hoechst 33258 stain, Exp. Cell. Res. 104:255-262.

26. M.L. Gefter, D.H. Margulies and M.D. Scharff, 1977, A simple method for polyethylene glycol-promoted hybridization of mouse myeloma cells, Somat. Cell Genet. 3:231-236.

27. R.H. Kennet, K.A. Denis, A.S. Tung and N.R. Kinman, 1978, Hybrid plasmocytoma production: Fusions with adult spleen cells, monoclonal spleen fragments, neonatal spleen cells and human spleen cells, Current Topics in Microbiol. and Immunol. 81:77-91.

28. W.J. Mellman, 1965, Human Peripheral blood leukocyte cultures. In Human Chromosome Methodology. Editor J.J. Yunis, Academic Press, NY, pp.21-49.

29. H. Subak-Sharpe, R.R. Burk and J.D. Pitts, 1969, Metabolic cooperation between biochemically marked mammalian cells in tissue culture, J. Cell Sci. 4:353-367.

30. R.P. Cox, M.R. Krauss, M.E. Balis and J. Dancis, 1972, Communication between normal and enzyme deficient cells in tissue culture, Exp. Cell. Res. 74:251-268.
31. S.M. Friedman, G. Thompson, J.P. Halper and D. Knowles, 1981, OT-CLL: A human T cell chronic lymphocytic leukemia that produces IL-2 in high titer, J. Immunol. 128:935-940.
32. W. Szybalski, E.H. Szybalska and G. Ragni, 1962, Genetic studies with human cell lines, Nat. Cancer Inst. Monograph. 7:75-88.
33. J.W. Littlefield, 1964, Selection of hybrids from matings of fibroblasts in vitro and their presumed recombinants, Science 145:709-710.
34. M. Dechamps, B.R. DeSaint-Vincent, C. Evrard, M. Sassi and G. Buttin, 1974, Studies in 1-β-D-arabinofuranosyl-cytosine (ARA-C) resistant mutants of Chinese hamsters' fibroblasts. II. High resistance to ARA-C as a genetic marker for cellular hybridization, Exp. Cell Res. 86:269-279.
35. L. Medrano and H. Green, 1974, A uridine kinase-deficient mutant of 3T3 and a selective method for cells containing the enzyme, Cell 1:23-26.
36. T. Chan, C. Long and H. Green, 1975, A human-mouse somatic hybrid line selected for human deoxycytidine deaminase, Somat. Cell Genet. 1:81-90.
37. M.B. Meyers, O.P. van Diggelen, M. van Diggelin and S. Shin, 1980, Isolation of somatic cell mutants with specified alterations in hypoxanthine phosphoribosyl transferase, Somat. Cell Genet. 6:299-306.
38. J.M. Boyle, J. Hopkins, M. Fox, J.D. Allen and R.H. Leach, 1981, Interference in hybrid clone selection caused by mycoplasma hyorhinis infection, Exp. Cell Res. 132:67-72.

CULTURED HUMAN T LYMPHOCYTE LINES AND CLONES
AS IMMUNOGENETIC TOOLS

Dolores J. Schendel and Rudolf Wank

Institute of Immunology
University of Munich
Munich, Federal Republic of Germany

ABSTRACT

Human T lymphocytes were sensitized in vitro to allogeneic determinants encoded by the HLA region and subsequently cloned by limiting dilution in the presence of human T cell growth factor. An in vitro priming protocol entailing multiple rounds of allostimulation in mixed lymphocyte culture, using a disparate stimulating to responding cell ratio (10:1), enabled a high frequency of functionally active clones to be isolated. The specificity of the proliferative and cytotoxic reactivities was ascertained directly by screening in an HLA characterized family.

INTRODUCTION

Human thymus-derived (T) lymphocytes are comprised of different subpopulations of immunocompetent cells that individually perform specialized functions in an immune response. Mixed lymphocyte cultures (MLC) of cells from two individuals can be used to obtain T lymphocytes sensitized to antigens controlled by the major histocompatibility complex (MHC). This in vitro system has allowed identification of several T cell subpopulations mediating different functions in the immune response to alloantigens, including those showing MHC specific proliferation, cytotoxicity, or suppression (reviewed in 1).

Recently it has been shown that MHC specific T lymphocytes displaying these different functions can be maintained in long term culture by addition of growth factors released by phytohemagglutinin-activated leukocytes (2-5). Furthermore, using limiting dilution or soft agar cloning techniques, it has been possible to clone human T cells (6-11). Such alloantigen specific T cell lines and clones are extremely valuable immunogenetic reagents that allow the roles of various MHC antigens in the induction of functionally distinct T cell subpopulations, and in the control of their interactions, to be studied at a fine specificity level.

Techniques for culturing and cloning human T lymphocytes are in developmental stages; thus much information remains anecdotal. In this paper we describe the methods developed in our laboratory to isolate functionally distinct T cell clones specific for antigens controlled by the human MHC.

MATERIALS AND METHODS

Cells. The lymphocytes used in the in vitro priming phase for sensitization of the clones, as feeder cells during their continued culture, and as stimulating test cells in the secondary MLC (2° MLC) and the cell-mediated lympholysis (CML) assays, were prepared from defibrinated blood separated by standard techniques (12).

Culture medium. RPMI 1640 culture medium (Gibco, Grand Island, N.Y.), supplemented with L-glutamine (2mM), penicillin (100 units/ml) and streptomycin (100 µg/ml) was used throughout.

Human serum. 15% human serum was added to the RPMI culture medium; it was autologous to the cloned responding cells, and was used during the *in vitro* sensitization phase, the cloning phase, and for the continued maintenance of the clones. It was prepared as discussed elsewhere (13).

T cell growth factor. TCGF was made and tested according to the methods described earlier for its production from leukocytes stimulated with phytohemagglutinin (13). For the single batch used in these studies a concentration of 20% of the crude supernatant was selected.

X-irradiation. Various doses of X-irradiation, delivered at 125 rads/min by a cesium 137 source, were selected for different parts of these studies. A dose of 2000 rads was used to inactivate the stimulating cells in the first two rounds of allosensitization prior to cloning and to inactivate the stimulating test cells used in the 2° MLC. A dose of 7500 rads was used in the final round of allosensitization prior to cloning and, thereafter, for all feeder cells added to the growing colonies. This dose, in our hands, was required to restrict outgrowth of these cells in culture medium containing TCGF.

RESULTS AND DISCUSSION

Isolation of Human T Lymphocyte Clones

Figure 1 shows the scheme that we have used for the isolation of MHC specific human T cell clones in high frequency. It can be considered in three phases: 1) *in vitro* sensitization of MHC reactive cells, 2) seeding by limiting dilution with subsequent clonal expansion, and 3) screening for functional activity. In these studies alloantigen specific proliferation and cytotoxicity were tested, using 2° MLC and CML assays.

To obtain enriched numbers of alloreactive cells specific for MHC antigens, the lymphocytes to be cloned were first sensitized *in vitro* by repeated stimulation with allogeneic cells as outlined previously (11). Mixed lymphocyte cultures using peripheral blood cells from two unrelated individuals were established on day 0. Cells from one party were X-irradiated with 2000 rads and 10×10^6 of these stimulating cells (2_x) were added per 1×10^6 responding cells from individual 1. After 9 days the primed cells from individual 1 were washed, counted, and recultured with fresh X-irradiated (2_x) stimulating cells. Here again 10×10^6 X-irradiated cells were added per 1×10^6 responding cells. After 7 days the cells from the MLC were restimulated again with fresh 2_x cells, added in a 10-fold amount. The dose of X-irradiation was increased at this point from 2000 to 7500 rads to prevent outgrowth of stimulating cells during cloning. After an additional three days of culture the MLC cells were washed, counted and suspended in fresh culture medium. They were then seeded into special cloning plates (704160, Greiner and Sons, Nürtingen, FRG) which contained 24 wells of 2 ml volume, with the base of each 2 ml well separated into 16 microwells of 20 μl volume. Cells were added to the

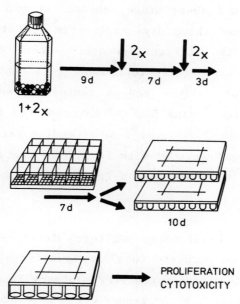

Fig. 1. Cloning of human T lymphocytes. Lymphocytes of individual 1 were sensitized in vitro to X-irradiated cells of individual 2, during three rounds of MLC. Thereafter, they were cloned by limiting dilution in a special cloning plate. After 7 days individual colonies were transferred to 200 µl microtiter plates and then after 10 days to 2 ml volume culture plates. When adequate numbers of cells were available, individual colonies were tested for proliferative and cytotoxic function.

larger wells in 0.5 ml of culture medium at various concentrations to yield 40, 10, 1 and 0.4 cells per 20 µl subdivision. These numbers were based on the total cell recoveries from the MLC, therefore, the actual responding cell number would be much lower since residual stimulating cells, which had been added three days earlier in a 10-fold amount, were included in the total cell counts. After 3 h the wells were examined microscopically to control cell number and then fed with 1.5 ml of culture medium containing an optimal concentration of TCGF, plus 5×10^4 X-irradiated feeder cells per 2 ml well. Four days later 1 ml of medium was removed from each larger well and replaced with fresh TCGF containing medium; at this time 5×10^4 additional X-irradiated feeder cells were added.

Emergence of colonies was monitored daily, and by day 7 adequate growth had occurred to allow transfer of individual colonies to 200 µl volume microtiter plates with round-shaped wells. Approximately 5×10^3 feeder cells were added per microtiter well. The colonies were expanded in these plates and 100 µl of culture medium was replaced with fresh TCGF containing medium every third day. Additional feeder cells (5×10^3) were added after 5 days. On day 10 contents of single microtiter wells were transferred to individual 2 ml volume culture wells (3524 Costar, Cambridge, MA) containing 2 ml of TCGF culture medium and 1×10^6 X-irradiated feeder cells. Thereafter, the cells were maintained in these 2 ml wells which were split and fed with fresh TCGF containing medium every 2nd or 3rd day, and with addition of feeder cells ($5-10 \times 10^5$ cells/well) every 7th-10th day.

Analysis of Functional Activity and Genetic Specificity

Clones showing specific memory proliferation, as detected

in 2° MLC, or cytotoxicity, as measured in CML, were of primary interest in these studies. In order to easily select such MHC specific clones at an early stage, microassays of cellular proliferation and cytotoxicity were refined for use with limited numbers of cloned cells. Prior to functional testing, the colonies were deprived of fresh feeder cells for at least three days and of TCGF for 12 h. Proliferative responses in 2° MLC were measured using procedures adapted from those described elsewhere (14), enabling specific reactivity to be analyzed with as few as 1000 cloned cells per test well. Basically, $1-5 \times 10^3$ cloned cells were distributed in 20 µl volumes in microtiter wells with V-shaped wells. Twenty-five $\times 10^4$ X-irradiated stimulating cells were added, also in 20 µl volumes. To determine the haplotype specificity, as discussed below, responses to stimulating cells from all members of an HLA-typed family were analyzed. These culture plates were incubated at 38.5°C in a humidified incubator with 6% CO_2. After 30 h the cultures were labeled with 2 µCi/well of tritiated thymidine of high specific activity (60-80 Ci/mmol, New England Nuclear, Boston, MA). The thymidine pulse was continued for 6 h giving a total culture time in the 2° MLC of 36 h.

Cellular cytotoxicity was measured using minor modifications of the CML assay described elsewhere in detail (15). Between 200 and 500 cloned cells were incubated with 2×10^3 labeled target cells. Cloned cells and target cells were incubated in an 80 µl culture volume during the 4 h CML test and, thereafter, 100 µl of medium was added to each test well prior to collection of supernatants. For screening in CML each individual colony was tested against six familial target cells so that HLA segregating activity, as discussed below, could be ascertained immediately.

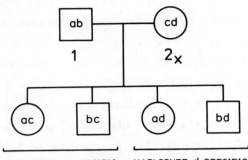

Fig. 2. Screening for HLA haplotype specific responses. Members of an HLA-typed family were used to screen the specificity of developing clones. Paternal cells, having the HLA a and b haplotypes, were used as the responding cells (1) in the in vitro sensitization phase prior to cloning. Maternal cells, with the HLA c and d haplotypes, were used as the specific stimulating cells (2_x). Following cloning, those colonies reacting to cells of the mother (cd) and ac and bc offspring would indicate specificity for a determinant encoded by the HLA c haplotype. Alternatively, reactivity to the mother (cd) and ad and bd offspring would indicate specificity for the HLA d haplotype.

Figure 2 illustrates the genetic constellation that was selected to allow rapid screening of clones specific for an HLA haplotype. The unrelated cells used in the in vitro sensitization phase were obtained from parents of an HLA typed family. Paternal cells, having the HLA haplotypes designated a and b, were sensitized to X-irradiated maternal cells with the HLA haplotypes c and d. In this case distinguishable responding cells become sensitized to antigens encoded by both the c and d haplotypes (16). Following cloning, proliferative cells specific for antigens encoded by the c haplotype incorporate radioactive thymidine only after stimulation with X-irradiated cells of the mother or ac and bc children; whereas clones specific for the d haplotype respond only to cells of the mother and children of ad and bd types. Colonies responding to all offspring would indicate that clonality had not been achieved, although it is possible that the c and d haplotypes might share an epitope that is recognized by a single clone. These possibilities can be distinguished through recloning of such colonies.

An experiment combining these technical and genetic approaches is shown in Table 1. Less than optimal growth was observed in the wells seeded at 40 cells per 20 µl well. Only one colony was selected for further transfer and no specific proliferation or cytotoxicity was seen with this colony, instead only weak responses to cells of the mother and all offspring. Sixty-five percent of the wells seeded at 10 cells per well contained colonies and 2 were randomly chosen for further study. Interestingly, both colonies showed HLA haplotype specific proliferation: one specific for an antigen encoded by the c haplotype and the other specific for an antigen of the d haplotype. Very strong cytotoxicity was mediated by one colony that was directed only against the family target cells having the c haplotype, in concordance with the specificity of 2° MLC proliferation. Many active colonies were found at the seeding of 1 cell per well. Of 13 randomly selected colonies from this group, 12 showed adequate growth after transfer to 2 ml wells. 2° MLC testing on day 29 revealed that 8 of these 12 showed

Table 1. Isolation and functional characterization of HLA specific human T cell clones

Number of cells seeded per well	Percentage* of positive wells	Numbers tested	HLA specific 2° MLC		HLA specific CML	
			c haplotype	d haplotype	c haplotype	d haplotype
40 cells	27.1%	1	0	0	0	0
10 cells	64.6%	2	1	1	1**	0
1 cells	28.1%	12	6	2	3***	0
0.4 cell	2.1%	1	0	1	0	0

* 96 wells of 20 μl were seeded at each cell number.

** Cytotoxicity ranged between 19 and 30% on the three HLA-c targets at an effector to target cell ratio of 0.1:1.

*** Cytotoxicity was only weak (7-10% specific lysis) with the HLA-c haplotype target cells.

proliferative responses specific for HLA. That is, 6 responded with thymidine uptake only when restimulated with cells carrying the HLA c haplotype, while 2 responded to restimulation only by cells expressing antigens of the d haplotype. In this group three colonies were found that mediated weak cytotoxicity. This CML was specific for an HLA haplotype and corresponded to the specificity of proliferation detected in the 2° MLC.

Several points in this protocol seem particularly important. First, the in vitro priming using three rounds of stimulation by allogeneic cells, added in a disparate ratio to the responding cells (i.e. 10:1), increases the numbers of MHC reactive cells, as determined by limiting dilution analysis (RW, unpublished observations) and by the frequency of MHC reactive clones that can be isolated. In these examples, 75% of the randomly selected colonies seeded at 1 cell per well were specific for an HLA haplotype. Second, it has been possible to retain these cells in culture for more than three months without any apparent crisis of growth and with retention of specificity. The cells were first sensitized to alloantigen and only upon cloning were they exposed to TCGF. Thereafter, they were maintained in the presence of TCGF, autologous serum, and with frequent addition of specific feeder cells, autologous to the stimulating cells used in the original in vitro priming. This constant readdition of specific allogeneic cells may be required to maintain the expression of the receptors for specific growth factors (supplied by the TCGF) on the clonal progeny. A stable source of TCGF is also critical. In this case it was prepared without the addition of extra drugs, such as indomethacin, which might have long term detrimental effects, and without addition of lymphoblastoid cell lines, which are potential sources of infective agents, such as mycoplasma, that could be transferred to the developing clones. Third, the miniaturized and optimized

2° MLC and CML assays require very few cells; thus, functionally active clones can be identified very early in their culture lives. Using the genetic constellation of the family, their MHC haplotype specificity can be ascertained immediately. This then allows rapid selection of clones that are of interest for further studies or that are important to reclone immediately.

Combined, these technicial and genetic considerations provide a simple approach to obtain multiple clones responsive to antigens encoded by the human MHC. Any interesting genetic combination can be selected for study since there is no requirement for previous in vivo sensitization. Related groups of MHC specific clones, such as the six specific for the HLA c haplotype isolated from the seeding at one cell/well, will enable fine specificity analysis of the roles of various HLA gene products in the induction and regulation of functionally distinct human T cell subsets, and analysis of surface markers that may delineate their pathways of differentiation.

ACKNOWLEDGMENT

This work was supported by grants from the Deutsche Forschungsgemeinschaft, SFB 37 (B3 and B4).

REFERENCES

1. van Rood, J.J., de Vries, R.R.P. and Bradley, B.A. 1981. Genetics and biology of the HLA system. In: The Role of the Major Histocompatibility Complex in Immunobiology (M.E. Dorf, Ed.) pp. 59-113. Garland STPM Press, New York.

2. Morgan, D.A., Ruscetti, F.J., and Gallo, R. 1976. Selective in vitro growth of T lymphocytes from normal human bone marrows. Science 193:1007.

3. Ruscetti, F.W., Morgan, D.A., and Gallo, R. 1977. Functional and morphologic characteristics of human T cells continuously grown in vitro. J. Immunol. 119:131.

4. Bonnard, G.D., Schendel, D.J., West, W.H., Alvarez, J.M., Maca, R.D., Yasaka, K., Fine, R.L., Herberman, R.B., de Landazuri, M.O., and Morgan, D.A. 1978. Continuous growth of normal human T lymphocytes in culture with retention of important functions. In: Human Lymphocyte Differentiation. Its Application to Cancer. (B. Serrou and C. Rosenthal, Eds.) pp. 319-326. Elsevier/North Holland, Amsterdam.

5. Maca, R.D., Bonnard, G.D., and Herberman, R.B. 1979. The suppression of mitogen and alloantigen stimulated peripheral blood lymphocytes by cultured human T lymphocytes. J. Immunol. 123:246.

6. Bach, F.H., Inouye, H., Hank, J.A., and Alter, B.J. 1979. Human T lymphocyte clones reactive in primed lymphocyte typing and cytotoxicity. Nature 281:307.

7. Kornbluth, J. and Dupont, B. 1980. Cloning and functional characterization of primary alloreactive human T lymphocytes. J. Exp. Med. 152:164s.

8. Goulmy, E., Blokland, E., van Rood, J.J., Charmot, D., Malissen, B., and Mawas, C. 1980. Production, expansion, and clonal analysis of T cells with specific HLA-restricted male lysis. J. Exp. Med. 152:182s.

9. Pawelec, G., Rehbein, A., Alter, B.J., and Wernet, P. 1980. Retention of specific proliferative capacity of cloned human T cells alloactivated against HLA-D alleles. Immunogenetics 11:527.

10. Malissen, B., Charmot, D., and Mawas, C. 1981. Expansion of human lymphocyte populations expressing specific immune reactivities. III. Specific colonies, either cytotoxic or proliferative, obtained from a population of responder cells primed in vitro. Preliminary immunogenetic analysis. Human Immunology 2:1.

11. Schendel, D.J., and Wank, R. 1981. Isolation of HLA-haplotype specific human T lymphocyte clones. Immunobiology 160:367.

12. Böyum, A. 1968. Separation of leukocytes from blood and bone marrow. Scand. J. Clin. Lab. Invest. Suppl. 97:21.

13. Schendel, D.J., and Wank, R. 1981. Production of human T cell growth factor. Human Immunology 2:325.

14. Wank, R. 1982. Rapid responses of lymphoctes in an optimized mixed lymphocyte culture. Tissue Antigens 19:323.

15. European CML Study Group. 1980. Human histocompatibility testing by T cell-mediated lympholysis: A European standard CML technique. Tissue Antigens 16:335.

16. Wank, R., Schendel, D.J., and Dupont, B. 1978. Typing for homozygous determinants with lymphocytes primed to two-haplotype differences. Transplant. Proc. 10:763.

LONG-TERM PERSISTANCE IN EXPERIMENTAL ANIMALS OF COMPONENTS
OF SKIN-EQUIVALENT GRAFTS FABRICATED IN THE LABORATORY

        Euguene Bell, Stephanie Ellsworth Sher,
        Barbara E. Hull and Robert L. Sarber

        Massachusetts Institute of Technology
        Department of Biology
        Cambridge, Massachusetts

        Seymour Rosen

        Beth Isreal Hospital
        Department of Pathology
        Boston, Massachusetts

Living skin equivalent grafts have been fabricated in the laboratory and applied to experimental animals in which they persist indefinately. In this paper we review the technology developed for tissue fabrication, give a short account of animal experiments and present evidence that dermal fibroblasts incorporated into grafted tissues proliferate and remain in situ in rat hosts for periods of at least 13 months[b].

---

[b] Grafts have been in place on experimental animals for as long as 24 months but have not been biopsied for karyotyping.

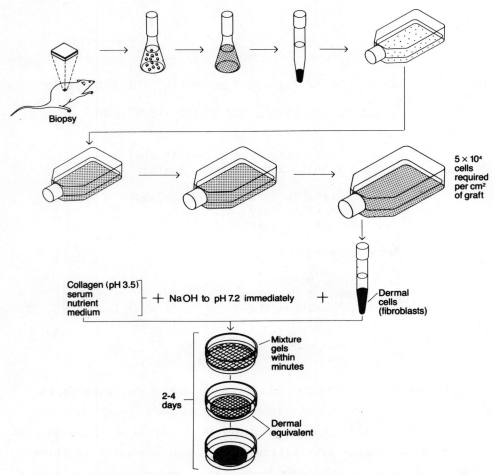

Fig. 1   Procedures for fabricating the dermal equivalent component of a skin-equivalent graft fabricated with cells from a biopsy taken from a potential graft recipient. Fibroblasts recovered from a biopsy are cut into fragments and digested with enzymes which degrade the dermal matrix. Cells are plated out and grown in flasks to the number needed for the graft. Components required for the lattice are combined with cells as shown and poured into a casting dish in which they gel and are contracted by the cells.

In practice to make a skin equivalent model or a graft to an experimental animal a cell strain is used or a biopsy is taken from a potential graft recipient. Cells are recovered from the biopsy and grown up in culture to a number commensurate with the size of the graft required (Fig. 1). The media used are McCoy's 5A with either 10 or 20% fetal bovine serum in a 5% CO_2 atmosphere for human cells, or Dulbecco's Modified Eagle's medium with 10% fetal bovine serum in a 10% CO_2 atmosphere for rat or rabbit cells. Cells are combined with collagen and other components of the tissue equivalent lattice as described in detail elsewhere (1,2).

When collagen, tissue culture medium and serum are combined at neutral pH with fibroblasts, a gelled collagen lattice forms almost immediately (1,3,4). The cells in the lattice extend processes which attach to the collagen fibrils to which they adhere. The substrate is compliant and therefore yields to the stresses generated by constituent cells. The action of the cells results in fluid loss from the lattice and the formation of a tissue fabric which has tensile strength and is graftable. Tensile strength tests with 4 day old lattices of 1.8 cm in diameter which contain 1.3×10^5 cells and 2.3 mg of rat tail collagen have been made. Such a lattice can withstand a weight of about 7.0 grams for 5.0 minutes before beginning to tear. We have called the tissue a dermal-equivalent. It can be cast in a dish or a pan of any size and undergoes a reduction in thickness and area. It can also be cast as a cylinder which forms a sheet when cut open.

The lattice contracts maximally after about four days. The rate of the contraction and the ultimate lattice dimensions are proportional to cell number and inversely proportional to the amount of collagen present (1).

After the lattice has contracted it is ready to receive a suspension of epidermal cells from a second biopsy (Fig. 2). Trypsin-dissociated cells (5) are applied to the surface of the dermal equivalent, attach and proliferate. Islands of basal cells grow on the substrate to form a continuous epidermal sheet which undergoes differentiation and keratinizes (6,7).

What has been fabricated then is a tissue or organ equivalent consisting of two distinct cell types incorporated into separate strata which are related to each other as they are _in vivo_. We are studying such tissue-equivalent constructs as a model system to assess how well they simulate tissues and organs _in vivo_ particularly with respect to the preservation of differentiated functions. In addition to skin equivalents we have fabricated model blood vessels consisting of three distinct layers[c]: First matrix ingredients, including smooth muscle cells are poured or cast in a cylindrical vessel having an axial rod. The tissue equivalent contracts around the core and the expressed fluid can then be removed. The components of a second lattice are then poured into the space between the tissue equivalent and the outer cylinder. The cells of the second layer are pericytes. When the second lattice has contracted around the first layer the double layered cylinder is slipped off the axial rod and the lumen of the tubular tissue equivalent is injected with endothelial cells which attach to the walls and line it. Other cell types such as cardiac muscle cells have also been used to construct tubes.

---

[c] Weinberg, C. and Bell, E., unpublished results.

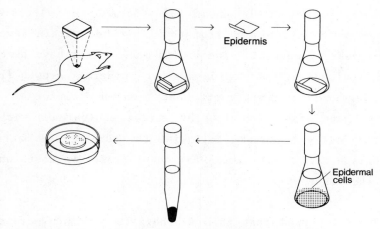

Fig. 2. To form the two-layered skin equivalent, a second biopsy is taken from the animal. Using a combination of trypsin and EDTA, the epidermal sheet is separated from the dermis and the sheet is then dissociated to yield individual epidermal cells. These epidermal cells are then spun down to form a pellet which is spread over the surface of the collagen lattice.

Connective-tissue equivalents can also be fabricated without cells.[d] Blood platelets are a satisfactory substitute to promote contaction of the collagen lattice. With one unit of

---

[d] Bell, E., Church, D., Sher, S., Hull, B., Sarber, R. and Soranno, T. Fabrication of a connective tissue substitute with platelets and collagen. Submitted for publication.

blood it is possible to fabricate a sheet 2500 cm² in area. A principal difference is that tissue equivalents cast with platelets contract in the thickness dimension only. Grafts to rabbits made up with autologous platelet dermal equivalents are infiltrated with fibroblasts from the graft bed within seven days and appear to persist. The grafts made so far have been of dermal equivalents without epidermis but epidermal growth from the periphery covers a graft of about 3 cm in diameter in a month. We have assembled platelet dermal equivalents with ³H-labelled collagen and are following the turnover of matrix autoradiographically after implantation of grafts in hosts which provided the platelets.

We have reported that when fibroblasts are incorporated into dermal-equivalent tissues *in vitro* they cease to divide[e]. This is true whether the cells are of low or high PDL. Thus the tissue equivalent provides a useful model system in which cells are homogeneously out of cycle. It is a connective tissue model whose constituents can be manipulated to study their mitogenic potential, or a model for testing the capacity of cells to respond to mitogenic stimuli *in vivo* by transplanting them in a tissue equivalent to an appropriate host.

We have followed, *in vivo*, the mitotic responses of cells incorporated into skin equivalent grafts implananted into syngeneic hosts or into hosts which donated the cells incorporated in the grafts. With syngeneic animals grafts are made up with female cells and grafted to a male host.

---

[e] Sarber, R., Hull, B., Merrill, C., Soranno, T. and Bell, E. 1981. Regulation of proliferation of fibroblasts of low and high population doubling levels grown in collagen lattices. Mechanism of Ageing and Development, 17:107-117.

GRAFTS FABRICATED IN THE LABORATORY

Cells are incorporated into dermal equivalents to be used for grafts at a density roughly equivalent to that of fibroblasts in dermis. Under ordinary culture conditions <u>in vitro</u>, cells in a dermal equivalent are out of cycle. For five days after grafting cell density in the dermal equivalent <u>in vivo</u> is unchanged (Fig. 3). By 9 days however the density of cells in the graft increases dramatically (Fig. 4); we have counted a five-to-seven-fold increase in cell number per unit area. This increase in cell density is largely a result of the proliferation of the original cells of the graft, as will be shown by the karyotyping data presented later. By 9 days grafts are extensively vascularized so that cells can receive signals from the host via a well developed circulation.

Although many grafts were sutured into place in early experiments they are now applied to the graft bed without suturing and allowed to overlap the surrounding host tissue. At the time of application the borders of the graft bed are tattoed, and the area of the graft tattoo to tattoo is measured during the life of the graft to monitor wound contraction. Under optimum conditions wound contraction is completely inhibited. An eight day and five month graft are shown in Figures 5 and 6. While the eight day graft is nearly transparent the five month graft has become opaque. The skin equivalent at five months is smooth, soft and matches the texture of surrounding skin except that it lacks secondary derivatives. The suture lines betwen the graft and host tissue are virtually invisible grossly. A histological section of a seven month graft shows that the hypervascularity, high fibroblast density and epidermal hypertrophy have all disappeared (Fig. 7). While the dense basketweave of the dermal matrix is not seen even after a graft has been on an animal one year, extensive remodeling is observed. We have reported that

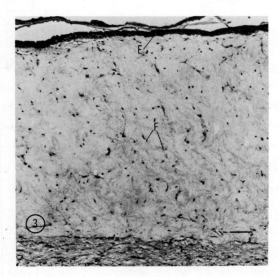

Fig. 3. This light micrograph shows a thick section (7 µm) of a paraffin-embedded graft removed from a rat five days after it was applied to the animal. Note the complete epidermal covering (E) and the sparse population of fibroblasts (F). The bar represents 0.3 mm.

there is a striking increase in birefringence of the dermis and a moderate increase in the thickness of collagen bundles (7). While early graft have little elastin, after 4 months elastic fibers have become visible. We can distinguish between the possibilities that the graft has been inflitrated by cells from the host or that dermal cells in the graft have multiplied, by sampling grafts and karyotyping cells.

To carry out such an experiment, grafts are made up with cells from female Fischer rats and transplanted to male hosts.

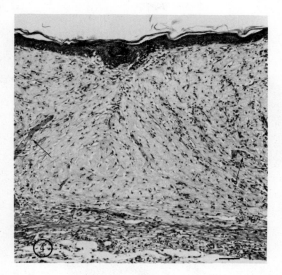

Fig. 4. This thick section illustrates that at nine days the density of fibroblasts has increased dramatically relative to that found at five days (Fig. 3), while the overall depth of the graft remains nearly constant. Note the appearance of blood vessels (arrows) within the dermis. The bar represents 0.3 mm.

When grafts are removed they are trimmed to adjacent tissue and cut up fragments are allowed to attach to a plastic petri plate. After fibroblasts grow out and are subcultivated in flasks, cultures are treated with colchicine and mitotic cells are shaken off. Cells are swollen in hypotonic medium, fixed, and dried down on glass slides.

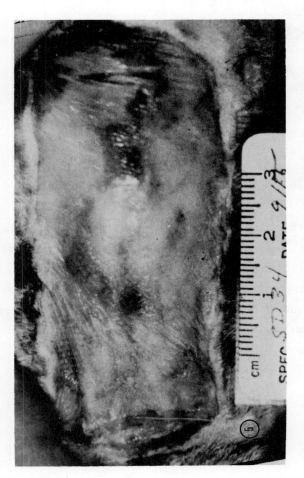

Fig. 5. An eight day graft on a male Spraque-Dawley rat. The area when grafted on September 9, 1981, was 7 x 5.3 cm. On September 16, 1981, when the rat was unbandaged the area was 7.7 x 3.8 cm. Contraction was 20%.

GRAFTS FABRICATED IN THE LABORATORY 429

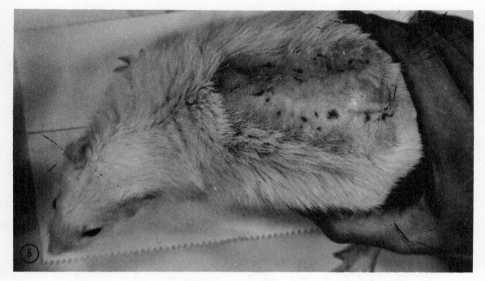

Fig. 6.   A five month old graft on a male Sprague-Dawley rat. The area when grafted on April 6, 1981, was 7.0 x 3.8 cm; when photographed September 14, 1981, the area measured 6.0 x 4.0 cm. Contraction was 10%.

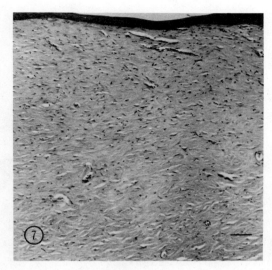

Fig. 7. This light micrograph shows a thick section (7 µm) of a seven month graft. The collagen bundles have become much thicker than at nine days (Fig. 4), indicating an extensive reorganization of the collagen network during the intervening months. The graft lacks both hair follicles and sweat glands, since these secondary derivatives are lost during our preparative procedures. The bar represents 0.3 mm.

We have studied the turnover of fibroblasts in grafts to syngenic rats over a period of thirteen months using the female karyotype as a genetic marker for the cells of the graft. Table 1 shows the trend over time. While initially over 90% of the cells of the graft are of donor genotype at seven and at thirteen months the number has dropped to about one half. By seven months the density of cells in the graft has returned to normal. The fact that so large a number of cells from the

Table 1

Age of graft at biopsy	# Female karyotypes as % of total	
9 days	91%	n* = 23
1 month	64%	n = 22
7 months	54%	n = 24
13 months	42%	n = 26
	*n = number of karyotypes analyzed.	

original graft remains is significant and provides evidence that skin equivalent grafts and the cells which consitute then become permanently integrated with the host skin. Since we do not yet know the degree of flux of fibroblasts in dermis normally, or the spatial distribution of male cells in the graft, or the degree of infiltration of host tissue by graft cells the significance of the dimunition of cells of female genotype with time in the graft cannot be fully evaluated. If fibroblasts of the dermis engage in moderate excursions normally, the mixing of graft and host tissues is to be expected. It is clear that the process in any event is one of long term and probably not in response to a wound healing stimulus. In grafts in which the dermal equivalent is made up with platelets rather than cells the invasive response of surrounding cells is almost immediate. The contrast suggests that the resident population of cells in dermal equivalents fabricated with cells is sensed by fibroblasts of the host. The signals are yet unknown to us as are the signals which mitotically activate a quiescent population of fibroblasts; so too as those which result in an ultimate reduction and return to normal of cell numbers in the dermal equivalent.

Although similar experiments with epidermal cells are not completed we have indirect evidence that epidermis persists as well. Histological sections of 3 cm diameter grafts five days after implantation show a complete epidermal covering which could not have formed by in-growth of epidermis from the periphery in so short a period.

While living tissue equivalents, as vehicles for returning cultured cells to organisms, have broad clinical possibilities, they also have potential value as model systems for examining basic problems in wound healing, tissue modeling, cell aging and cell differentiation.

ACKNOWLEDGEMENT

The expert technical assistant of Diane Church and Thomas Soranno, both of the Department of Biology, Massachusetts Institute of Technology, Cambridge, MA, is gratefully acknowledged.

REFERENCES

1. Bell, E., Ivarsson, B. and Merrill, C. 1979. Production of a tissue-like structure by contraction of collagen lattices by human fibroblasts of different proliferative potential in vitro. Proc. Nat'l. Acad. Sci. 76, No. 3: 1274-1278.

2. Bell, E., Ehrlich, H.P., Buttle, D.J. and Nakatsuji, T. 1981. Living tissue formed in vitro and accepted as skin-equivalent tissue of full thickness. Science 211: 1052-1054.

3. Ehrmann, R.L. and Gey, G.O. 1956. The growth of cells on a transparent gel of reconstituted rat-tail collagen. J. Nat'l. Cancer Inst. 16: 1375-1403.

4. Elsdale, T. and Bard, J. 1972. Collagen substrata for studies on cell behavior. J. Cell Biol. 54: 626-637.

5. McKeehan, W.L. 1977. The effect of temperature during trypsin treatment on viability and multiplication potential of single normal human and chicken fibroblasts. Cell Biol. Int. Rep. 1: 335.

6. Bell, E., Yanover, P., Levinstone, D. Merrill, C., Sher, S., Marek, L. and Young I. 1977. Cell division and cell motility in cell lineages. J. Cell Biol. 75, (2, Part 2): 6a.

7. Bell, E., Ehrlich, H.P., Sher, S., Merrill, C., Sarber, R., Hull, B., Nakatsuji, T., Church, D. and Buttle, D.J. 1981. Development and use of a living skin equivalent. J. Plastic and Recon. Surgery 67, No. 3:386-392.

ANGIOTENSIN RECEPTORS AND THE CONTROL OF Na^+ AND K^+ TRANSPORT IN CULTURED AORTIC SMOOTH MUSCLE AND BRAIN MICROVESSEL CELLS

Tommy A. Brock and Jeffrey B. Smith

Cardiovascular Research and Training Center and
Department of Pharmacology, Univerisity of
Alabama in Birmingham, Birmingham, Alabama   35294

ABSTRACT

We examined the effect of angiotensin on Na^+ and K^+ transport by cultures of aortic smooth muscle and brain microvessel cells. Angiotensin II (AII) and angiotensin III (AIII) stimulated net Na^+ uptake, which was assayed in the presence of ouabain to block Na^+ efflux via the Na^+-K^+ pump. The combination of saturating concentrations of AII and AIII produced no greater stimulation of net Na^+ uptake than either angiotensin by itself. AII also stimulated ouabain-sensitive $^{86}Rb^+$ uptake by cultured aortic smooth muscle and brain microvessel cells. AII was not as effective as monensin, a Na^+ ionophore, in stimulating the Na^+-K^+ pump. In the presence of monensin, AII had no effect on ouabain-sensitive $^{86}Rb^+$ uptake. These data are consistent with previous observations suggesting that AII stimulates the Na^+-K^+ pump by supplying it with more of its rate-limiting substrate, Na^+.

^{125}I-AII bound tightly to cultured aortic muscle cells. Approximately 5 nM unlabelled AII half-maximally inhibited ^{125}I-AII binding. Incubation of intact cultures with 1.0μM AII produced a time-dependent decrease in specific ^{125}I-AII binding. This slow loss of AII binding was prevented by methylamine, a known inhibitor of receptor clustering and internalization. We conclude (1) that angiotensin increases the Na^+ permeability of smooth muscle cells in culture, thereby stimulating the Na^+-K^+ pump; (2) that AII binds to a specific receptor on the cell surface; and (3) that the AII-receptor complex may be internalized by a methylamine-inhibited pathway.

INTRODUCTION

Vascular smooth muscle (VSM) contracts when the concentration of free Ca^{2+} in the cytoplasm increases (1,2,3). The octapeptide hormone angiotensin II (AII) contracts VSM directly by binding to a specific receptor on the muscle cell surface (4,5). Although AII-receptor interactions have been well-characterized in a variety of different cell types (4-9), the transmembrane signalling events which lead to the increase in internal Ca^{2+} remain to be clarified.

The Na^+ gradient in VSM has been postulated to be a major determinant of vascular contractility (10,11). Increasing the Na^+ gradient may augment relaxation, while decreasing the Na^+ gradient may augment contraction. There have been incongruous reports concerning AII and its effect on Na^+ movements in isolated smooth muscle. Angiotensin II has been reported to increase (12) or not change (13) tissue Na^+, or to stimulate Na^+ efflux (14) in isolated arterial tissue. We have recently reported that AII markedly stimulates Na^+ uptake and Na^+-K^+ pump activity in smooth muscle cells cultured from explants of rat aorta (15). Cultures of vascular smooth muscle provide a convenient model for studying the effects of vasoactive agents on Na^+ and K^+ transport because a homogeneous population of cells can be obtained which is free from neural influences and an extensive extracellular matrix.

In the present study, we extend our previous observations of the stimulation of Na^+ entry and Na^+-K^+ pump activity by AII to: (1) smooth muscle cells cultured from rat aorta prepared by enzymatic dispersion; and (2) microvessel cultures derived from explants of pig brain microvessels. The results of this study demonstrate that the stimulation of Na^+ entry by angiotensin is associated with the binding of AII to a high affinity receptor. We also present data suggesting that the AII-receptor complex may be slowly internalized in the aortic muscle cultures.

METHODS

Cell Culture - Aorta. Explants of the tunica media from rat aorta were cultured as previously described (15). We also cultured aortic smooth muscle cells after enzymically dispersing the tissue by the following procedure. Thoracic aortae were removed from 12 ether-anesthetized, Sprague-Dawley rats (300-350 g) under sterile conditions, and placed in Medium 199 (Gibco) containing 100 units/ml penicillin G and 100 µg/ml streptomycin. Following the removal of loose connective tissue, the vessels were placed in 5 ml of Medium 199 containing 0.5 mg/ml elastase (Sigma), 1 mg/ml soybean trypsin inhibitor (Sigma), 1 mg/ml defatted bovine serum albumin (BSA, Sigma), and 1.0 mg/ml collagenase (CLS IV, Worthington). After a 30 min incubation at 37°C without agitation the tunica adventitia was

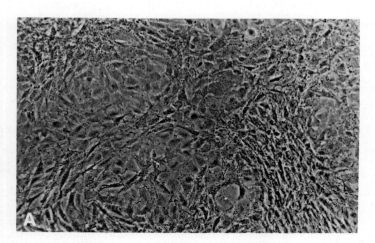

Fig. 1A. Light micrograph showing the characteristic appearance of smooth muscle cultures isolated from rat thoracic aorta by enzymic dispersion. (100x)

removed with forceps as an intact everted tube, thus leaving behind thin tubes of tunica media and intima. This tissue was thoroughly rinsed, in Medium 199 placed in 5 ml of fresh enzyme solution for an additional 30 minutes, and rinsed again. Then the vessels were minced with a scissors and allowed to digest for an additional 1.5 - 2 hours in 5 ml of the enzyme solution. At the end of this period, Medium 199 containing 10% fetal bovine serum (FBS) (Gibco) was added to the cell suspension, and the dispersed cells were washed twice by centrifugation at 1000 x g for 10 minutes. The pellet was suspended in Medium 199 containing 10% FBS, 10 mM HEPES buffer, penicillin G and streptomycin, and 1×10^7 cells were seeded into a 25 cm^2 culture flask (Costar). Stock cultures and culture dishes were prepared as previously described (15). Non-growing smooth

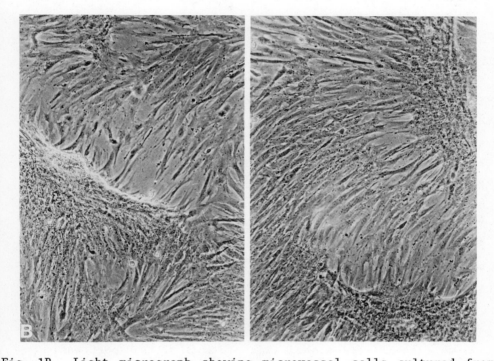

Fig. 1B. Light micrograph showing microvessel cells cultured from explants of blood vessels (100-250 µm) isolated from newborn pig brain. (100x)

Both cell types grow in hills and valleys which are characteristic of smooth muscle cells in culture. The rat aortic cells have a more irregular shape, whereas the pig microvessel cells are spindle-shaped.

muscle cells were obtained by placing confluent dishes in serum-free Medium 199 for 5-10 days or by allowing the cells to exhaust the growth medium of mitogenic factors. All cultures were used 10-16 days after the original plating date and were obtained from explants unless otherwise indicated. Figure 1A illustrates the light microscopic appearance of the cultured smooth muscle cells.

Cell Culture - Brain Microvessels. Microvessels were isolated and cultured from newborn pig brain as previously described (16,17). Individual microvessels (100-250 µm) were attached to the bottom of 35 mm culture dishes (Falcon) and covered with a 1:1 mixture of DME:F12 (Gibco) containing 20% FBS. After 2-3 days, cells began to migrate from these vessels onto the culture dish, which became confluent in about 7 days. The amount of FBS was then reduced to 10% and the cultures were handled as previously described (15). Figure 1B illustrates the appearance of pig microvessel cultures as viewed by the light microscope.

$^{86}Rb^{+}$ Uptake and Total Cell Na^{+} and K^{+}. The cultured cells (35 mm dishes) were washed twice with 37°C transport assay buffer (TAB), mM: NaCl, 120, KCl, 5; $CaCl_2$, 2; $MgCl_2$, 1; HEPES, 20; pH, 7.6 adjusted with tris-(hydroxymethyl) aminomethane. Then 1 ml of TAB was added and the cultures were allowed to incubate for 20 min ± 2 mM ouabain with either AII or monensin. $^{86}RbCl$ (New England Nuclear) was added (1×10^6 cpm) and the incubation continued for 10 min. The cultures were washed 6 times with cold, isotonic 0.1 M $MgCl_2$. Intracellular radioactivity and ions were extracted with 0.1 N HNO_3. Radioactivity was measured by Cerenkov radiation in a liquid scintillation counter. Total cell Na^{+} and K^{+} were determined from this same sample by atomic absorption spectrophotometry. Protein was measured by the method of Lowry et al. (18) using BSA as a standard.

^{125}I-Angiotensin binding. Total and non-specific ^{125}I-AII binding (1880 µCi/µg, New England Nuclear) were assayed on cultures of intact smooth muscle cells. The cultures were incubated at room temperature with 0.04 nM ^{125}I-AII in 1 ml of TAB containing 1 mg/ml BSA. Then the cultures were washed 6 times with TAB. The amount of cell bound ^{125}I-AII was measured in a 1.0 N NaOH extract (1 ml) of each culture. Non-specific binding which was estimated in the presence of 10 µM [Ile$_5$]-AII, was approximately 20% of the total.

RESULTS

Angiotensin increases net Na^+ uptake in the presence of ouabain. The effects of angiotensin II and III on net Na^+ uptake in smooth muscle cultured from rat aorta following enzymic-dispersion are shown in Table 1. Both AII and AIII stimulated net Na^+ uptake in the presence of ouabain. Ouabain was present in order to block Na^+ efflux via the Na^+-K^+ pump. In the absence of ouabain, neither AII nor AIII consistently increased total cell Na^+. AIII appeared to be just as effective as AII in stimulating net Na^+ uptake. The combination of saturating concentrations of AII and AIII (0.1 µM) produced no greater stimulation of net Na^+ uptake then either angiotensin by itself. In all cases, there was a concomitant loss in cell K^+ which was similar in magnitude to the gain in cell Na^+.

Angiotensin increases the activity of the Na^+-K^+ pump. Since the activity of the pump is normally limited by the level of Na^+ inside the cell (19,20), the increase in Na^+ permeability produced by angiotensin might be expected to increase Na^+-K^+ pump activity. Figure 2 shows the effects of AII and monensin, a Na^+ ionophore, on the rate of ouabain-sensitive $^{86}Rb^+$ uptake by non-growing muscle cultures. Angiotensin II stimulated pump

Table 1. Effects of AII and AIII on net Na^+ uptake by cultured smooth muscle cells.

Additions	Net Na^+ Uptake	
	(μmol/30 min/mg protein)	
None	0.083 ± 0.010	(1.00)
AII, 0.1 μM	0.121 ± 0.009	(1.47)
0.2 μM	0.115 ± 0.007	(1.39)
AIII, 0.1 μM	0.115 ± 0.012	(1.40)
0.2 μM	0.129 ± 0.009	(1.56)
AII + AIII*	0.129 ± 0.001	(1.56)

*Angiotensin II and III were present at equal concentrations, 0.1 μM. Non-growing cultures of smooth muscle obtained from rat aorta by enzymic dispersion were washed once with TAB, and total cell Na^+ was measured after a 30 min incubation in the presence of 2 mM ouabain. Values are means ± SEM of duplicate determinations.

activity by nearly 2.5-fold, but was much less effective than monensin. Ouabain-insensitive $^{86}Rb^+$ uptake was not affected by either AII or monensin (data not shown). The effects of AII and monensin on $^{86}Rb^+$ uptake were essentially the same whether cell growth was arrested in serum-free Medium 199 or in the original plating medium which contained 10% FBS (Figure 2).

Total cell Na^+ was 0.035 ± 0.004 and 0.271 ± 0 .005 μmol/mg protein in the absence and presence of monensin, respectively. If AII increases ouabain-sensitive $^{86}Rb^+$ uptake by supplying the pump with more of its rate-limiting substrate (Na^+), then AII would not be expected to increase pump activity in the monensin-treated cells. As can be seen in Figure 3, in the

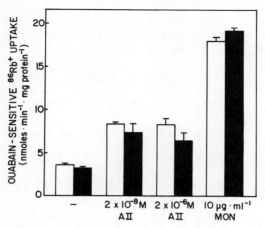

Fig. 2. Effects of AII and monensin on the rate of ouabain-sensitive $^{86}Rb^+$ uptake by non-growing smooth muscle cells cultured from rat aorta. Cultures were placed in serum-free Medium 199 for 7 days (white bars) or left in the original plating medium for 10 days (black bars). Values are means ± SEM for 2 to 10 identically treated cultures.

presence of monensin, AII had no effect on the rate of ouabain-sensitive $^{86}Rb^+$ uptake.

Table 2 shows the effects of AII and monensin on the rate of $^{86}Rb^+$ uptake by cultured cells from pig brain microvessels. Both AII and monensin stimulated pump activity in these cultures, although the stimulation of the Na^+-K^+ pump by AII was not as great as that seen in the rat aortic cultures. Ouabain-insensitive $^{86}Rb^+$ uptake was not increased by either agent (Table 2). Therefore, AII selectively increased the activity of the Na^+-K^+ pump without affecting the passive permeability of the cell to $^{86}Rb^+$. AII increased net Na^+ uptake by approximately 30% in the cultured microvessel cells (data not shown).

Table 2. Effect of angiotensin II on the rate of $^{86}Rb^+$ uptake by cultured microvessel cells from pig brain.

Additions	$^{86}Rb^+$ Uptake (nmol/min/mg protein)	
	Ouabain-sensitive	Ouabain-insensitive
None	3.10 ± 0.33 (1.00)	4.44 ± 0.27 (1.00)
AII, 0.2 µM	4.74 ± 0.25 (1.53)	4.42 ± 0.37 (0.99)
Monensin, 10 µg/ml	22.74 ± 0.31 (7.33)	2.67 ± 0.09 (0.60)

Values are mean ± SEM for 2 to 6 determinations in each of three independent experiments. Microvessel cultures were left for approximately 12 days in 10% FBS in DME/F12.

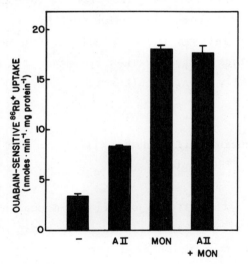

Fig. 3. Effect of AII on the rate of ouabain-sensitive $^{86}Rb^+$ uptake in the presence of monensin. AII and monensin were present as indicated at 0.1 µM and 10 µg/ml, respectively. Aortic muscle cultures were placed in serum-free Medium 199 for 7 days prior to the experiment. Values are means ± SEM for 2 to 6 identically treated cultures.

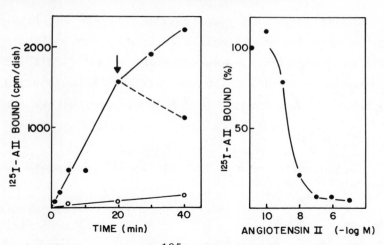

Fig. 4. A. Time course of ^{125}I-AII binding to cultured aortic smooth muscle cells. ^{125}I-AII binding was assayed as described in Methods. Each point represents the mean of duplicate determinations. Closed circles denote total ^{125}I-AII binding; open circles denote binding in the presence of 10 µM unlabelled AII. At the time indicated by the arrow, 10 µM [Ile$_5$]-AII was added to the cultures. The dashed line represents the displacement of ^{125}I-AII from the cells.
B. Inhibition of ^{125}I-AII binding by unlabelled AII. Each point represents the mean of duplicate determinations. Binding was allowed to proceed for 20 minutes at room temperatre.

Angiotensin II binding. Figure 4 shows the binding of ^{125}I-AII to cultures of aortic smooth muscle cells at 22°C. The rate of ^{125}I-AII binding was linear for the first 20 minutes, and then began to level off (Figure 4a). The amount of non-specific binding which occurred in the presence of 10 µM AII was less than 20% of the total binding. The addition of 10 µM unlabelled AII resulted in a slow displacement of bound ^{125}I-AII (dotted line) from the cells.

The ability of AII to compete with ^{125}I-AII for binding sites of the cultured smooth muscle cells is shown in Figure 4b. [Ile$_5$]-AII was very effective in competing with ^{125}I-AII for the binding sites on the cells. The concentration of unlabelled AII which reduced ^{125}-AII binding by 50% was approximately 5 nM.

Table 3. Angiotensin-induced decrease in AII binding to cultured aortic smooth muscle cells.

Additions	^{125}I-AII Binding (% Control)
None	100
AII, 1 µM, 2 hours	87
AII, 1 µM, 4 hours	52
Methylamine, 10 mM, 4 hours	99
AII + methylamine, 4 hours	89

Each value represents the mean of duplicate determinations. Non-growing smooth muscle cells were incubated at 37°C in 1 ml of Medium 199 containing 1 mg/ml BSA and the indicated additions. After 2 or 4 hrs the cells were washed 3 times with TAB, and then incubated for 40 minutes at 37°C in TAB containing 1 mg/ml BSA and ^{125}I-AII. Non-specific binding, which was measured in the presence of 10 µM unlabelled AII, was not significantly affected by any of the additions.

Incubation of the cultures with 1 µm unlabelled AII produced a time-dependent loss of ^{125}I-AII binding (Table 3). At 4 hours, there was a 48% decrease in the amount of ^{125}I-AII bound. The addition of methylamine, a known inhibitor of receptor clustering and internalization (26), during the 4 hour treatment with 1 µM AII, prevented most of the loss of ^{125}I-AII binding.

Incubation of the cells with methylamine alone had no significant effect on ^{125}I-AII binding (Table 3). These data suggest that the AII-receptor complex is slowly internalized in the aortic smooth muscle cultures.

DISCUSSION

Vascular smooth muscle in culture represents a potentially important system to study the mechanism by which certain vasoactive agents influence the transport and steady-state levels of cellular cations. Recently we (15) reported that angiotensin increases Na^+ and K^+ transport in smooth muscle cultured from explants of the tunica media from rat aorta. The increase in K^+ transport appears to be secondary to the effect of angiotensin on Na^+ permeability. We hypothesized that the binding of angiotensin to a specific receptor on the cell surface activates a Na^+ channel which is gated by the hormone. An increase in membrane Na^+ permeability would be expected to stimulate K^+ transport because cell Na^+ is normally rate-limiting for the Na^+-K^+ pump (19,20).

Here we presented similar findings for cultured cells from porcine brain microvessels and smooth muscle cells cultured from rat aorta by enzymic dispersion. In both types of cultured cells, angiotensin increased passive Na^+ entry, as well as Na^+-K^+ pump activity. AII and AIII appear to act via the same receptor in these cultures since the combination of saturating concentrations of AII and AIII produced no greater stimulation of net Na^+ uptake than either peptide by itself. Increasing cell Na^+ augments the contraction of VSM *in vivo* (10,11). In addition, the secondary stimulation of the Na^+-K^+ pump would help to re-establish the Na^+ gradient and possibly limit the contractile response.

The presence of an electrogenic Na^+-K^+ pump in vascular smooth muscle is well-documented (21,22). The following observations suggest that AII stimulates the pump by increasing its supply of Na^+: 1) Monensin, a Na^+ ionophore, stimulated ouabain-sensitive $^{86}Rb^+$ uptake, aortic smooth muscle and brain microvessels; 2) In the presence of monensin, AII had no effect on Na^+-K^+ pump activity; 3) AII did not stimulate pump activity in cells incubated in a Na^+-free medium (unpublished data) or in cells loaded with Na^+ by incubation in a K^+-free medium (15); and 4) AII had no effect on (Na^+, K^+)-ATPase activity in membranes isolated from aortic smooth muscle cultures (unpublished data). Of interest are our observations that angiotensin does not increase total cell Na^+ unless ouabain is present to block Na^+ efflux via the pump. Our previous studies with monensin (15,20) suggest that Na^+-K^+ pump activity is a sensitive indicator of the Na^+ level in the vicinity of the cytoplasmic surface of the plasma membrane. Thus, AII may preferentially increase Na^+ in this region and thereby increase pump activity without producing detectable increases in total cell Na^+. It should be noted that the methods used to arrest cell growth in this study had no apparent effect on the ability of AII and monensin to stimulate the pump.

In the present study, we have shown that the stimulation of Na^+ entry by angiotensin is associated with the binding of AII to a high affinity receptor on the surface of the cultured muscle cells from rat aorta. The specific binding of ^{125}I-AII to the cultured smooth muscle cells was time-dependent and partially reversible. Similar high affinity binding sites for AII have been characterized in rabbit aorta (5,23), guinea pig aorta (24) and rat mesenteric artery (9).

Considerable evidence now exists indicating that several different hormone-receptor complexes are internalized and degraded after the hormone binds to a receptor on the cell surface (25,26). In cultures of rat mesenteric artery, it has recently been shown that treatment of cells with AII results in a time-dependent loss of ^{125}I-AII binding capacity (27). Recovery of the full capacity to bind AII was blocked by cyclohexamide, thus indicating that protein synthesis was required. In the present study, treatment of intact smooth muscle cells with AII resulted in a slow decrease in ^{125}I-AII binding. The transglutaminase inhibitor, methylamine, prevented most of the loss of ^{125}I-AII binding. Methylamine is known to inhibit receptor clustering and the internalization of hormone-receptor complexes (28). These data suggest that the AII-receptor complex may be slowly internalized and subsequently degraded by lysosomal enzymes. We do not yet know how the regulation of Na^+ and K^+ transport is affected by this process in the cultured smooth muscle cells.

The mechanism by which angiotensin increases free, cytoplasmic Ca^{2+} levels and causes contraction is unknown. Our data suggest that the stimulation of Na^+ entry into the smooth muscle cell may be of primary importance in this process. We previously suggested that the angiotensin receptor is coupled to a Na^+ channel (15). Increasing Na^+ entry by the activation of this channel would be expected to depolarize the membrane, thus opening voltage-dependent Ca^{2+} channels (29). Electrophysiological evidence suggests that AII does, in fact, depolarize smooth muscle (30-33), perhaps by a Na^+-dependent mechanism (32,33). An increase in Na^+-permeability may also elevate intracellular Ca^{2+} by decreasing Ca^{2+} exit via a Na^+-Ca^{2+} antiporter (10,11). Alternatively, AII may directly activate a specific calcium channel (29). Angiotensin II has been shown to stimulate the release of calcium from a microsomal preparation of rabbit aorta (34). Recently, AII was also demonstrated to alter Ca^{2+} fluxes in adrenal glomerulosa cells

(35). If the stimulation of Ca^{2+} entry is the primary mode of action of AII, then the increased entry of Na^+ must be closely linked to this event.

Angiotensin is known to stimulate aldosterone biosynthesis by adrenocortical cells (36) in addition to its ability to contract smooth muscle. It has recently been shown that angiotensin also stimulates growth in Swiss 3T3 cells (37) and adrenal glomerulosa cells in culture (38). Both sodium and calcium ions have been implicated in steroidogenesis (36,39) and the control of cell proliferation (20,40,41). Therefore, it appears that angiotensin initiates a series of similar ionic events in different cell types which produce a variety of biological responses. We are currently examining the effects of angiotensin on isotopic sodium and calcium fluxes and growth in cultured aortic smooth muscle cells.

ACKNOWLEDGEMENTS

The authors wish to thank Cindy Smith for preparing the smooth muscle and microvessel cultures, Anne Brock for drawing the figures, and Suellen Walker for typing the manuscript.

REFERENCES

1. Filo, R.S., Bohr, D.F., and Ruegg, J.C. 1974. Glycerinated skeletal and smooth muscle: calcium and magnesium dependence. Science 147:1581-1583.
2. Ruegg, J.G. 1971. Smooth muscle tone. Physiol. Rev. 51:201-248.
3. Johansson, B. 1978. Processes involved in vascular smooth muscle contraction and relaxation. Circ. Res. 43:I14-I20.

4. Devynck, M.A. and Meyer, P. 1976. Angiotensin receptors in vascular tissue. Am. J. Med. 61:758-767.
5. Schultz, G.A., Galardy, R.W., and Jamieson, J.D. 1981. Biological activity of an angiotensin II-ferritin conjugate on rabbit aortic smooth muscle. Biochem. 20:3412-3418.
6. Glossman, H., Baukal, A.J., and Catt, K.J. 1974. Properties of angiotensin II receptors in the bovine and rat adrenal cortex. J. Biol. Chem. 249:825-834.
7. Sraer, J.D., Sraer, J., Ardaillous, R., and Minoune, O. 1974. Evidence for renal glomerular receptors for angiotensin II. Kidney Int. 6:241-246.
8. Bennett, J.P., Jr. and Snyder, S.H. 1976. Angiotensin II binding to mammalian brain membranes. J. Biol. Chem. 251:7423-7430.
9. Gunther, S., Gimbrone, M.A., Jr. and Alexander, R.W. 1980. Identification and characterization of the high affinity vascular angiotensin II receptor in rat mesenteric artery. Circ. Res. 47:278-286.
10. Van Breeman, C., Aaronson, P. and Loutzenhiser, R. 1979. Sodium-calcium interaction in mammalian smooth muscle. Pharmacol. Rev. 30:167-208.
11. Blaustein, M.P. 1977. Sodium ions, calcium ions, blood pressure regulation, and hypertension: a reassessment and a hypothesis. Am. J. Physiol. 232:C165-C173.
12. Friedman, S.M. and Allardyce, D.B. 1962. Sodium and tension in an artery segment. Circ. Res. 11:84-89.
13. Guignard, J.P., and Friedman, S.M. 1971. Vascular ionic effects of angiotensin II in the rat. Proc. Soc. Exp. Biol. Med. 137:157-160.
14. Turker, R.K., Page, I.H., and Khairallah, P.A. 1967. Angiotensin alteration of sodium fluxes in smooth muscle. Arch. Int. Pharmacodyn. Ther. 165:394-404.

15. Brock, T.A., Lewis, L.J., and Smith, J.B. 1982. Angiotensin increases Na^+ entry and Na^+-K^+ pump activity in cultures of smooth muscle from rat aorta. Proc. Natl. Acad. Sci. U.S.A. 79:1438-1442.
16. Brendel, K., Meezan, E. and Carlson, E.C. 1974. Isolated brain microvessels: a purified, metabolically active preparation from bovine cerebral cortex. Science 185:953-955.
17. DeBault, L.E., Kahn, L.E., Frommes, S.P. and Cancilla, P.A. 1979. Cerebral microvessels and derived cells in tissue culture: isolation and preliminary characterization. In Vitro 15:473-487.
18. Lowry, O.H., Rosebrough, N.J., Farr, A.L., and Randall, R.J. 1951. Protein measurement with the folin phenol reagent. J. Biol. Chem. 193:265-275.
19. Thomas, R.C. 1972. Electrogenic sodium pump in nerve and muscle cells. Physiol. Rev. 52:563-594.
20. Smith, J.B. and E. Rozengurt. 1978. Serum stimulates the Na^+-K^+ pump in quiescent fibroblasts by increasing Na^+ entry. Proc. Natl. Acad. Sci. U.S.A. 75:5560-5564.
21. Anderson, D.K. 1976. Cell potential and the sodium-potassium pump in vascular smooth muscle. Fed. Proc. 35:1293-1297.
22. Fleming, W.W. 1980. The electrogenic Na^+,K^+- pump in smooth muscle: physiological and pharmacological significance. Ann. Rev. Pharmacol. Toxicol. 20:129.
23. Lin, S.Y., and Goodfriend, T.S. 1970. Angiotensin receptors. Am. J. Physiol. 218:1319-1328.
24. LeMorvan, P., and Palaic, D. 1975. Characterization of the angiotensin receptor in guinea pig aorta. J. Pharmacol. Exp. Ther. 195:167-175.
25. Kaplan, J. 1981. Polypeptide-binding membrane receptors: analysis and classification. Science 212:14-20.

26. Pastan, I.H. and Willingham, M.C. 1981. Receptor mediated endocytosis of hormones in cultured cells. Ann. Rev. Physiol. 43:239-250.
27. Alexander, R.W., Hyman, S., Atkinson, W. and Gimbrone, M.A., Jr. 1980. Regulation of angiotensin II receptors in cultured vascular smooth muscle cells. Circulation 62:(Suppl. III):90 (abs.).
28. Maxfield, F.R., Willingham, M.C., Davies, P.J.A. and Pastan, I. 1979. Amines inhibit the clustering of α_2-macroglobulin on the fibroblast cell surface. Nature (London) 277:661-663.
29. Bolton, T.B. 1979. Mechanisms of action of transmitters and other substances in smooth muscle. Physiol. Rev. 59:606-718.
30. Ohasi, H., Nonomura, Y. and Ohaga, A. 1976. Effects of angiotensin, bradykinin, and ocytocin on electrical and mechanical activities in the Taenia coli of the guinea pig. Jap. J. Pharmacol. 17:247-257.
31. Somlyo, A.P., and Somlyo, A.V. 1971. Electrophysiological correlates of the inequality of maximal vascular smooth muscle contractions elicited by drugs. In: Proc. Symp. Physiol. Pharmacol. Vascular Neuroeffector systems. eds. Bevan, J.A., Furchgott, R.F., Maxwell, R.A., and Somlyo, A.P. Karger, Basel, pp. 216-228.
32. Hamon, G. and Worcel, M. 1979. Electrophysiological study of the action of angiotensin II on the rat myometrium. Circ. Res. 45:234-243.
33. Zelcer, E. and Sperelakis, N. 1981. Angiotensin induction of active responses in cultured reaggregates of rat aortic smooth muscle cells. Blood Vessels, In press.
34. Baudoin, M., Meyer, P., Fesmandijian, S. and Morgat, J.L. 1972. Calcium release induced by interaction of angiotensin with its receptors in smooth muscle cell microsomes. Nature 235:336-338.

35. Elliot, M.E., and T.L. Goodfriend. 1981. Angiotensin alters $^{45}Ca^{2+}$ fluxes in bovine adrenal glomerulosa cells. Proc. Natl. Acad. Sci. U.S.A. 78:3044-3048.
36. Peach, M.J. 1977. Renin-angiotensin system: biochemistry and mechanisms of action. Physiol. Rev. 57:313-370.
37. Schelling, P., Ganten, D., Speck, G. and Fisher, H. 1979. Effects of angiotensin II and angiotensin II antagonist saralasin on cell growth and renin in 3T3 and SV3T3 cells. J. Gen. Physiol. 98:503-514.
38. Gill, G.N., Ill, C.R., and Simonian, M.H. 1977. Angiotensin stimulation of bovine adrenocortical cell growth. Proc. Natl. Acad. Sci. U.S.A. 74:5569-5573.
39. Schiffrin, E.L., Lis, M., Gutkowska, J. and Genest, J. 1981. Role of Ca^{2+} in response of adrenal glomerulosa cells to angiotensin II, ACTH, K^+, and ouabain. Am. J. Physiol. 241:E42-E46.
40. Leffert, L.H., Editor. 1980. Growth regulation by ion fluxes. Ann. N.Y. Aca. Sci. 339;1-335.
41. Smith, J.B. and Rozengurt, E. 1978: Lithium transport by fibroblastic mouse cells: characterization and stimulation by serum and purified growth factors in quiescent cultures. J. Cell. Physiol. 97:441-449.
42. Smith, J.B. 1980. The Na^+ transporter in normal and transformed 3T3 cells: Kinetics and modulation by Ca ions. Dev. Biochem. 14:101-102.

ANTIGENIC EXPRESSION OF HUMAN MELANOMA CELLS IN SERUM-FREE MEDIUM

Thomas F. Bumol, John R. Harper, Darwin O. Chee and
Ralph A. Reisfeld

Scripps Clinic and Research Foundation
La Jolla, CA 92037

ABSTRACT

A human melanoma cell line, M14, adapted to grow in serum free synthetic media was examined for its expression and secretion of several serologically defined melanoma associated antigens (MAA) previously described in this laboratory. Melanoma associated antigen expression and secretion was identical to that of M14 cells grown in parallel in serum supplemented medium. Spent synthetic media was found to be an enriched serum free source for the initial isolation of 100 kilodalton secreted glycoprotein MAA. M14 melanoma cells grown in synthetic media were also shown to be adaptable to the double agar clonogenic assay facilitating the examination of clonal heterogeneity in functional studies of MAA in melanoma tumor biology.

This work was supported by grants CA28420 from the National Institutes of Health and ACS IM-218 from the American Cancer Society. This is publication No. 2576 of Scripps Clinic and Research Foundation.

Abbreviations: MAA - melanoma associated antigens; SDS-PAGE - sodium dodecylsulfate polyacrylamide gel electrophoresis

Recent investigations from this laboratory have focused on characterizing human melanoma associated antigens (MAA) found either as secreted or cell surface associated glycoproteins in human melanoma cell lines.[1,2,3] In these studies, monoclonal and polyclonal antiserums to melanoma cell components have been developed to specifically identify these MAAs immunochemically and provide a means to study the structural biochemistry of these determinants.[1,2,3] At this time we have identified two antigens on which our research efforts are targeted: 1) a 100,000 dalton secreted glycoprotein (100K) common to melanoma, sarcoma and neuroblastoma tumor cell lines, and 2) a 250,000 dalton-high molecular weight component glycoprotein-proteoglycan complex which is thus far restricted to melanoma cells.

The ultimate goal of our efforts is two-fold. Initially, we hope to develop schemes to isolate these melanoma associated antigens in sufficient quantities to obtain detailed structural information on these molecules, and secondly, we wish to implicate these glycoproteins in functional aspects of the biology of metastatic human melanoma in vitro. In both of these regards we have been characterizing a human melanoma cell line, M14, which has been adapted to grow in chemically defined serum free synthetic medium, for its expression and secretion of several serologically defined MAAs and for its ability for anchorage independent growth in soft agar. We will demonstrate the feasibility and advantages of this system for the study of human melanoma antigens and the biology of malignant melanoma.

MATERIALS AND METHODS

Cells and Culture

The M14 melanoma cell line was originally derived from a metastatic lesion and adapted to grow in 10% fetal calf serum supplemented RPMI 1640 media by Morton and colleagues at UCLA. The M14 cell line was adapted to grow in serum free CDM media by Chee et al[4] with 2 mM L-glutamine and 50 µg/ml gentamycin sulfate. This same cell line was then adapted to grow in this laboratory in Iscoves modified D-MEM serum free media formulated by Iscove and Melchers[5] and commercially available from Centaurus, Inc. (Irvine, CA), as Synmed-I. The formulation of Synmed-I is listed in Table 1. The L14 cell line is an autologous lymphoblastoid cell line derived from the same patient as the M14 melanoma cell line.

Antisera

The 9.2.27 monoclonal antibody was developed against a 4M urea extract of cultured melanoma cells (M21) and recognizes a melanoma cell glycoprotein-proteoglycan complex consisting of a 250,000 dalton N-linked glycoprotein and a high molecular weight component (HMW-C) greater than M_r 500,000.[2] 250/165, obtained from Drs. Ferrone and Imai, Research Institute of Scripps Clinic and F11 and E4, originally developed by Chee (personal communication), are monoclonal antibodies which recognize a 100,000 d glycoprotein secreted from melanoma cells. Polyclonal antisera 7501 R5, 1450 R4 and 3497 were raised in this laboratory against whole human melanoma cells, purified human plasma fibronectin, and immunoadsorbents containing the 9.2.27 monoclonal antibody defined antigen, respectively.

Table 1. Composition of Synmed-I*

Inorganic Salts	mg/l
$CaCl_2$	165
$FeCl_3 \cdot 6H_2O$	0.002
KCl	330
KNO_3	0.076
$MgSO_4$	97.67
NaCl	4505
$NaHCO_3$	3024
$NaH_2PO_4 \cdot H_2O$	125
$Na_2SeO_3 \cdot 5H_2O$	0.0173

Amino Acids	
L-alanine	25
L-asparagine·H_2O	28.4
L-arginine·HCl	84
L-aspartic acid	30
L-cystine·2HCl	91.24
L-glutamic acid	75
Glycine	30
L-histidine·HCl·H_2O	42
L-isoleucine	105
L-leucine	105
L-lysine·HCl	146
L-methionine	30
L-phenylalanine	66
L-proline	40
L-serine	42
L-threonine	95
L-tryptophan	16
L-tyrosine·(Na_2)	104.2
L-valine	94

Vitamins	mg/l
Biotin	0.013
D-calcium pentophenate	4.0
Choline	4.0
Folic acid	4.0
i-inositol	7.2
Nicotinamide	4.0
Pyridoxal·HCl	4.0
Riboflavin	0.4
Thiamine·HCl	4.0
Vitamin B_{12}	0.013

Additional Components	
Transferrin	1.0
Bovine serum albumin	400
Soybean lipid	50-100
Cholesterol	12-25
D-glucose	4500
Phenol red	15
HEPES	5958
Sodium pyruvate	110

* Centaurus Biological Corporation, Anaheim, CA.

Radioimmunometric Binding Analysis

Cell surface antigens and their secretion into spent tissue culture media were assayed utilizing a ^{125}I-protein radioimmunometric binding assay developed in this laboratory for solid phase target antigens.[6] The antigenic preparation used for solid phase radioimmunometric binding analysis of spent media antigens was a 30 X concentrate of spent Synmed media obtained by concentration on an Amicon Hollowfiber Dialyzer Concentrator DC2A apparatus (Amicon, Lexington, MA).

Analysis of Indirect Immunoprecipitation by SDS-PAGE

M14 cells growing in Synmed-I were biosynthetically labelled with 2 mCi ^{3}H-leucine (42 Ci/mmole, New England Nuclear) for 24-48 hours at 37°C. Spent tissue culture media was obtained by centrifugation, dialyzed and then utilized for indirect immunoprecipitation analysis. Cell pellets were extracted in RIPA lysis buffer[7] and these detergent extracts used for indirect immunoprecipitation analysis on a protein A-Sepharose immunoadsorbent as previously described.[1,2] SDS-PAGE on immunoprecipitated antigens was followed by fluorography according to previously published procedures.[1,2]

Chromatofocusing Procedure

Chromatofocusing of MAA in concentrated spent Synmed media was carried out on PBE resin (Pharmacia) with an eluting pH gradient ranging from pH 5.0 to pH 8.0 in Tris-acetate buffer. All procedures for elution of components at their isoelectric point were according to the manufacturer's procedure (Pharmacia, Inc.).

Double-Agar Clonogenic Assay

Cultured human malignant melanoma cells (M14) grown in serum-free chemically-defined medium (Synmed-I) were harvested from tissue culture flasks by vigorous agitation and dispersed into a single cell suspension by repeated pipetting. These cells were then plated in a double-agar clonogenic assay slightly modified from that previously described by Salmon and colleagues (8,9). Briefly, underlayers were prepared using Synmed-I supplemented with 2 mM L-glutamine, 50 mg/l gentamycin sulfate and 0.6% agar (Difco). One-half milliliter of underlayer was set in each well of a 24- well tissue culture plate (Costar). The single cell suspensions were finally resuspended in Synmed (glutamine-gentamycin), brought to 0.3% agar (5×10^4 cells/ml) and 2.5×10^4 cells were plated over each precast underlayer. Cultures were maintained at 30°C in a 5% CO_2 atmosphere with 100% humidity until colony enumeration and morphological examination on days 5-10.

Results

The initial experiments were designed to compare the expression of MAAs on M14 cells grown in serum containing medium to that of M14 cells grown in synthetic media. Cells were harvested in log growth phase by EDTA extraction and exceeded 90% viability by trypan blue exclusion in both cases. The L14 lymphoblastoid cell line served as a specificity control and the results of a typical cell surface radioimmunometric binding assay with several monoclonal and polyclonal antiserums is summarized in Figure 1. The M14-Synmed cells demonstrated equivalent or slightly elevated cell surface expression of the antigenic determinants recognized by the 9.2.27 and F11 monoclonal antibodies and the 3497 R1 polyclonal antiserum when analyzed in parallel with the M14 cells grown in FCS. In contrast to the binding of antibodies to melanoma cells, the parallel experiment on the autologous L14 lymphoblastoid cell line showed no significant cell surface binding of these monoclonal antibodies recognizing MAA (Figure 1). Thus, M14 melanoma cells grown in synthetic media for extended periods of time (in this case greater than two years) have equivalent antigenic expression as their counterpart cell line maintained in serum supplemented media.

We then examined the spent tissue culture fluid of the M14-Synmed cell line for a variety of antigens which are secreted and/or shed from human melanoma cells.[1,2,3] The results of a

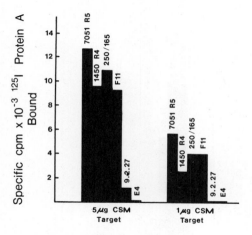

Figure 1

Comparison of the cell surface expression of serologically defined MAA on the M14 cell line grown in parallel under serum free conditions (Synmed) and in serum supplemented RPMI 1640 medium (fetal calf serum, FCS). Hybridoma spent media (SM), isolated IgG fractions from hybridoma ascites and straight polyclonal antisera were assayed for binding to the cell surfaces of the M14 melanoma cell line and a control autologous lymphoblastoid cell line, L14, in a ^{125}I-protein A-Sepharose radioimmunometric binding assay previously described.[7] The results depicted represent the average of duplicate determinations on 2×10^5 viable target cells with background subtraction of irrelevant antisera controls, including normal rabbit sera and spent media from the secreting myeloma cell line used for hybridoma production.

typical radioimmunometric binding analysis using a 30 X concentrate of spent synthetic media as a target is shown in Figure 2. Both polyclonal antiserums (7051R5, 1450R4) and monoclonal 250/165 and F11 antibodies bind significantly in this experiment; however, monoclonal antibodies directed mainly against cell surface determinants (9.2.27 and E4) bind poorly. The predominant glycoprotein recognized by the polyclonal antisera 1450R4 is the extracellular matrix protein fibronectin, demonstrating that melanoma cells grown in synthetic media produce this glycoprotein. Since monoclonal antibodies 250/165 and F11 both recognize a 100K glycoprotein in spent media of biosynthetically labelled melanoma cell lines grown in serum supplemented media, we examined the molecular

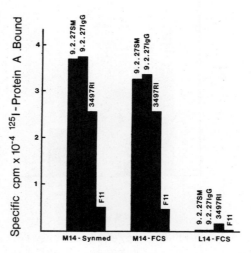

Figure 2

Radioimmunometric binding analysis of MAA in spent synthetic media from the M14 melanoma cell line. A 30 X concentrate of spent synthetic media (CSM) was applied to flexible 96-well plates at 5 µg and 1 µg protein/well and allowed to dry at 40°C. Polyclonal and monoclonal antisera at dilutions ranging from 1:10 to 1:20 were assayed in ^{125}I-protein A radioimmunometric binding assay at these two target protein concentrations. The results are the average of duplicate determinations with background subtraction of irrelevant control antisera as described in Figure 1.

profile of these antigens in synthetic media. The results of an experiment determining immunoprecipitated ^{3}H-leucine labelled antigens by SDS-PAGE and fluorography is shown in Figure 3. The molecular profile of the 9.2.27 monoclonal antibody defined antigens in detergent lysates of M14-Synmed cells was found identical to that previously reported on cells grown in media with serum supplements.[2] In addition, the secreted 100K antigen found previously in melanoma lines grown in serum containing media was found in immunoprecipitates obtained with both monoclonal (F11) and polyclonal (7051) antisera from ^{3}H-leucine labelled synthetic spent media (Figure 3). Thus, the cell surface expression, secretion and molecular profiles of MAA in synthetic media grown M14

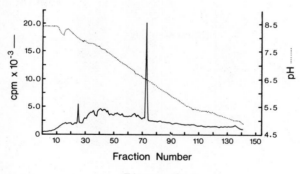

Figure 3

Indirect immunoprecipitation analysis of serologically defined MAA. ^{3}H-leucine labeled M14-Synmed cells, detergent extracts and spent media served as a source of biosynthetically labeled antigens on monoclonal and polyclonal antisera immunoadsorbents. The immunoprecipitated antigens were eluted in SDS-PAGE sample buffer,[1,2] analyzed on 6% acrylamide gels and subjected to fluorography. Molecular weight markers include myosin (200 Kd), phosphorylase B (92.5 Kd) and ovalbumin (43 Kd). DF designates the dye front and this fluorograph represents a 5 day exposure.

melanoma cells were identical to those previously described for cell lines grown in serum supplemented media.[1,2,3]

The M14 cells propagated in synthetic media consequently provide an excellent model system to study MAA expression on the cell surface and secreted into spent culture media. In addition, the low protein content of Synmed-I (bovine serum albumin is the major protein species at 400 mg liter) facilitates the design of effective protocols for isolation of secreted MAA in the absence of massive amounts of contaminating serum proteins found in serum supplemented spent

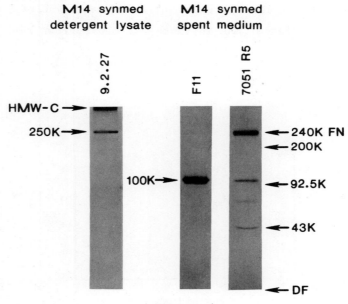

Figure 4

Chromatofocusing analysis of the 100 kilodalton melanoma associated antigen defined by F11 monoclonal antibody. A 2 ml fraction of spent concentrated synthetic medium was applied to a 15 cm x 1 cm column of PBE 94 (Pharmacia, Inc.) on 0.025 M Tris-acetate buffer, pH 8.3. The eluent buffer which generated the self-forming pH gradient (----) on this column is a pH 5.0 polybuffer 96 (30%) + polybuffer 74 (70%) (Pharmacia, Inc.). Antigenic activity determined by the F11 antibody was monitored by solid phase radioimmunometric binding analysis of column fractions and is represented by the specific ^{125}I-protein A cpm bound per column fraction.

media. At this time we have developed initial procedures to purify the 100K antigen from a 30 X concentrated spent synthetic media preparation by taking advantage of the apparent isoelectric point (pH 6.3) of this protein as determined by two-dimensional isoelectrofocusing-SDS-PAGE analysis.[10] An example of an analysis with a chromatofocusing column that elutes proteins according to their pI from complex mixtures is shown in Figure 4. By combining this fractionation technique with our solid phase radioimmunometric binding assay for the 100K antigen we were able to determine that the F11 defined antigens elute at a pH of 6.4, agreeing closely with our previous data of an apparent pI of 6.3.[10] We currently are scaling up this procedure to isolate the 100K antigen in a combination with CM-cellulose fractionation and high performance liquid chromatography molecular sieving techniques.

In addition to being an excellent model system for the immunochemical study of human MAA, the chemically defined synthetic media also provides an *in vitro* test system to analyze several biological properties of metastatic melanoma under controlled culture conditions. In this regard we successfully established the double agar clonogenic assay developed by Salmon and colleagues (8,9) to study anchorage independent growth of M14 melanoma cells in synthetic media (J.R. Harper, T.F. Bumol and R.A. Reisfeld, in preparation). An example of the results obtained with this method is shown in Figure 5. Specifically, single cell suspension of M14 cells can establish colonies in soft agar-Synmed-I combinations, permitting the selection of colony morphology and size of these cells from those simultaneously established in soft agar. Further studies examining the role of our serologically defined MAA in this system, including adhesion and growth in the presence of monoclonal antibodies, are currently in progress. In addition, these novel conditions established for this melanoma cell line are being applied to adapt fresh surgical melanoma specimens in the double-agar clonogenic assay to study tumor cell heterogeneity.

DISCUSSION

The data presented in this report indicate that the M14 human melanoma line grown in synthetic media expresses and secretes several serologically defined melanoma associated antigens that are identical to those previously described on melanoma cell lines grown in serum supplemented media. Our data also demonstrate that we established that this tissue culture system, free of serum, is suitable to study the immunochemical and biochemical nature of several monoclonal antibody defined antigens and to isolate these components.

Our initial attempts to isolate the 100K glycoprotein MAA defined by the F11 monoclonal antibody demonstrate that concentrated spent synthetic media can serve as an excellent source for this antigen. A combination of procedures, i.e., chromatofocusing of concentrated spent synthetic media and use of a radioimmunometric binding assay utilizing monoclonal antibody F11, demonstrates the elution of the 100K antigen at its isoelectric point of pH 6.3 (Figure 4). Similar studies on a large scale involving this technique and others are currently ongoing to provide sufficient material for amino acid sequence analysis of this interesting secreted melanoma associated glycoprotein.

In addition to the advantages provided by a serum free tissue culture system for the characterization of melanoma associated antigens, we have used the double agar clonogenic assay to demonstrate that the M14 melanoma cell line can be adapted to anchorage independent growth in soft agar (Figure 5). The critical

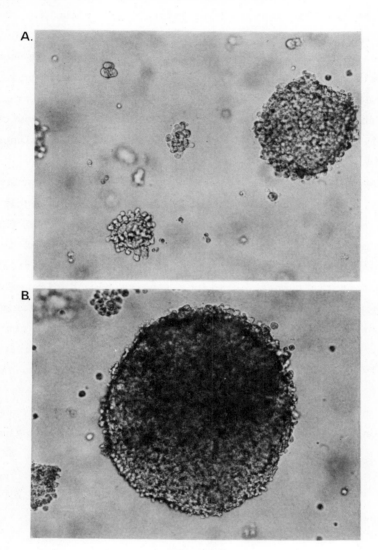

Figure 5

Double agar clongenic assay of M14 melanoma cells grown in synthetic media. Figure 5A depicts a field of cloned M14 melanoma cells demonstrating heterogeneity in colony size and morphology after 6 days of growth (100X magnification). Figure 5B represents a large colony of M14 melanoma cells in a tight colony formation which is growing into the Synmed-agar combination (100X magnification). At this stage the colony can be easily manipulated and transferred for further studies on this clonal population of M14 human melanoma cells.

requirements for this important in vitro property can now be examined by manipulation of this chemically defined system, including studies on tumor cell line heterogeneity in antigenic expression as well as growth and in vitro correlates to metastatic potential. This information should provide the technology required to apply this assay to fresh surgically explanted melanoma cells to directly study the cell biology of metastatic human melanoma.

The overall importance of using synthetic media to study all cell lines, and in particular human tumor lines, has recently been discussed by Barnes and Sato.[11] It is their contention that synthetic media provide the optimum tissue culture system for a critical study of disorders of proliferation and differentiation such as metastatic tumors. Previous reports have indicated that certain human tumors grow significantly better in synthetic media than in conventional, serum supplemented media and that certain unique differentiation characteristics were specifically expressed exclusively under culture conditions in synthetic, serum free medium.[12,13] Thus, the cell culture system described here for human melanoma cells may well prove to be optimal for a careful exploration of the functional roles of MAAs in melanoma tumor biology.

Acknowledgements

The technical assistance of Mr. John Brock, Ms. Vicky McCarthy and Ms. Kathleen Rocker is gratefully acknowledged. We also thank Ms. Dee Davidson for typing of the manuscript.

T.F. Bumol is a recipient of fellowship DRG-HHH-F-2 from the Damon Runyon-Walter Winchell Cancer Fund.

REFERENCES

1. D.R. Galloway, R.P. McCabe, M.A. Pellegrino, S. Ferrone and R.A. Reisfeld, Human melanoma associated antigens: immunochemical characterization with xenoantisera, J. Immunol. 126:62-66, 1981.
2. A.C. Morgan, D.R. Galloway and R.A. Reisfeld, Production and characterization of a monoclonal antibody to a melanoma associated glycoprotein, Hybridoma, in press, 1981.
3. A.C. Morgan, D.R. Galloway, F.C. Jensen, B.C. Giovanella and R.A. Reisfeld, Human melanoma associated antigens: presence on cultured normal fetal melanocytes, Proc. Natl. Acad. Sci. USA 78:3834-3838, 1981.
4. D.O. Chee, A.W. Boddie, J.A. Roth, E.C. Holmes and D. Martin, Production of melanoma associated antigens by a defined malignant melanoma cell strain grown in chemically defined medium, Cancer Res. 36:1503-1509, 1976.
5. N.N. Iscove and F. Melchers, Complete replacement of serum by albumin, transferrin and soybean lipid in cultures of lipopolysaccharide-reactive B lymphocytes, J. Exp. Med. 147:923-933, 1978.
6. A.C. Morgan, D.R. Galloway, B.S. Wilson and R.A. Reisfeld, Human melanoma associated antigens: a solid phase assay for detection of specific antibody, J. Immunol. Meth. 39:233-246, 1980.
7. Z. Gilead, Y. Jeng, S.W.M. Wold, K. Sugawara, H.M. Rho, M.L. Harter and M. Green, Immunological identification of two adenovirus 2 induced early proteins possible involved in cell transformation, Nature 264:263-266, 1976.
8. A. Hamburger and S.E. Salmon, Primary bioassay of human malignant stem cells, Science 197:461-463, 1977.

9. F.L. Meyskens and S.E. Salmon, Inhibition of human melanoma colony formation by retinoids, Cancer Res. 39:4055-4057, 1979.
10. R.A. Reisfeld, D.R. Galloway, A.C. Morgan, L.E. Walker and T.F. Bumol, Human melanoma associated antigens: an immunochemical and molecular profile, Cancer Bull. 33:211-218, 1982.
11. D. Barnes and G. Sato, Serum-free cell culture: a unifying approach, Cell 22:649-655, 1980.
12. D. Barnes, J. van der Bosch, H. Masui, K. Miyazaki and G. Sato, The culture of human tumor cells in serum-free medium. In: "Interferons: A Volume of Methods in Enzymology." (S. Pestka, ed.) Academic Press, New York, in press.
13. H. Murakami and M. Masui, Hormonal control of human colon carcinoma growth in serum-free medium, Proc. Natl. Acad. Sci. USA 77:3464-3468, 1980.

USE OF CELL CULTURE TO IDENTIFY HUMAN PRECANCER

Henry C. Lyko and James X. Hartmann

Department of Biological Sciences
Florida Atlantic University
Boca Raton, Florida 33431

ABSTRACT

In the United States, colon cancer is the most common form of internal cancer in both sexes. Prevention of the disease depends on early diagnosis of polyps or pre-cancerous lesions. The response of normal human colon fibroblasts (CRL1459) was used to identify individuals with clinical pre-cancer. Their plasma induced transformation associated morphology characterized by the retraction of cellular processes, cell rounding and eventual detachment from the vessel surface. Those plasma samples which induced a transformation associated morphology contained significantly increased levels of protease as shown by casein hydrolysis (Bio-Rad, CA). We are using hyperproteinasemia as a biomarker to identify individuals

---

Abbreviations: ACR - adenomatosis of the colon and rectum; FPC - familial polyposis coli; GS - Gardner syndrome; VFPC - variant familial polyposis coli; TM - transformed morphology; PPA - Plasma protease activity; CEA - carcinoembryonic antigen; MNNG - N'methyl N' Nitroso N' Nitrosoguanidine; TPA - 12-0-tetradecanoyl phorbal--13 acetate; PA - plasminogen activator; LETS - Large External Transformation Sensitive protein; C-AMP - 3'5' cyclic adenosine monophosphate; C-GMP - 3'5' cyclic guanosine monophosphate.

with polyps who have hereditary adenomatosis of the colon and rectum (ACR). We are currently evaluating cell cultures versus biochemical assays as a means for early detection of precancerous tumors in the general population. The findings of a tumor associated protease in clinical precancer, and its effect on cell cultures support our proposal that protease activity promotes tumor progression in ACR and may represent the gene defect in this hereditary disease.

INTRODUCTION

One of twenty Americans will develop colorectal cancer resulting in 120,000 new cases in 1981 with 54,900 deaths.[1] The development of this malignancy is believed to be preceeded in many instances by the benign neoplasm, the adenomatous polyp.[2] We are interested in the human hereditary disorder, adenomatosis of the colon and rectum (ACR), clinically known as familial polyposis coli and Gardner syndrome (GS). ACR is an autosomal dominant disorder that invariably progresses to cancer by midlife. The phenotype is expressed in stages. ACR Gene carriers are born with a mucosa void of neoplasia. This mucosa is progressively replaced by hundreds of adenomas until malignant transformation of one or more polyps is the ultimate and invariable fate of affected individuals. Polyposis precedes carcinoma and develops in the second to third decade of life. Two thirds of the proband group have one or more primary carcinomas of the bowel by 42.5 years of age.[2] Hence, this disease predisposes to cancer twenty years earlier than the general population counterpart to this malignancy. ACR is widely accepted as the human model for studying the development of bowel cancer.[3]

A gene determined propensity to adenomatosis and cancer could result from an abnormal population of intestinal cells, or could arise in a population of intestinal cells which are exposed to an abnormal metabolite such as a bile acid. It was suspected that an abnormal metabolite induced transfomration in ACR, notably because of the wide variety of extracolonic tumors in diverse anatomical locations associated with the Gardner syndrome.[4] In addition to multiple polyposis of the large bowel, GS has extracolonic manifestations including polyposis in the gastric and periampullary region, periampullary carcinoma, epidermoid cysts of the skin, osteomatosis of the jaw, face and large bones, desmoids of the messentary and extramessentary, and other fibromatous tumors. We tested the hypothesis that an abnormal metabolite predisposed to ACR by determining if ACR plasma or serum could alter the morphology of normal cells in culture.

RESULTS

Plasma samples were obtained from ACR individuals with neoplasia, progeny at risk for inheriting the ACR gene, ACR individuals who have undergone total colectomy, proctectomy and have an ileostomy, and apparently healthy controls.[a,b] Plasma from nine of twelve individuals (75%) affected with ACR who had benign neoplasms,

---

[a] Whole blood was drawn on EDTA, centrifuged at 2000 g, removed by pipetting and frozen at -70°C in 1 ml aliquots. Cells (a normal human colon cell culture CRL1459 from the American Type Culture Collection) were subcultivated every 3-4 days by routine trypsin procedure. For experiments, cells ($3 \times 10^4/0.5$ml) were plated into 16mm sterile wells (Corning # 35425). Plasma was filter sterilized using 0.2μm filters, and added to minimum essential medium (MEM-C) containing 10% fetal bovine serum, Hepes buffer and antibiotics at 1:6 ratio. Wells seeded 24 hours or 7 days previously were used. Human control plasma (from blood types A, B, and O) was obtained locally and from Baltimore. All experiments were assayed in triplicate and repeated three times.

transformed the normal morphology of normal human colon fibroblasts in culture.[a] The transformed morphology (TM) included retraction of cell processes, a spherical morphology, detachment of cells from the surface of the culture flask, and homologous cell agglutination (see Figure 1). These observations were also seen when normal human dermal fibroblasts were used (unpublished results) and therefore, are not specific to the cell line reported here. The effect of ACR plasma on cells was (in nine instances) comparable to the effects wrought by 5% trypsin. Plasma from three ACR individuals, progeny at risk, and controls did not have a demonstrable effect on cells in culture.

The morphological alterations were reversible when cells were replated in control plasma alone or in control plasma/culture medium or in culture medium alone.[a] ACR serum samples were all negative for having a demonstrable effect on cells in culture.

We have been engaged in a blind study[b] to determine if plasma protease activity (PPA) is a biomarker of neoplasia in ACR. Plasma from nine of twelve ACR individuals with polyps had an average 100 fold increase ($p>0.001$) in protease activity compared to controls.[c] The enzyme activity and clinical histories are presented in Table One. PPA was detected in a modified casein agar plate assay using a Protease Detection Kit (Biorad, Richmond, CA).[c] The nine plasma

---

[b] The 75 subjects included 12 ACR individuals with benigh neoplasms; 40 progeny at risk for inheriting the ACR gene; 22 control subjects without a family history of ACR, 3 ACR individuals who have undergone total coloprotectomy. Plasma samples were obtained from Anne J. Krush, M.S., Assistant Professor of Medicine, Johns Hopkins Hospital (JHH). Genetic analysis an dclinical histories were provided by Dr. Edmond A. Murphy, M.D., Professor of Medicine, JHH. Clinical evaluation (colonoscopy and biopsies) were conducted by the Division of Gastroenterology, Department of Medicine, under the supervision of Dr. Thomas Hendrix, M.D., JHH. ACR patients are part of the ongoing Polyposis Study, JHH Baltimore, MD.

CELL CULTURE TO IDENTIFY HUMAN PRECANCER 475

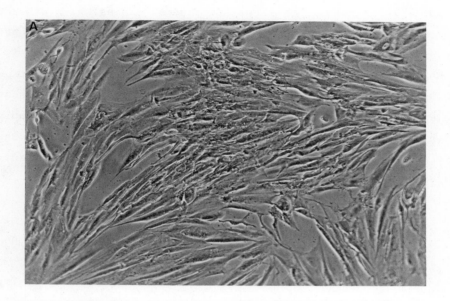

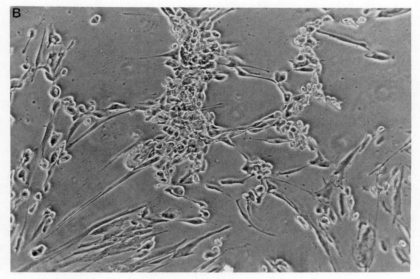

Figure 1. Photomicrograph of normal human colon cells treated with plasma from patients with familial polyposis.

A. Monolayer prior to the addition of plasma. B. Retraction of cell processes, rounding, and agglutination of rounded cells at 12 min. of treatment.

samples with increased PPA were the same samples which caused a TM of cells in culture. The three ACR plasma samples which did not cause a TM also were negative for increased PPA compared to controls. One individual at risk for ACR had increased PPA although this person is clinically negative for bowel neoplasia. This female has benign neoplasms of the breast (see Table One). The highest PPA was found in an individual who had benign adenomas and a polyp which histologically approached carcinoma in situ. This was the only documented possible cancer in this series. This individual had variant FPC as seen in Table One.

The age range for the ACR group with increased PPA was 15 to 50. The age range for the ACR group negative for increased PPA was 14 to 30. FPC, variant FPC and GS phenotypes were identified by increased PPA and TM, whereas GS phenotypes were negative for TM or PPA in three cases (Table One).

All serum samples were negative for protease activity in the casein agar plate assay. This was observed for serum samples obtained from donors that showed increased protease activity in plasma.

---

[c] Samples from Baltimore were air-shipped on dry ice and arrived frozen. Blood from all participants was obtained prior to the administration of any drugs and prior to colonoscopy or sigmoidoscopy. Caseinolytic activity of whole plasma was assayed using a protease detection kit (Biorad, Richmond, CA) with the following modifications: Ten ml of blood was obtained via venipuncture in vacutainer tubes containing EDTA and centrifuged at 2000g; plasma was removed by pipetting, coded and frozen at -70°C in 1 ml aliquots. Following thawing of samples, 50 microliters of plasma was placed into 9mm wells in casein agar plates made with 2 casein-agar tablets per 10ml of deionized water. All positive samples assayed gave maximum lysis zones at 2 hours. A dilution series of standard plasmin (fibrinolysin which dissolves casein) was used in the casein agar plate assay to quantitate the enzyme activity in ACR plasma. A student t test was used for data analysis.

TABLE ONE

A. ACR Group with Neoplasia

AGE	DISEASE	PROTEASE LEVEL (Std. plasmin CTA units/ml$\times 10^{-1}$)	CLINICAL HISTORY OF POLYPOSIS
39	GS	.4(I.O.C.)*	tubular adenomas
49	V.F.R.C.	80(I.O.C.)	tubular adenomas and villous adenoma that approached carcinoma in situ.
15	FPC	.4(I.O.C.)	tubular adenomas
40	FPC	70(I.O.C.)	aggressive adenoma formation
50	GS	.4(I.O.C.)	hyperplastic polyp, previous tubular adenomas
30	FPC	.4(I.O.C.)	tubular adenomas
U.K.	GS	.6(I.O.C.)	tubular adenomas
14	GS	0.02(W.C.R.)**	tubular adenomas
30	GS	0.00(W.C.R.)	tubular adenomas
25	GS	0.02(W.C.R.)	tubular adenomas

B. Progeny at Risk Clinically Negative

50	--	.8(I.O.C.)	This female has benign neoplasms of the breast. One of her progeny has GS. Her mother died from colon cancer

The remaining progeny at risk, 3 ACR individuals with total coloproctomy, and the controls had protease activity within control range (0.0 to 0.1 std. plasmin CTA units/ml$\times 10^{-1}$).
* increased over controls significantly, p>.001.
** within control range

DISCUSSION

A majority (75%) of the ACR individuals studied were identified by increased PPA and exhibited plasma that caused a marked effect on cells in culture. Cancer embryonic antigen (CEA) is useful in identifying only 20% of the ACR group with benign neoplasms.[5] CEA is not a biomarker which can be used to identify people with early cancer (hence curable cancer) in the general population.[6] Assay of occult blood in stool will identify 3 to 5% of the people with colorectal cancer.[7] The assay methods reported in this communication suggest that a majority of ACR individuals with clinical precancer can be identified by increased PPA or cell culture. If so, then these methods may be adaptable to mass screening an asymptomatic population who have premalignant neoplasms of the bowel. One of the ACR individual identified was only 15 years of age. The average age for the onset of cancer in ACR is 42.5 years.[2] Autosomal dominant disorders are notorious for variability in expression from family to family and even case to case. Therefore, this young man may have been a candidate for early cancer. If not, increased PPA may have been useful as a biomarker of neoplasia, 30 years before cancer could be expected to develop. This point suggests another interesting possibility. Because increased PPA may exist for many years in ACR (suggested by the age range of ACR individuals identified, 15-50), it is possible that plasma protease promotes the progression to cancer in this disease?

There is evidence that protease is causally related to malignant transformation. Therefore, a cellular model for neoplastic development in ACR is presented.

Cellular Model for Development of Neoplasia in ACR

Step 1. Initiation

Transformation is believed to result from a multi-stage process. In ACR, germinal mutation is presumed to initiate this process. In our model (see Figure 2 for the remainder of the discussion) step one initiation is represented by a gene defect in DNA. This is compatible with germinal mutation or with the integration of virus into the genome. Evidence in support of the gene defect resulting in an initiated state comes from two experiments using ACR subepidermal fibroblasts.[8,9] In the first report cells were exposed to an initiating chemical carcinogen (N-methyl-N'-nitro'N'nitrosoguanidine, or MNNG). Cells exposed to MNNG expressed morphological alterations, growth to high saturation

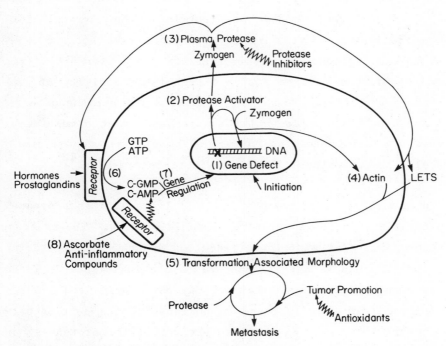

Figure 2. Cellular Model of Neoplastic Development in ACR.

density, formation of large cell aggregates above an agar base, and formation of colonies in semisolid medium. However, the cells did not form tumors in athymic mice, a final step in complete transformation.[10] When ACR cells were exposed to a tumor promoter (12-O-tetradecanoyl phorbal-13 acetate, or TPA) alone, the cells were transformed and grew in athymic mice. The conclusion is that

ACR cells are in an initiated state, requiring tumor promotion to induce transformation.[9]

Step 2. Intracellular Protease Activator

It has been reported that ACR subepidermal cells produce increased (2 to 4 fold) amounts of plasminogen activator in cell culture.[11] This activator of the proenzyme zymogen has been correlated with chemical, viral and irradiation induced transformation.[10] In our model we have not restricted the intracellular protease activator to plasminogen activator (PA), because ACR plasma is caseinolytic but not fibrinolytic in a standard fibrin plate (unpublished results). The fibrinolytic activity of PA reported in the literature may in fact represent a class of unrelated enzymes. It has recently been reported that a membrane-enriched fraction containing PA (49.000 dalton) may contribute to extracellular proteolysis, important in transformation.[12]

In our model we propose a role for intracellular protease activator (IPA) in the alteration of nucleic acids. The protease inhibitor (antipain) inhibits irradiation induced transformation after (only) a single day of exposure to the inhibitor.[13] These authors suggested that the inhibitor may have influenced DNA repair enzymes. If this is found to be true, then a detrimental influence of intracellular protease activators (or proteases) on DNA repair enzymes may be considered. Increased tetraploidy is observed in some ACR (GS) cells in culture,[14] however, not in FPC cell cultures. Therefore, a 2 to 4 fold increase in PA is probably not related to an increased number of genes, a result of tetraploidy. More studies are needed on the ACR PA. Determination of the cellular distribution and molecular weight of this protease activator will result in a better understanding of transformation in ACR.

Step 3. Plasma Protease

We have observed increased plasma protease activity in ACR plasma, and suggested this enzyme is involved in tumor promotion in this heritable disorder.[15,16,17] Extracellular proteolysis is believed to be an important step in tumor invasion and metastasis.[18,19] Extracellular proteolysis transforms the morphology of normal 3T3 cells and initiates growth in confluent quiescent cell population.[20] Also, extracellular proteolysis may be related to confluency, growth and transformation in 3T3 cells. Confluent cells have an intracellular PA that is not released extracellularly.[21] Most of the PA is found in a heavy membrane fraction. In growing and transformed cells, the PA is released and is found in a lighter plasma membrane-enriched fraction(s). Extracellular proteolysis has been associated with human cancer.[22] In our report we have suggested an association with human precancer. Although we have not shown directly that protease in ACR plasma transformed the morphology of cells in culture, we have demonstrated that plasma with increased protease activity caused this effect. Although the morphological alterations observed could be ascribable to the effect of protease,[17,20] there is evidence to suggest an indirect effect was involved. Since an earlier report[17] we have been concerned that ACR plasma transformed the morphology of cells in the presence of fetal bovine serum, rich in protease inhibitors which should compete with cell surface receptors for protease active sites. In our model (step 5) the transformation associated morphology may be the result of an alteration of intracellular actin, or an alteration in cell surface fibronectin (Large External Transformation Sensitive, or LETS) protein.

Step 4. Actin

ACR skin cell cultures have been examined by immunofluorescence.[23] It was reported that there is an altered distribution of intracellular actin. Intracellular PA may alter the cytoskeleton in a plasminogen independent process.[24] Therefore, the observation of altered actin distribution in ACR cells may be the result of increased PA. However, extracellular proteolysis can also affect the cytoskeleton[25] and plasminogen enhances TPA induced alterations in the intracellular matrix.[24] Therefore, in our model we suggest that intracellular PA affects actin distribution in ACR, an effect that would be enhanced in the presence of ACR plasma which is high in protease.

Step 5. Transformation Associated Morphology

We offer the following hypothesis to account for the observation that ACR plasma, high in protease, can alter the morphology of cells in the presence of fetal bovine serum. Fibronectin (LETS in our model) is digested by plasmin protease to yield protease active fragments that enhance virally transformed morphological characteristics, in the presence of fetal bovine serum.[26,27] It was hypothesized[27] that the fragments maintain an equal or increased affinity for fibronectin interaction sites on the cell surface, selectively binding the fragments to the cell surface in the presence of the fetal serum. ACR plasma containing increased protease activity may already have active fibronectin fragments (in vivo) when added to cells. Alternatively, the plasma protease may be digesting fibronectin in vitro in the cell culture model to yield the same result. This would suggest an indirect role for protease in ACR plasma in inducing a transfomration associated morphology of cells in culture.

Step 6. Cyclic Nucleotides

Protease lowers the intracellular level of 3'5' cyclic adenosine monophosphate and stimulates DNA synthesis.[28] C-AMP is a second messenger in hormone response which is proposed as a major growth regulator.[29] C-AMP and 3'5' cyclic guanosine monophosphate (C-GMP) are proposed to have opposing effects on cell growth.[29]

Adenomas are permissive for continuous DNA synthesis.[30] In an animal model for colon cancer, a fall in intracellular C-AMP and C-AMP-dependent protein kinase activity (following exposure to a rodent colon carcinogen) was proposed for permissive DNA synthesis.[31] We have previously reviewed the etiology and prevention of bowel cancer[32] and proposed a role for C-AMP in adenoma regression (step 8 in our model).

Step 7. Gene Regulation

It is generally agreed that initiation in ACR is a mutational event, compatible with a multistage progression to cancer. What is not agreed upon is the mechanisms of action for the promotional and latent stages. Our model is compatible with an epigenetic promotional stage, but not restricted to it. The pathway proposed could result from a second somatic mutation.

A theory has been advanced that involves a gene-amplification model of carcinogenesis.[33] In this model initiation occurs on a recessive gene that results in a single tandem duplication of a proto-oncogene. The result is an increase in gene product insufficient to cause transformation. Promotion would involve sister chromatid exchange through several rounds of cell division, to increase the frequencies of duplication, until enough gene product is produced to cause transformation. This model has considerable biological plausibility for some forms of cancer. However, ACR is an autosomal dominant disorder. Transformation should result soon after birth if the gene was dominant in producing copies of a gene product responsible for transformation. However, a resolution to this problem may be suggested. It may be possible that the ACR gene is inhibited at birth by a modifying allele. A somatic mutation of the modifying allele would remove the restriction on the ACR gene for producing a gene product. This first somatic mutation would be the promotional stage in ACR, the development of adenomatosis. Additional promotional events would be any factor capable of stimulating sister chromatid exchange via cell division, although this stage would affect latency, and not represent an additional mutation. Therefore, this model would suggest a two stage process in ACR. The requirement of only one somatic mutation (on a modifying allele of a dominant oncogene) would follow the clinical outcome, an invariable progression to transformation. It is attractive (and biologically plausible) to suggest the gene product produced is intracellular protease activator. Initiation (inhibited by a modifying allele) would result in the production of a small amount of the gene product (such as a 2-4 fold increase in PA). Promotion would follow the somatic mutation of the inhibiting allele allowing transformation to occur when enough gene product (protease activator) was produced to cross a threshold for full transformation.

ACKNOWLEDGEMENTS

The authors would like to thank Anne J. Krush, M.S.; Edmond A. Murphy, M.D., D.Sc.; Thomas Hendrix, M.D. and the members of the Gastroenterology Division of the Department of Medicine, Johns Hopkins Hospital, Baltimore, MD.; M. Clare Wilson, R.N. and the members of the Florida Atlantic University Health Services; James Baron, M.D., Boca Raton, FL, for their cooperation with this project. We thank Anamarie Di Nunzio for assistance in preparation of the manuscript.

REFERENCES

1. American Cancer Society Facts and Figures, 1981.
2. H.J.R. Bussey, "Familial Polyposi Coli," Johns Hopkins University Press, Baltimore, MD (1975).

3. C.R. Sachatello and W.O. Griffen, Familial polyposis coli, Am. J. Surg. 128:198, 1975.
4. V.A. McKusick, Genetics and colon cancer: A review, Digest Dis. 19(10):954, 1974.
5. T. Alm and B. Wahren, Carcinoembryonic antigen in hereditary adenomatosis of the colon and rectum. Scand. J. Gastroenterol. 19(8):875, 1975.
6. A.V. Jubert, T.M. Talbott and T.M. Maycroft, Characteristics of adenocarcinomas of the colorectum with low levels of pre-operative plasma carcinoembryonic antigen (CEA), Cancer 42:635, 1978.
7. J.K. Isley, Jr. and R.B. Akin, A community-based colon and rectal cancer screening program. J. Florida M.A. July:501, 1981.
8. J.S. Rhim, R.J. Huebner, P. Arnstein and L. Kopelovich, Chemical transformation of cultured human skin fibroboasts derived from individuals with hereditary adenomatosis of the colon and rectum, Int. J. Cancer 16(5):565, 1980.
9. L. Kopelovich, N.E. Bias and L. Helson, Tumour promotors alone induces neoplastic transformation of fibroblasts from human genetically predisposed to cancer, Nature 282:619, 1979.
10. R. Pollack, et al., Production of plasminogen activator and colonial growth in semisolid medium are in vitro correlates of tumorigeneicity in the immune-deficient nude mouse. In: "Proteases and Control, Vol. 2," (Cold Springs Harbor Conference on Cell Proliferation), (E. Reich, D.B. Rifkin and E. Shaw, eds.) Cold Springs Harbor Laboratory, NY (1975).
11. L. Kopelovich, Hereditary adenomatosis of the colon and rectum. A model of tumor progression. In: "Cancer Invasion and Metastasis: Biological Mechanisms and Therapy," (B.S. Day et al., eds.), Raven Press, NY, pp. 383-95 (1977).
12. S. Jaken and P.H. Black, Regulation of plasminogen activator

in 3T3 cells: Effect of phorbol myristate acetate on subcellular distribution and molecular weight. J. Cell Biol. 90:727, 1981.

13. A.R. Kennedy and J.B. Little, Effects of protease inhibitors on radiation transformation in vitro. Cancer Res. 41(6):2103, 1981.

14. B.S. Danes and E.J. Gardner, The Gardner Syndrome: A cell culture study on kindred 109. J. Med. Genet. 15:346, 1978.

15. H.C. Lyko and J.X. Hartmann, Detection of plasma protease in heritable adenomatosis coli. Abstr. In: "Cancer Research Proceedings," Waverly Press, Inc., Baltimore, MD 21:81, 1980.

16. H.C. Lyko and J.X. Hartmann, Plasma protease and inhibitor activity identifies patients with a precancerous condition. Dancer Detection and Prevention 3(1):326, 1980.

17. H.C. Lyko and J.X. Hartman, Familial polyposis coli plasma causes a transformation associated morphology of cells in vitro: Hyperproteinasemia and colorectal polyps, Cancer Detection and Prevention 4:401-405, 1981.

18. A. Vaheri, et al., Fibronectin and proteases in tumor invasion. European Organization for Research on Treatment of Cancer (EORTC) Monograph Series, In: "Proteinases and Tumor Invasion," (P. Straeuli, A.J. Barrett and A. Baici, eds.) Raven Press, NY, pp. 49-58 (1980).

19. J.B. Boyd, et al., Production and secretion of proteolytic enzymes by normal and neoplastic cells. J. Surg. Oncol. 11(3):275, 1979.

20. P. Whur, J.J. Silcox, J.A. Boston and D.C. Williams, Plasminogen activation transforms the morphology of quiescent 3T3 cell monolayers and initiates growth, Br. J. Cancer 39(6):718, 1979.

21. S. Jaken and P.H. Black, Differences in intracellular distribution of plasminogen activator in growing, confluent, and

transformed 3T3 cells. Proc. Natl. Acad. Sci. 76(1):246, 1979.
22. E. Wilmes, O.L. Schonberger and K. Hochstrasser, Studies for proteolysis in malignant tumours. Laryngol. Rhinol Otol. 58(11):861, 1979.
23. L. Kopelvich, et al., Organization of actin containing cables in cultured skin fibroblasts from individuals at high risk of colon cancer. Int. J. Cancer 26(3):301, 1980.
24. D.B. Rifkin, R.M. Crowe and R. Pollack, Tumor promotors induce changes in the chick embryo fibroblast cytoskeleton, Cell 18:361, 1979.
25. R. Pollack and D. Rifkin, Actin-containing cables within anchorage-dependent rat embryo cells are dissociated by plasmin and trypsin, Cell 6:495, 1975.
26. C. Kryceve-Martinerie, et al., Transformation-enhancing factor(s) released from chicken Rous sarcoma cells: Effect on some transformation parameters. Virol. 112(2):436, 1981.
27. G. De Petro, S. Bartali, T. Vartio and A. Vaheri, Transforming-enhancing activity of gelatin-binding fragments of fibronectin. Proc. Natl. Acad. Sci. 78(8):4965, 1981.
28. M.M. Burger, et al., Growth control and cyclic alterations of cyclic AMP in the cell cycle. Nature New Biol. 239:161, 1972.
29. F.R. DeRubertis, R. Chayoth and J.B. Field, The content and metabolism of cyclic adenosine 3'5' monophosphate and cyclic guanosine 3'5' monophosphate in adenocarcinoma of the human colon, J. Clin. Invest. 57:641, 1976.
30. M. Lipkin, Cell kinetics: Summary of recent findings in studies of gastro intestinal disease in man. J. Envir. Pathol. Toxicol. 2(1):9, 1978.
31. F.R. DeRubertis and P.A. Craven, Early alterations in rat colonic mucosal cyclic nucleotides metabolism and protein

kinase activity induced by 1,2,-Dimethylhydrazine, Cancer Res. 40:4589, 1980.

32. H.C. Lyko and J.X. Hartmann, Ascorbate, cyclic nucleotides, citrus and a model for preventing large bowel cancer. J. Theor. Biol. 83:675, 1980.

33. M.L. Pall, Gene-amplification model of carcinogenesis. Proc. Natl. Acad. Sci. 78:2465, 1981.

USES OF TISSUE CULTURE AND CRYOPRESERVATION IN
PANCREATIC ISLET TRANSPLANTATION

Collin J. Weber*, F. Xavier Pi-Sunyer**, Earl
Zimmerman***, Gajanan Nilaver***, Michael Kazim*,
Orion Hegre**** and Keith Reemtsma*

Departments of Surgery*, Medicine** and Neurology***
Columbia University College of Physicians & Surgeons
New York, New York
The Department of Anatomy
University of Minnesota
Minneapolis, Minnesota****

INTRODUCTION

Diabetes mellitus remains a major health problem, with several million effected patients in this country. It is the fifth leading cause of death, the most common cause of blindness, and a major etiologic factor in renal failure (1-4).

It is generally acknowledged that the long-term course of insulin-dependent diabetes mellitus is favorably affected by careful control of blood glucose.[3-5] Unfortunately, even well controlled diabetic patients develop progressive microvascular disease, particularly retinopathy and nephropathy.[4-6]

Studies of pancreatic islet transplantation have been stimulated by the premise that islet transplants might be a physiologic approach to insulin replacement therapy in diabetic patients. Results from a large number of experiments published in recent years have supported the potential efficacy of islet transplantation in man.[7-14] In addition to prompt reversal of experimental hyperglycemia,[10-13] islet isotransplants in diabetic animals have been associated with marked improvement of some experimental diabetes-related metabolic,[15-18] and renal functional and morphologic abnormalities.[19-22] Experimental pancreatic islet grafts have been well-tolerated in most implant sites, including the portal vein, peritoneal cavity, spleen, muscle, and renal subcapsule.[8,10,11-15,23-25,27] In addition, neonatal and fetal islets have been shown to be advantageous for transplantation, in part because of the relative ease of islet isolation and viability.[8,11,25-30] The general conclusion drawn from most experimental studies of islet transplantation is that the procedure is both safe and effective.[7-9] The metabolic effects of islet transplantation,[15,16,23] and optimal means for isolation and storage of donor islets[26-31] remain important areas of current investigation in many laboratories.

The appeal of pancreatic islet transplantation in treatment of insulin-dependent diabetes mellitus has been enhanced by recent demonstrations of long-term function of islets grafted into relatively safe (intra-muscular and subcutaneous) sites,[5,32,33] successful reversal of spontaneous diabetes with experimental islet grafts,[34] beneficial effects of islet grafts on experimental renal functional and morphologic abnormalities,[20-22] limitations as well as advantages of insulin infusion pump therapy in human diabetics,[5,35] encouraging results with human pancreatic islet autografts,[36,37] and prolongation of experimental islet allograft and xenograft function using relatively non-toxic means of treatment of donor islets.[38-40]

Isolation of large numbers of human pancreatic islets for transplantation has proven to be one of the most difficult problems related to the islet transplant effort.[41] Not only is islet isolation from the adult human pancreas extremely difficult for technical reasons,[27,42-44] but adult human islets almost certainly will remain in relatively short supply in the foreseeable future. Fetal or neonatal mammalian pancreas may be an ideal source of donor islet tissue.[28,30,45-47] Unfortunately, these sources also will continue to be limited by both logistic and ethical considerations.

The Rationale for Islet Culture

In the last few years, tissue culture has emerged as a potential method for solution of several problems hindering human islet transplantation. Although disadvantages of islet culture (Table 1) presently limit its clinical application, distinct advantages (Table 2) suggest that islet culture deserves further experimental study.

Table 1

DISADVANTAGES OF ISLET CULTURE

1) LABORIOUS AND EXPENSIVE
2) POORLY SUITED TO EMERGENCY SURGERY
3) ISLET YIELD PRESENTLY LOW
4) RISK OF CONTAMINATION
5) ? PHYSIOLOGIC ALTERATION OF ISLETS

Table 2

ADVANTAGES OF ISLET CULTURE

1) ISLET PURIFICATION / ISOLATION
2) ISLET PRESERVATION
3) PHYSIOLOGIC STUDY OF DONOR ISLETS
4) MORPHOLOGY / PATHOLOGY STUDY
5) STERILITY CHECK
6) ? IMMUNOLOGIC ALTERATION OF ISLETS

Culture for Islet Isolation & Preservation

First, short-term culture of late fetal or neonatal pancreatic fragments is a remarkably efficient means of serendipitous isolation of islets. Lazarow[13] was the first to observe that exocrine pancreatic cells autolyse rapidly, leaving a relatively purified preparation of islets after 4-8 days _in vitro_. This finding has been confirmed.[8,26,30,31,41,44,48-51] Our laboratory has documented both isolation and preservation of rodent and human islets in Petri dish, and in millipore and grid organ culture systems.

Tissue Preparation for Islet Culture-Isolation

Procedures and Sacrifice of Experimental Animals

Pancreas excisions and islet implantations are done using deep general anesthesia and sterile technique, with subsequent humane sacrifice by inhalation ether or intravenous barbiturate, according to the principles set forth in the "Guide for the Care and Use of Laboratory Animals," Institute of Laboratory Animal Resources, National Research Council, DHEW, Publ. #(NIH)74-23.

Sources of Human Pancreas

To date, out laboratory has had experience in culture isolation of islets from the human pancreas in 27 experiments. Much of this data has been published.[27,44,51] These specimens were obtained as part of institutional organ donor retrieval in 16 instances (donor age three weeks-to-40 years). An additional 11 specimens have been obtained from operative pancreatic resections, as part of an islet salvage protocol.[76]

Donor Management and Tissue Fragmentation

Fasting of pancreas donor animals and pilocarpine treatment may be helpful in enhancing reproducibility of islet isolation.[26] Rapid pancreas excision may be followed by wet weight measurement, gland injection with 4°C Hank's BSS (HBSS) (Gibco), followed by coarse & then fine fragmentation.

Fine, uniform, relatively atraumatic pancreatic fragmentation ($1mm^3$) is clearly important in islet preparation. We have obtained excellent results with use of the McIlwain (Brinkman Instr.) chopper in manipulation of donor pancreas, at 4°C in HBSS.

Enzymatic Dispersion of Pancreatic Fragments

The pioneering work of Moskalewski,[52] Hellman,[53] Hellerstrom,[54] and Lazarow[13] firmly established the usefulness of enzymatic dissociation of pancreatic fragments with collagenase. Problems persist in standardization of collagenase efficacy and purity.[55] Nevertheless, brief exposure of pancreatic fragments to a solution of collagenase (Type IV, with low tryptic activity) remains the standard method for dispersion of minced tissue. Additional components of the islet isolation solution with which several authors have had success in islet isolation are listed in Table 3.[13,16,26,28,30,31,40-44,47,49-60]

For most dissociations of rodent and human pancreatic fragments, we have used 30-50 collagenase per gram of tissue, with 37°C waterbath agitation for 20-30 minutes.[12,44] This method regularly achieves uniform dissociation of the neonatal rat pancreas (Figures 1,2).

Table 3

ISLET ISOLATION SOLUTION
(per liter) (37°C)***

Hank's BSS (Gibco)
Fatty acid-free albumin 0.2mg/ml
  (Sigma)
Collagenase Type IV- 6-10 mg/ml &
  Worthington
DNase - 0.04% (Sigma)
Hyaluronidase- 1.8 mg/ml (Sigma)
Trypsin* - 0.25%(wt/v) (Sigma)
Dispase -II (Boehringer)-5g/liter
Cephalothin- 500 mg/L (alternately, cloxacillin - 50 mg/L + gentamicin
                      20 mg/L or Penicillin-G 100 U/ml + streptomycin -
                      100 ug/ml).
Fungizone 0 0.25 ug/ml
Essential amino acid solution (Gibco)- 1.% (v/v)
Essential vitamin solution (Gibco)- 1.% (v/v)
NaHCO3 (7.5% solution), to adjust pH to 7.3 - 7.4
Osmolality adjusted (NaCl) to 337 mosmol/L

Glucose- 5.6 mM
EGTA*- 0.02%(wt/v)(Sigma)
HEPES**- 24.5 mM (Sigma)
Solucortef- 125mg/L
Na-heparin- 100 units/L

---

*   Trypsin activity is optimal in a Ca++ Mg++ - free solution, with
      EGTA
      1,2, di(2 aminoethoxy)ethane-N,N,N'-tetra-acetic acid).
**   N-2-hydroxyethylpiperazine-N'-2-ethane'sulfonic acid
*** Dispersion terminated by lowering the temperature to 4-5°C, by
      addition of the above solution, minus collagenase, trypsin and
      hyaluronidase, plus soybean trypsin inhibitor 0.01% (wt/v) (Sigma),
      5% donor serum or 10 mg/ml albumin, 0.5 mM $CaCl_2$ and 1.0 mM $MgCl_2$,
      and 500 IU/ml, Trasylol (Delbay).
      Approximately 20-50. mg per gram (or ml volume) of pancreas
      fragments.
      Collagenase requires calcium for activity.

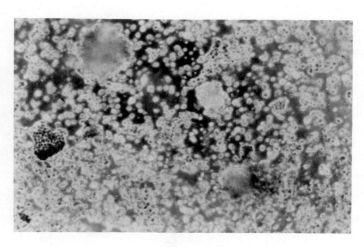

Figure 1

Neonatal rat pancreas, dissociated with collagenase. Phase contrast, trypan blue exclusion. Orig. mag. X125. Note three islets and multiple single exocrine cells.

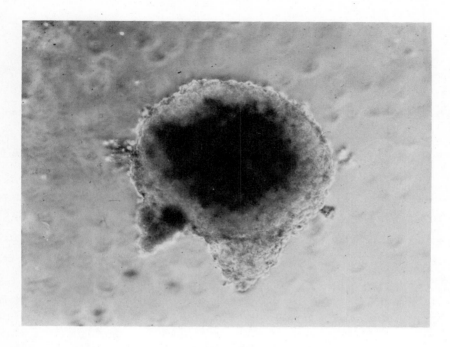

Figure 2

Same as Figure 1. Large isolated islet, with attached exocrine fragments, immediately after collagenase dissociation.

Separation of human pancreatic islets by collagenase digestion is much less reproducible than is the case for the rodent pancreas. The reasons for this limitation are largely unknown, although the more dense stroma of the human pancreas has been cited as a possible explanation.[26] Addition of DNase, trypsin and Dispase to the dissociation solution have not been found useful to date. More vigorous agitation,[61] intraparenchymal injections of enzymes and differential sedimentation through ficoll or percoll[26,58,62] have yielded modest improvements in yield. Because of the evident difficulties in adequate isolation of islets from the higher mammalian pancreas, we have pursued the observation of Lazarow,[13] investigating the usefulness of short-term culture for islet isolation from pancreatic fragments.

Islet Culture Methodology

In recent years, much has been written concerning the methodology of pancreatic islet culture. Based on the data of Nielson[31] and Andersson[71], optional culture medium may be RPMI. Optimal medium pH is probably 7.2-7.3.[72] Islets may be cultured in fetal calf serum or homologous species-specific serum. However, fetal calf serum promotes disruption of the islet capsule and monolayer outgrowth, whereas homologous serum allows maintenance of islet integrity.[63] Classic protein concentration is 10%, and some protein is essential for islet viability in vitro.[68] After the first few days, protein concentration may be reduced to 0.2%.[42,61] Islets may be cultured on millipore or on stainless steel grids as well as in classic Petri dishes. However, use of bacteriologic Petri dishes may be optimal in terms of long term maintenance of intact islets.[42]

Methodology of Islet Culture-Isolation

In our laboratory, fine mechanical mincing of the pancreatic specimen in Hank's balanced salt solution, pH 7.3-7.4, buffered with 25 mMol HEPES, is followed by incubation of 500-1000 mg aliquots of tissue each, in 5 ml of Hank's balanced salt solution, containing 30 mg of Type IV collagenase (Worthington), at 37°, with agitation (80-90 cycles/min) for 30 minutes. Isolated islets evident under the dissecting microscope are collected immediately using a braking pipette, and are placed in Petri dish culture. Remaining fragments are placed in separate culture dishes. Plastic Petri dishes (Nunc-bacteriologic type) which do not allow adherence of islets are used in all cases. Cultures are maintained at 37°, in a humidified atmosphere of 95% air/5% CO_2. Culture medium consists of RPMI 1640 (Gibco) (other media are used interchangeably) (MEM, #199, CMRL-1066); (Gibco) supplemented with

10% human serum albumin; (Abbott or Armour) 10,000 units/liter of penicillin; 100 mg/l of streptomycin; 1% essential amino acid solution; 1% essential vitamin solution; HEPES buffer, 25 mMol; Fungizone, 0.25 g/ml; and glucose, 11 mMol (all reagents Gibco). Culture medium is changed daily for the first four days, and then twice weekly. The methods have been modified from the studies of Nielsen,[31,61] Andersson,[42,70,71] Goldman,[63] and Hellerstrom.[30,54] Islet number is quantified weekly, using an inverted phase microscope; and this quantification is corroborated by sampling of cultured islets, which are fixed (formalin-paraffin embedding) and stained immunohistochemically for insulin, glucagon, and somatostatin.[65,73,77] Functional assessment of islet viability in culture is ascertained by sequential or simultaneous two-hour

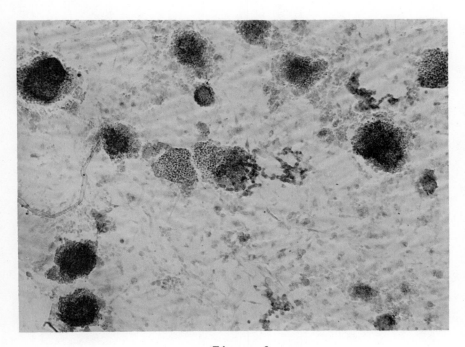

Figure 3

Neonatal rat pancreas, collagenase dissociated and Petri dish cultured X 24 hours. Aldehyde-fuchsin, orig. mag. X80. Note insulin staining of intact islets and chains of beta cells, adherent to dish. (Courtesy of M. Gagarin).

incubations of known numbers of islets in 0, 50 and 500 mg/dl glucose, -containing medium as well as 20 mMol theophylline.[65]

Data Analysis and Statistics

In vitro experiments and sampling are done in duplicate or triplicate whenever possible, with all appropriate controls. In vitro control and experimental data obtained from assays of coded samples are expressed as Mean SEM, or SD and are compared for significance of differences by use of Student's "t"-test, and for correllation using linear least square regression analysis.

Culture Isolation of Islet: Results

Twenty-four hour Petri dish culture of neonatal rat pancreas fragments first subjected to collagenase results in marked

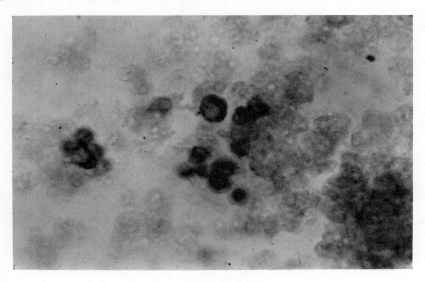

Figure 4

Same as Figure 3.. Orig. mag. X400. Note aldehyde fuchsin positive beta cells in culture. (M. Gagarin).

purification of both intact islets and islet fragments (Figures 3,4). Use of fetal calf serum and tissue culture Petri dishes yields adherent islets and chains of islet cells, as noted by Goldman.[63] Interestingly, simple pancreatic fragmentation (without collagenase) may be combined with millipore-grid "organ" culture[48,65] to yield isolation of functional neonatal rat and neonatal mouse islets after 4-6 days in vitro.[51,65] (Figures 5,6,7) (Table 4).

Recent studies in our laboratories have documented application of tissue culture isolation methodology to human pancreas specimens as well.[27,44,51] Intact, functional human islets may be identified as early as day #3 in vitro (Figures 8,9).

Working with human as well as murine pancreas, we have observed islet attachment to standard Petri dishes (Falcon) when culture-isolated in medium MEM, #199, CMRL or RPMI (all Gibco)

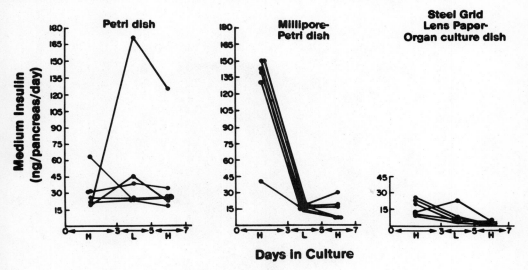

Figure 5

Neonatal mouse pancreatic fragment culture. Comparison of medium insulin in Petri dish, millipore grid and stainless steel grid organ culture methods. (H) = medium glucose 300 mg/dl. (L) = 100 mg/dl.

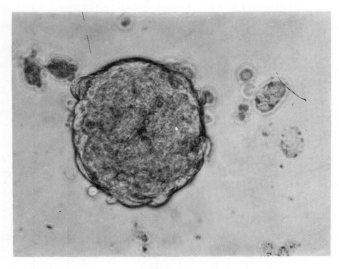

Figure 6

Neonatal mouse islet, culture-isolated, day #6 <u>in vitro</u>. Phase contrast microscopy. Orig. mag. X125. Note lack of adherence to Nunc dish and intact capsule.

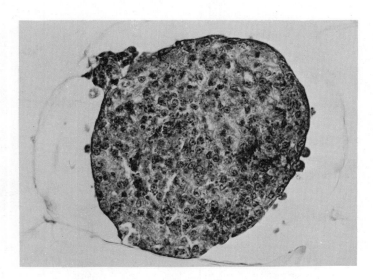

Figure 7

Same as Figure 6. Formalin fixation, hematoxylin-eosin. Original mag. X250.

Table 4

Donor Age	N	Medium Insulin		
		Glucose= 100 mg/dl	Glucose= 300 mg/dl	Glucose=300 mg/dl + 10 mM Theophyllin
3-4 days	7	1.1 ± .2	6.3 ± 1.1	7.7 ± 1.7
11-12 days	4	0.9 ± .1	1.0 ± .1	0.5 ± .1

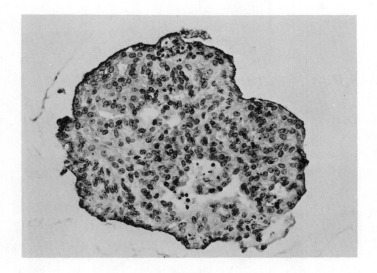

Figure 8

Human islet, culture isolation, day #3-4 in vitro. Clinical diagnosis = hereditary pancreatitis. Permanent fixation, hematoxylin-eosin. Orig. mag. X250.

supplemented with 10% fetal calf serum (BBL or Gibco), confirming data of Goldman,[63] Nielsen,[31] Hellerstrom[30], and Andersson,[70,71] (see Figure 10).[44,51] Substitution of homologous human serum albumin, 10% v/v (Abbott or Armour), and culture-isolation in bacteriologic Petri dishes (Nunc) results in maintenance of islet capsules, non-adherence and functional integrity, as evidence by a glucose-and theophylline-stimulated insulin release in vitro for up to 16 days (see Figures 9,11,12; Table 4), and by demonstration

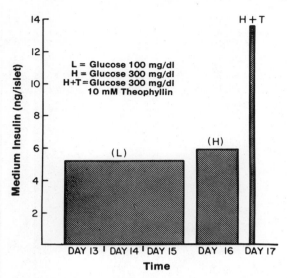

Figure 9

Medium insulin versus time in vitro for islets shown in Figure 8.

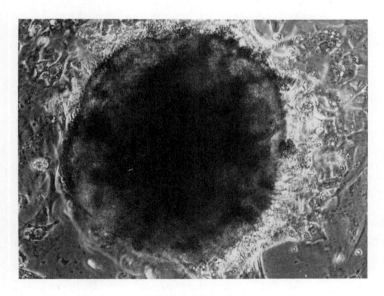

Figure 10

Human cadaveric islet, culture isolation in standard Petri dishes, day #8 in vitro. Note adherence and initiation of monolayer outgrowth. Phase contrast, orig. mag. X 125.

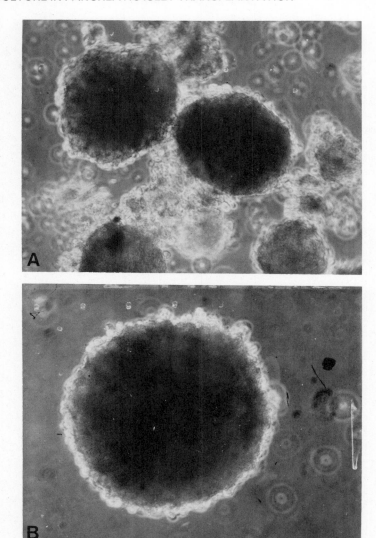

Figure 11

11A Human neonatal cadaver islets, culture isolation in Nunc dishes, day #5-6 in vitro. Note persistence of some chains of cells and nonadherence of islets. Phase contrast, orig. mag. X 125.

11B Human islet, culture-isolated, in culture, day #16 in vitro. Phase contrast, orig. mag. X 125. Note nonadherence when bacteriologic (Nunc) dishes used.

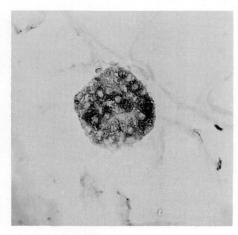

Figure 12
Neonatal mouse inslet, culture-isolated. Permanent fixation and immunoperoxidase staining for insulin (G. Nilaver and E. Zimmerman). Orig. mag. X 250.

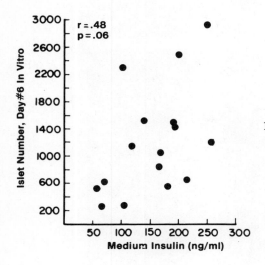

Figure 13
Correlation of medium insulin with culture-isolated neonatal murine islet number (day 6).

of beta cell insulin using peroxidase immunohistochemistry. (Figure 13)[77]

Extended studies of culture isolation of neonatal murine islets have correllated medium insulin[74] content with islet number (Figure 14). Similar data has been obtained correlating both medium insulin[74] and glucagon[78] with culture isolation islet number in three recent human experiments. In addition, comparison

of insulin release per murine islet prepared by percoll
sedimentation-overnight culture versus six day culture-isolation
using fetal calf serum without, or with heat inactivation (of FCS)
has shown the latter method to be advantageous. (Figure 15),
confirming the data of Andersson.[71]

Transplantation of Culture-Isolated Rat Islets

Several laboratories have documented reversal of experimental
(streptozotocin or alloxan) diabetes by isotransplantation of
isolated-cultured or culture-isolated rat islets
(8,12,13,25,27,32, 38,48,69,79). We have shown prompt reversal of
hyperglycemia, with normal weight gain in diabetic rats treated
with 24 hour cultured islets intraperitoneally (Figures 16,17), or
with 6-day culture-isolated islets implanted intramuscularly.[32,65]
Immunohistochemical studies have demonstrated the presence of
insulin, glucagon and somatostatin-containing cells in culture
isolated islet isografts biopsied from rectus muscle sites four
months after successful implantation. (See Figures 18-21)[73]

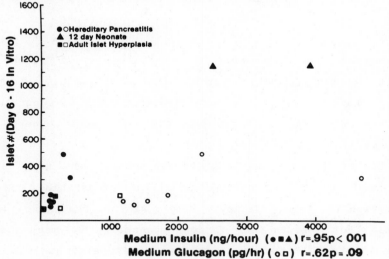

Figure 14
Correlation of medium insulin and glucagon with culture-isolated
human islet number.

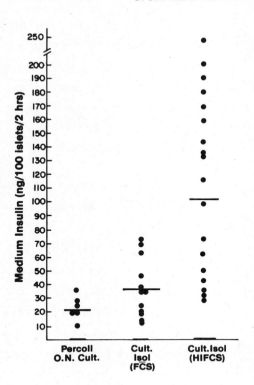

Figure 15

Medium insulin production per 100 islets, expressed as ng/2 hours in three techniques: collagenase-percoll-overnight cultures; collanase-culture X 6 days in medium containing 10% FCS; and collagenase-culture X 6 days in medium containing heat inactivated FCS.

Long-term functional studies of cultured versus "fresh" islet isografts have suggested comparable functional results. Feasibility of implantation of cultured islets in man is supported by these data, with obvious advantages vis a'vis human experimentation as outlined in Table 2.

Anticipating eventual islet transplants in man, what must be known about a given technique of islet preparation and preservation is its efficiency. Simply stated, the basic question is this: "For a given mass of donor pancreas, how much insulin may its islets be expected to produce after being subjected to separation and preservation?" The numerator is hormone production, and the denominator is donor tissue mass (wet weight of tissue).

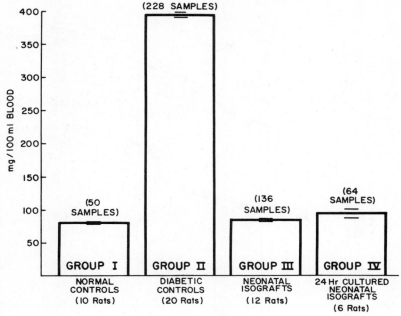

Figure 16

Fasting blood glucose in normal, diabetic and islet isografted rats. Mean ± SEM for data collected over six months. Note similarity of data for recipients of fresh and cultured islets.

The opportunity to quantify a donor islet preparation prior to transplantation, and the ability predict subsequent in vivo metabolic effects of the islet graft on the basis of in vitro functional data are the chief advantages offered by culture to the islet transplant effort. In vitro-in vivo correlates have been documented in rodent islet isograft models.[27,48,65] Additional advantages of islet purification in culture prior to transplantation include enhanced engraftment in intramuscular and subcutaneous sites, by virtue of avoidance of close contact with damaging exocrine enzymes,[33,65] and the opportunity to perform the graft electively.

Another potential advantage of donor islet culture may be the ability to alter graft immunogenicity in vitro. Encouraging data from Lacy and associates[38-40] await independent confirmation.

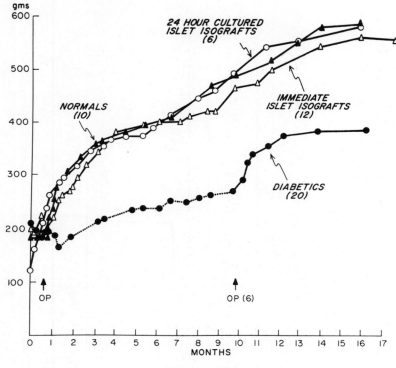

Figure 17

Body weight over seventeen months for normal, diabetic and islet isografted rats. Note similarity of weight gain for recipients of fresh and cultured islets. Note also impaired weight gain of animals given islets (with prompt restoration of normoglycemia - not shown) after 10-11 months of diabetes.

This aspect of islet culture deserves much further investigation in the near future.

Islet Cryopreservation

Cryopreservation is an attractive and logical alternative to culture for long-term islet storage or "banking". Storage of islets from both human and animal sources by means of tissue culture has been limited to 1-3 weeks in most instances. Growth factors and optimal in vitro conditions remain to be defined fully. On the other hand, cryopreservation may allow islet storage for up to 12-24 months. Several laboratories have

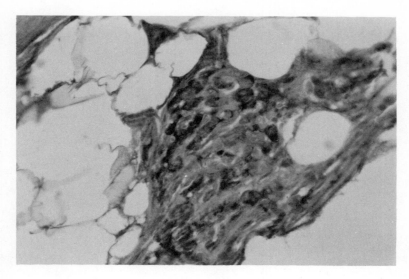

Figure 18

Neonatal rat islet, culture-isolated and isografted intramuscularly (note rectus muscle at corner of photo) and biopsied four months after restoration of normoglycemia. Permanent fixation and aldehyde-fuchsin stain. (O.D. Hegre). Orig. mag. X 250.

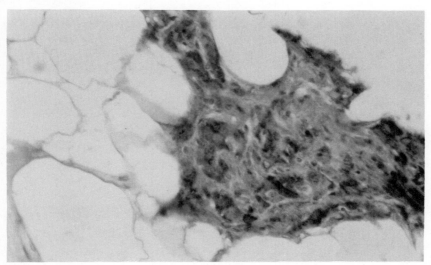

Figure 19

Adjacent section to that shown in Figure 17. Immunohistochemistry for insulin (O.D. Hegre).

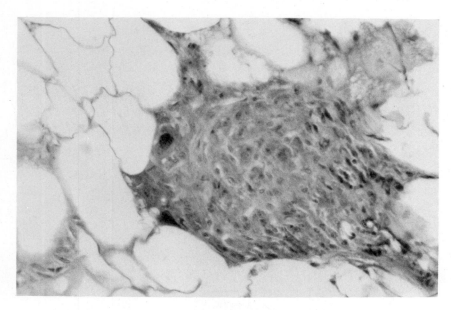

Figure 20

Adjacent section, immunohistochemistry for glucagon. (O.D. Hegre).

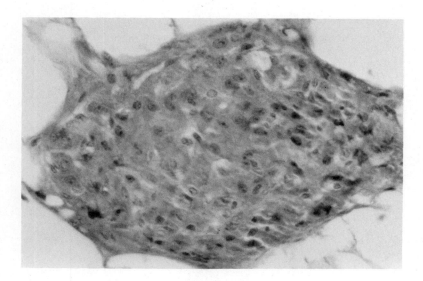

Figure 21

Adjacent section, immunohistochemistry for somatostatin. Note peripherally located somatostatin and glucagon-containing cells. (O.D. Hegre).

reported success in cryopreservation of rodent[80-82] and human[83-85] islets. In addition, reversal of experimental diabetes has been achieved by isotransplantation of freeze-thawed rat islets.[86,87]

Cryopreservation Methodology

Studies in our laboratory[85] have employed cryopreservation methods similar to those of Kemp, et. al[83] Islets are prepared for cryopreservation by transfer to RPMI 1640 containing 10% fetal calf serum or human serum albumin and 1.5 Mol dimethylsulfoxide, at 4°C for 60 minutes; followed by controlled rate freezing (Cryo-Med) at 0.25°C/min to -80°C, with subsequent storage in the vapor phase of liquid nitrogen at greater than -100°C, as

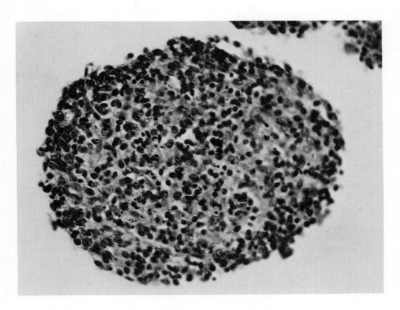

Figure 22

Human neonatal cadaveric islet, culture-isolated, cryopreserved X 70 days and freeze-thawed. Permanent fixation and hematoxylin-eosin staining. Orig. mag. X 250.

previously described. Islets are thawed by transfer to a 37°C water bath, until just before lysis of the last ice crystals, followed by replacement in tissue culture medium, 4°C, as described above, for functional and histological assessment of viability.

It has been possible to demonstrate histologic viability of human islets after freeze-thawing in several instances involving chronic pancreatitis[85] and neonatal cadaver specimens (see Figure 22). With both mouse and human islets, recovery has been approximately 20-40 percent, as asessed by pre- versus post-cryopreservation islet number (Figure 23), and 10%, when investigated by functional studies (Figure 24).

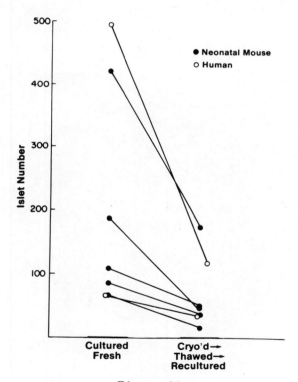

Figure 23

Islet number before and after cryopreservation: human and murine islet experiments.

These results may be improved in the future; certainly much further investigation will be required in order to optimize islet cryopreservation methods. Factors which deserve study include freezing rate, thawing rate, medium additives, tissue size (fragments-versus-islets-versus-islet cells), and type of manipulation prior to cryopreservation. It has been suggested, for example, that cell membrane injury induced by collagenase isolation may be repaired during a period of culture prior to cryopreservation.[88]

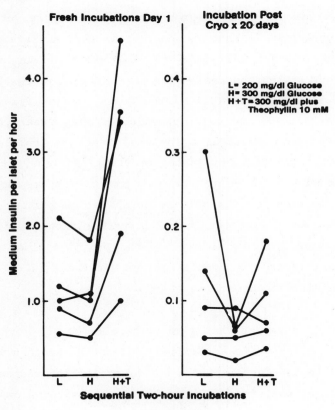

Figure 24

Medium insulin in sequential incubations with 200, 300 and 300 mg glucose plus 10 mM theophyllin, for islets before and after cryopreservation. Each islet sample studied pre- and post-freeze-thawing.

The importance of <u>cooling rate</u> and <u>additives</u> have been emphasized by Mazur[89] and Farrant.[90] Mazur's hypothesis predicts that the survival of frozen-thawed cells is importantly influenced by the cooling rate, and that the larger the cell, the more slowly it must be cooled in order to avoid intracellular ice formation. His data suggest that cells exhibit maximal survival at an intermediate cooling rate, dependent on the cell type, additives present, and the warming rate. Cell death at rapid cooling rates results from formation of intracellular ice crystals during cooling and their growth to damaging size during warming. Cells damaged by being cooled too slowly are injured by as yet poorly understood mechanisms. What is apparent is that <u>additives</u> such as glycerol and dimethyl sulfoxide (ME_2SO), which do not protect against injury at supraoptimal rates of cooling, are essential for survival of slowly frozen (i.e. large cells such as islet cells) mammalian cells. The two major theories of slow freezing injury suggest either that protection (by additives) is 'colligative' (not well substantiated), or that protection requires presence of the additive within the cell. The evidence on requirement of cell membrane permeation is increasingly negative, suggesting that it may be sufficient to protect the cell surface alone. Additional elements in slow freezing injury are thought to include altered intracellular electrolyte concentrations and toxic (i.e. osmotic) effects of additives themselves.

It also is apparent that mammalian cells do not survive slow freezing to a significant extent unless a protective non-electrolyte is added; up to a point, protection increases with concentration of additive; additives protect against slow freezing injury but not against rapid freezing injury; and finally, in some cases protection may be achieved without passage of additive into cells.

A variety of cryopreservation 'additives' have been used, including sucrose, glycerol, polyvinylpyrrolidone (PVP), hypertonic saline, ME_2SO. As summarized by Mazur[89], Leibo[91] and Farrant[90], each 'additive' has advantages and disadvantages, ME_2SO being slightly superior in most systems tested.

As Leibo[91] has stated so elegantly, ultimate survival of frozen mammalian cells also is dependent upon the <u>rate</u> at which they are warmed. Basically, the faster the rewarming, the better. Moderate rates of cooling may allow formation of small, but potentially damaging ice crystals. Rapid warming can obviate the consequences of growth of these crystals during thawing.

Finally, <u>the cell surface</u> plays a critical role in cryopreservation. Its permeability to water is a major factor determining the cooling rate that produced intracellular ice and cell death.[92] Its permeability to solutes determines both the extent to which protective additives permeate cells prior to and during freezing, and the osmotic response of cells during thawing and subsequent dilution and manipulation. Cell and orangelle surface membranes also are the chief targets of injury.[89]

Each of these factors may vary with tissue origin; future studies addressing definition of optimal islet cryopreservation conditions are needed.

CONCLUSIONS

Pancreatic fragment, islet and islet cell culture are established methods of short-term islet preservation. Culture of late fetal and neonatal pancreatic fragments is a remarkably efficient means of accidental isolation of islets, since in most culture systems, exocrine pancreatic cells autolyze rapidly.

Cultured animal islets will reverse experimental diabetes. Insulin, glucagon and somatostatin-containing cells can be demonstrated in long term isografts of culture-isolated islets. In addition, in vitro evaluation of donor islets is predictive of in vivo islet graft function.

Human islets may be isolated and stored in culture, but are much more difficult to isolate and maintain. Islet cryopreservation is a relatively new area of investigation. It is of clear potential value in long-term islet storage. In light of the considerable documentation now available attesting to the general safety and efficacy of experimental islet transplantation, much further work in islet culture, isolation and cryopreservation is warranted.

ACKNOWLEDGEMENTS

The authors are indepted to Ms. Sandra Rivera, Mrs. Yim Dam and Ms. Connie Le Peter for technical assistance; Ms. Ida Nathan for photomicrography; and Ms. Wilma Garcia for manuscript preparation. The authors are grateful, for support in cryopreservation methodology, to: The Transplantation Unit, Walter Reed Army Medical Center, Washington, D.C., COL. Jimmy Light, M.D., Director; and the Tissue Bank, Naval Medical Research Institute, Bethesda, Maryland, CMDR. D.M. Strong, Ph.D., Director.

REFERENCES

1. O.B. Crofford, Report of the National Commission on Diabetes. U.S. DHEW Publication. #(NIH) 76:1018. Washington, D.C.; Government Printing Office, 1975.

2. E. Takazakura, Y. Nakamofo, H. Hayakawa, K. Kawai, S. Muramoto, K. Yosidu, M. Shimizu, A. Shinoda and J. Takeuchi, Onset and progression of diabetic glomerulosclerosis: A prospective study based on serial renal biopsies. Diabetes 24:1, 1975.

3. R. Engerman, J. Bloodworth, Jr. and S. Nelson, Relationship of microvascular disease in diabetes to metabolic control. Diabetes 26:760, 1977.

4. J. Pirart, J. Lauvaux and C. Eisendrath, Diabetic retinopathy, nephropathy, neuropathy. Relation to duration and control: A statistical study of 4,400 diabetics. Diabetologia 11:370, 1975.

5. R.A. Rizza, J.E. Gerich, M.W. Haymond, R.E. Westland, L.D. Hall, A.H. Clemens and J.S. Service, Control of blood sugar in insulin-dependent diabetes: Comparison of an artificial endocrine pancreas, continuous subcutaneous insulin infusion, and intensified conventional insulin therapy. NEJM 303:1313, 1980.

6. H. Lestradet, L. Papoz, C.L. Hellouin de Menibus, F. Levavaseur, J. Blesse, L. Billaud, F. Battistelli, P.H. Tric and F. Lestradet, Longterm study of mortality and vascular complication in juvenile-onset (Type I) diabetes. Diabetes 30:175-179, 1981.

7. K. Reemtsma and C. Weber, Pancreas and pancreatic islet transplantation. In Sabiston D Jr (ED) The Sabiston Textbook of Surgery: The Biologic Basis of Surgical Practice, 12th Edition of the David-Christopher Textbook of Surgery, WB Saunder, Phila, 1981, p. 537.

8. O. Hegre and A. Lazarow, Islet transplantation. In Volk BW and Wellman K (Eds) The Diabetic Pancreas. Plenum, New York, p. 517.

9. D. Sutherland, A. Matas and J. Najarian, Pancreatic islet cell transplantation. Surg Cl N Am 58:365, 1978.

10. D. Scharp, C. Kemp, M. Knight, J. Murphy, W. Newton, W. Ballinger and P. Lacy, Long term results of protan vein islet isografts and allografts in the treatment of streptozotocin-induced diabetes. Diabetes 23:359, 1974.

11. J. Brown, W.R. Clark, R.K. Makoff, H. Weisman, J.A. Kemp and Y. Mullen, Pancreas transplantation for diabetes mellitus. Ann Int Med 89:951, 1978.

12. C. Weber, R. Weil, R. McIntosh and K. Reemtsma, Transplantation of pancreatic islets from neonatal to adult rats. Transplantation 19:442, 1975.

13. A. Lazarow, L. Wells, A.M. Carpenter, O. Hegre, R. Leonard and R. McEvoy, Islet differentiation, organ culture and transplantation. Diabetes 22:877, 1973.

14. C. Reckard, M. Ziegler and C. Barger, Physiological and immunological consequences of transplanting isolated pancreatic islets. Surgery 74:91, 1973.

15. M. Pipeleers-Marichal, D. Pipeleers, J. Cutler, et al., Metabolic and morphologic studies in intraportan islet transplanted rats. Diabetes 25:1041, 1976.

16. E. Trimble, E. Siegel, H. Berthoud and A. Renold, Intraportan islet transplantation: Functional assessment in conscious unretrained rats. Endocrinol 106:791, 1980.

17. C. Weber, R. Lerner, P. Felig, M. Hardy and K. Reemtsma, Hyperinsulinemia and hyperglucagonemia following pancreatic islet transplantation in diabetic rats. Diabetes 25:944, 1976.

18. C. Weber, M. Hardy, G. Williams, R. Lerner and K. Reemtsma, Pancreatic islet transplantation and hyperparathyroidism in diabetic rats. Surg Forum 27:317, 1976.

19. S.M. Mauer, M.W. Steffes, D.E.R. Sutherland, J.S. Najarian, A.F. Michael and D.M. Brown, Studies of the rate of regression of the glomerular lesion in diabetic rats treated with pancreatic islet transplantation. Diabetes 24:280, 1975.

20. M. Steffes, D. Brown, J. Basgen, A. Metan and S. Mauer, Glomerular basement membrane thickness following islet transplantation in the diabetic rat. Lab Invest 41:116, 1979.

21. C. Weber, F. Silva, M. Hardy, C. Pirani and K. Reemtsma, Effects of islet transplantation on renal function and morphology of short- and long-term diabetic rats. Transplant Proc 11:549, 1979.

22. C. Weber, F. Silva, M. Hardy, C. Pirani and K. Reemtsma, Islet transplants and nephropathy. In Friedkman E and L'Esperance F (Eds) Diabetic Renal-Retinal Syndrome. Grune and Stratton Inc, New York, 1980, p 373.

23. J. Brown, Y. Mullen, W. Clark, I. Molnar and D. Heininger, Importance of hepatic portal circulation for insulin action in streptozotocin diabetic rats transplanted with fetal pancreases. J Clin Invest 64:1688, 1979.

24. A. Andersson, Islet implantation normalizes hyperglycemia caused by streptozotocin-induced insulitis: Experiments in mice. Lancet 1:582-584, 1979.

25. F. Pi-Sunyer, R. Woo, C. Weber, M. Hardy and K. Reemtsma, Selection of a practical method of pancreatic islet transplantation in the rat. Surg. Forum 30:301, 1979.

26. D.W. Scharp, R. Downing, R. Merrell and M. Greider, Isolating the elusive islet. Diabetes 29(Suppl 1):19, 1980.

27. K. Reemtsma, C. Weber, F. Pi-Sunyer, R. Lerner, E. Zimmerman and M. Hardy, Alternatives in pancreatic islet transplantation: Tissue culture studies. Diabetes 29(Suppl 1):45, 1980.

28. A. Agren, A. Andersson, C. Bjorken, C. Groth and R. Gunnarsson, Hellerstrom pancreas: Culture and function in vitro. Diabetes 29:64, 1980.

29. J. Brown, J. Kemp, S. Hurt and W. Clark, Cryopreservation of human fetal pancreas. Diabetes 29:70, 1980.

30. C. Hellerstrom, N. Lewis, H. Borg, R. Johnson and N. Freinkel, Method for large-scale isolation of pancreatic islets by tissue culture of fetal rat pancreas. Diabetes 28:769, 1979.
31. J. Nielsen, J. Brundstedt, A. Andersson and C. Frimodt-Moller, Preservation of beta cell function in adult human pancreatic islets for several months in vitro. Diabetologia 16:97, 1979.
32. K.V. Axen and F.X. Pi-Sunyer, Long-term reversal of streptozotocin-induced diabetes in rats by intramuscular islet implantation. Transplantation 31:439-441, 1981.
33. R.C. Kramp and I.M. Burr, Subcutaneous, isogeneic transplantation of duct-ligated pancreas in streptozotocin-diabetic mice: relationships between carbohydrate tolerance and hormone content in transplantation or host pancreas. Diabetes 30:857, 1981.
34. A. Naji, W.K. Silvers, S.A. Plotkin, D. Dafoe and C.F. Barker, Successful islet transplantation in spontaneous diabetes. Surgery 86:218, 1979.
35. W.V. Tamborlane, R.L. Hnitz, M. Bergman, M. Genel, P. Felig and R.S. Sherwin, Insulin-infusion-pump treatment of diabetes: Insluence of improved metabolic control on plasma somatomedin levels. NEJM 305:303-307, 1981.
36. D.E.R. Sutherland, Pancreas and islet transplantation. II. Clinical trials. Diabetologia 20:435, 1981.
37. J.S. Najarian, D.E.R. Sutherland, D. Baumbartner, B. Burke, J.J. Rynasiewicz, A.J. Matas and F.C. Boetz, Total or near total pancreatectomy and islet autotransplantation for treatment of chronic pancreatitis. Ann Surg 192:526, 1980.
38. P.E. Lacy, J.M. Davie, E.H. Finke and D.W. Scharp, Prolongation of islet allograft survival. Transplantation 27:171, 1979.
39. P.E. Lacy, J.M. Davie and E.H. Finke, Prolongation of islet xenograft survival (rat to mouse). Diabetes 30:285, 1981.

40. D. Faustman, V. Hauptfeld, P.E. Lacy and J.M. David, Prolongation of islet allograft survival by pretreatment of islets with antibody directed to Ia determinants. Proc. Natl Acad Sci 78:5156, 1981.

41. D. Scharp, R. Merrell, S. Feldman, R. Downing, P. Lacy and W. Ballinger, The use of gyrotation culture for the preservation of isolated islets. Surg For 29:200, 1978.

42. A. Andersson and C. Hellerstrom, Explant culture: Pancreatic islets. In Methods in Cell Biology, Vol 21B, Academic Press, New York, 1980, 135-151.

43. G. Lundgren, A. Andersson, H. Borg, K. Buschard, C. Groth, R. Gunnarsson, C. Hellerstrom, B. Petersson and J. Ostman, Structural and functional integrity of isolated human islets of Langerhans maintained in tissue culture for 2-3 weeks. Transpl Proc 9:237, 1977.

44. C. Weber, M. Hardy, R. Lerne and K. Reemtsma, Tissue culture isolation and preservation of human cadaveric pancreatic islets. Surgery 81:270, 1977.

45. J. Brown, D. Heininger, J. Kuret and Y. Mullen, Islet cells grow after transplantation of fetal pancreas and control of diabetes. Diabetes 30:9, 1981.

46. R.C. McEvoy and O.D. Hegre, Syngeneic transplantation of fetal rat pancreas. II. Effect of insulin treatment on the growth and differentiation of pancreatic implants fifteen days after transplantation. Diabetes 27:988, 1978.

47. I. Swenne, A. Bone, S.L. Howell and C. Hellerstrom, Effects of glucose and amino acids on the biosynthesis of DNA and insulin in fetal rat islets maintained in tissue culture. Diabetes 29:686, 1980.

48. O. Hegre, R. McEvoy, V. Bachelder and A. Lazarow, Fetal rat pancreas. Differentiation of the islet cell component in vivo and in vitro. Diabetes 22:577, 1973.

49. W. Chick, A. Like and V. Lauris, Pancreatic beta cell culture: preparation of purified monolayers. Endocrinology 96:637, 1975.

50. A. Matas, D. Sutherland, M. Steffes and J. Najarian, Short-term culturek of adult pancreatic fragments for purification and transplantation of islets of Langerhans. Surgery 80:1976.

51. K. Reemtsma, C. Weber, F. Pi-Sunyer, R. Lerner, E. Zimmerman and M. Hardy, Organ culture studies for pancreatic islet transplantation. Transplant Proc XI:1002, 1979.

52. S. Moskalewski, Isolation and culture of islets of Langerhans of the guinea pig. Gen Comp Endocrin 5:342, 1965.

53. B. Hellman, Methodological approaches to studies on the pancreatic islets. Diabetologia 6:110, 1970.

54. C. Hellerstrom, A method for microdissection of intact pancreatic islets in mammals. Acta Endocrin 45:122, 1964.

55. H. Hahn, M. Ziegler, H. Jahr, R. Butler and K. Kohnert, Investigations on isolated islets of Langerhans in vitro XIII. Experiments concerning the preparation conditions with collagenase. Endokrinologie 67:67, 1976.

56. A. Lernmark, The preparation of, and studies on, free cell suspensions from mouse pancreatic islets. Diabetologia 20:432-438, 1974.

57. A. Andersson, H. Borg, C. Groth, R. Gunnarsson, C. Hellerstrom, G. Lundgren, Westman and J. Ostman, Survival of isolated human islets of Langerhans maintained in tissue culture. J Clin Invest 57:1295, 1976.

58. A. Shibata, C. Lugvigsen, S. Naber, M. McDaniel and P. Lacy, Standardization of a digestion-filtration method for isolation of pancreatic islets. Diabetes 25:667, 1976.

59. J. Ono, R. Takaki, H. Okano and M. Fukuma, Long-term culture of pancreatic islet cells with special reference to the B-cell function. In Vitro 15:95-102, 1979.

60. H. Kromann, M. Christy, J. Egeberg, A. Lernmark, J. Nerup and H. Richter-Olesen, Direct streptozotocin toxicity on dispersed mouse islet cells determined by 51 Cr-release, Med Biol 58:322-328, 1980.

61. J.H. Nielsen, Beta-cell function in isolated human pancreatic islets in long-term tissue culture. Acta Biol Med Germ 40:55, 1981.

62. J. Brunstedt, Rapid isolation of functionally intact pancreatic islets from mice and rats by Percoll gradient centrifugation. Diabetes & Metabolism 6:87, 1980.

63. H. Goldman, E. Colle and P. Braseau, The investigation of the human pancreatic endocrine cells in long-term culture. Diabetes 24:421, 1975.

64. G. Gomori, Observations with differential stains on human islets of Langerhans. Am J Pathol 171:395, 1941.

65. C. Weber, M. Greenwood, A. Zatriqi, M. Hardy, R. Lerner and K. Reemtsma, Effects of islet transplantation on growth and glucose homeostasis in diabetic rats. Diabetes 24:419, 1975.

66. P.E. Lacy and R.L. Gingerich, Approaches to culturing beta cells. In E. Von Wasielewski, W. Chick (Eds). Pancreatic beta cell culture. Excerpta Medica, Amsterdam, 1977, p.37.

67. W. Chick, D. King and V. Lauris, Techniques for the preparation and maintenance of pancreatic beta cell monolayer cultures. Ibid p. 85.

68. A.F. Nakhooda, C.B. Wolheim, B. Blondel and E. Marliss, Secretary function of endocrine pancreatic cells in monolayer culture: characterization and applications. Ibid p. 119.

69. O.D. Hegre, R.C. McEvoy and R.V. Schmitt, Isotransplantation of organ-cultured pancreas: reversal of alloxan diabetes. Ibid. p. 71.

70. A. Andersson and C. Hellerstrom, Isolated pancreatic islets in tissue culture: An investigative tool for studies of islet metabolism and hormone production. Ibid p.55.

71. A. Andersson, Isolated mouse pancreatic islets in culture: effects of serum and different culture media on the insulin production of the islets. Diabetologia 14:397, 1978.
72. J. Brunstedt and J.H. Nielsen, Long-term effect of Ph on B-cell function in isolated islets of Langerhans in tissue culture. Diabetologia 15:181, 1978.
73. S. Erlandsen, O. Hegre, J. Parsons, R. McEvoy and R. Elde, Pancreatic islet cell hormones, distribution of cell types in the islet, and evidence for the presence of somatostatin and gastrin within the D cell. J Hist Cyto 24:883, 1976.
74. V. Herbert, K. Lau, C. Gottlieb and S. Bleicher, Coated charcoal immunoassay of insulin. J Clin Endoc Metab 25:1375, 1965.
75. G. Snedecor and W. Cochran, Statistical Methods, Iowa State University Press, Ames, Iowa, 6th Edition, 1972.
76. C. Weber, R. Hirata, D. Strong, A. Gamez, D. Wilkinson, F. Pi-Sunyer, M. Hardy and K. Reemtsma, Quality control in human islet autotransplantation. Transpl Proc 12(2):199, 1980.
77. G. Nilaver, E. Zimmerman, R. Defendini, A. Liotta, D. Krieger and M. Brownstein, Adrenocorticotropin and b-lipoprotein in the hypothalamis.
78. E. Aguilar-Parada, A.M. Eisentraut and R.H. Unger, Effects of starvation on plasma pancreatic gulcagon in normal man. Diabetes 18:717, 1969.
79. H. Selawry, J. Harrison, M. Paripa and D. Mintz, Pancreatic islet isotransplantation: effects of age and organ culture of donor islets on reversal of diabetes in rats. Diabetes 27:625, 1978.
80. J. Kemp, P. Mazur, Y. Mullen, R. Miller, W. Clark and J. Brown, Reversal of experimental diabetes by fetal rat pancreas. I. Survival and function of fetal rat pancreas frozen to -196°C, Transpl Proc 9:325, 1977.

81. W. Payne, D. Sutherland, A. Matas and J. Najarian, Cryopreservation of neonatal rat islet tissue. Surg For 29:347, 1978.
82. R. Rajotte, H. Stewart, W. Voss, T. Shnitka and J. Dossetor, Viability studies on frozen-thawed rat islets of Langerhans. Cryobiology 14:116, 1977.
83. J.A. Kemp, S.N. Hurt, J. Brown and W.R. Clark, Recovery and function of human fetal pancreas frozen to -196°C. Transplantation 32:10, 1981.
84. J. Brown, J. Kemp, S. Hurt and W. Clark, Cryopreservation of human fetal pancreas. Diabetes 29:70, 1980.
85. C. Weber, D. Strong, R. Hirata, J. Collins, J. Light, J. Nicholl, H. Press, A. Gomez, D. Wilkinson, R. Budd, L. Jones and K. Reemtsma, New approaches to islet salvage: Cryopreservation, culture and perfusion of pancreatic fragments. Transpl Proc 11(2):195, 1980.
86. R. Rajotte, D. Scharp, R. Downing, G. Molnar and W. Ballinger, The transplantation of frozen-thawed rat islets transported between centers. Diabetes 28:377, 1979.
87. G. Nakagawara, Y. Kojima, T. Mizukami, S. Ono and I. Miyazaki, Transplantation of cryopreserved pancreatic islets into the portal vein. Transpl Proc XIII:1503, 1981.
88. J. Brown, W. Clark, R. Makoff, H. Weisman, J. Kemp and Y. Mullen, Pancreas transplantation for diabetes mellitus. Ann Int Med 89:951, 1978.
89. P. Mazur, Mechanisms of injury and protecting cells and tissues at low temperatures. INSERM 62:37, 1976.
90. J. Farrant, C. Walter and S. Knight, Cryopreservation and selection of cells. INSERM 62:61, 1976.
91. S. Leibo, Preservation of mammalian cells and embryos by freezing. INSERM 61:311, 1976.
92. G. Mansoori, Kinetics of water loss from cells at subzero centigrade temperatures. Cryobiology 12:34, 1975.

MECHANISMS OF MEDIATOR RELEASE FROM NEUTROPHILS

Gerald Weissmann, Charles Serhan[+], Helen M. Korchak[*]
and James E. Smolen[*]

Division of Rheumatology
Department of Medicine
New York University School of Medicine
550 First Avenue
New York, NY 10016

ABSTRACT

The encounter of neutrophils with immune complexes and complement components - in the bulk phase or on a surface - leads to their secretion of lysosomal hydrolases, especially neutral proteases, which provoke tissue injury. Secretion of lysosomal enzymes and generation of reactive oxygen species (e.g., $O_2^{\bar{\cdot}}$ generation are stimulus-specific and can be dissected to establish cause and effect relationships by means of: a) kinetic analysis, b) variations in the stimulus, and c) use of impermeant reagents to block discrete responses. Neutrophils also generate products of 11-cyclooxygenase (e.g., PGE_2, TxA_2) and of the 5- and 15-lipoxygenases (mono-, di-, and tri-HETEs, LTB_4, and their isomers). But the cyclooxygenase products (save TxA_2) are not phlogistic by themselves: they inhibit the functions of neutrophils, platelets, macrophages, and mast cells. The most

---

[+]Recipient of Predoctoral Fellowship CAO-9161, Department of Pathology, New York University School of Medicine
[*]Fellows of the Arthritis Foundation

potent pro-inflammatory agent yet identified as a product of arachidonate is LTB_4. LTB_4 is a potent Ca ionophore, constricts airways, is a potent chemoattractant, and induces local inflammation.

Neutrophils, Immune Complexes and Rheumatoid Arthritis (RA)

Neutrophils constitute over 90% of cells found in synovial fluid of RA patients (1). Since such fluids also contain immune complexes (IgG-IgG, IgG-IgM rheumatoid factors) and complement split products (C5a, C5a desarg, C3b, etc.) all of the reactants are present of the local Arthus lesion (2,3,4). Moreover, neutrophils from RA patients endocytose these complexes and complement (C) in vivo (5) and in vitro (6). In consequence, we (7) and others (8,9) have suggested that lysosomal enzymes, and other mediators of inflammation released by neutrophils after uptake of immune complexes account, at least in part, for rheumatoid inflammation. But RA fluids also contain abundant anti-proteases, such as $alpha_1$-antitrypsin, and it has recently been argued that, since the level of lysosomal enzymes in RA fluids does not correlate with erosive arthritis (10), release of proteases into the fluid cannot account for tissue injury. However, the neutrophils of synovial fluid may simply be markers of an Arthus reaction in the synoxium itself, or at the surface of cartilage, which permits entry of lysosomal proteases, but not antiproteases (11). Immune complexes are trapped by articular cartilage in RA (12,13) and reactions between neutrophils and cartilage containing such complexes are the basis of an in vitro model established by Henson (14) and ourselves (15), in which it can be demonstrated that immune complexes disposed on a surface are better stimuli to the secretion of granule contents than the same complexes presented in suspension (14,15,16). Immune electron microscopy reveals that neutrophils degranulate when they encounter cartilage with immune complexes entrapped: neutrophils first degrade, and then invade the tissue. In addition, extensive studies by Mohr and coworkers (17,18,19) - contradicting older views (20) - have shown that neutrophils are abundant in inflamed RA synovium and at the interface of cartilage with pannus. Release of neutrophil-derived elastase into cartilage subjacent to pannus and in inflamed synovium was demonstrated by cyto- and immunocytochemical techniques (17). In confirmation, Saklatvala and Barrett (21) found that at least half the neutral protease activity of rheumatoid synovium can be attributed to neutrophil-derived catepsin G and elastase. In experimental arthritis, pannus rich in neutrophils was also shown capable of invading living cartilage (18), corresponding to the observations of Ziff's group (13) that immune complexes were removed from the pannus-cartilage interface, from which proteoglycans are leached.

Stimulus-Secretion Coupling in the Human Neutrophil

Since our original observations (22,23) that lysosomal enzymes are regurgitated from viable neutrophils in the course of phagocytosis, this process - now extensively documented (14,24,25, 26) - has been placed into the broader context of stimulus-secretion coupling (27). Cells of pancreas (28), salivary gland (29), adrenal (29) as well as basophils (29) and macrophages (31), secrete stored materials in consequence of ligand-receptor interactions or exhibition of Ca ionophores. Indeed, enzyme secretion from neutrophils is provoked not only by ingestion of immune complexes or Ig-coated particles via Fc receptors (14,15), or by C3b-opsonized particles via C-receptors (26,32), but by chemotactic peptides (C5a, FMLP), by lectins (concanavalin A), by the tumor promoter phorbol myristate acetate (PMA), and by lipid- or water-soluble ionohpores (A23187, inomycin, PGB_x), by arachidonate, and by leukotriene B_4 (33,34,35,36,37). Secretion of lysosomal enzymes in response to the secretagogues listed above is accompanied by release into the medium of other mediators of inflammation, most important of which are reactive oxygen species ($O_2^{\top}$, H_2O_2, $OH^\cdot$ and singely oxygen) and lipid products derived from membrane phospholipids (prostaglandins, leukotrienes, and plateletactivating factor) (38,39,40).

Stimulus-secretion coupling is, however, only one aspect of neutrophil activation, first described by Fleck (41). The activated neutrophil aggregates (42), adheres avidly to surfaces (43), responds to gradients of secretagogues by chemotaxis (24), and fuses its two major types of granules (azurophil and specific) with the phagocytic vacuole or the plasma membrane (25,44).

Neutrophil activation is probably designed to assure the ingestion and killing of offending microorganisms, in the course of which granule constitutents are introjected into the phagocytic vacuole: "Covert secretion" (45). However, when the particle/cell ratio is great, "overt secretion" results: granule contents are secreted into the medium (23). Overt secretion is conveniently studied in cells treated with cytochalasin B. When cytochalasin B is added to neutrophils, particles or soluble secretagogues still engage appropriate receptors but are not endocytosed (37,46). Enzyme secretion, and release of $O_2^{\top}$ and various lipid mediators (e.g., prostaglandins, thromboxane A_2) is enhanced (38,39,46). Indeed, Henson has pointed out that the cytochalasin B-treated neutrophil in suspension behaves as if it were a normal cell that had engaged surface-bound ligands such as immune complexes trapped in cartilage (16).

Activation of the neutrophil has become amenable to rapid kinetic analysis (47). By means of continuous recording-, stop/flow-, and rapid centrifugation-techniques, several laboratories, including our own, have timed the early events of stimulus-secretion coupling (48,49,50). Each secretagogue provokes a somewhat different pattern of response, but, in general, the following sequence is observed. The earliest (0-5 sec) responses to ligand receptor interactions are a change in membrane potential, $\Delta\Psi$ (51,52), loss of membrane-associated Ca at the site of interaction (53,54), degradation of

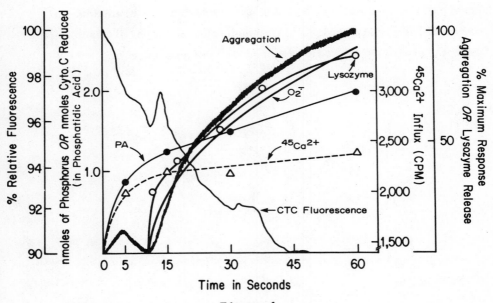

Figure 1

Example of kinetic analysis of neutrophil activation (see text for references). Neutrophils - aliquots of the same suspension from one donor - exposed to 10^{-7} F-met-leu-phe were studied with respect to 1. CTC fluorescence (= mobilization of membrane - associated calcium), 2. formation of phosphatidic acid (PA) (two-dimensional TLC), 3. uptake of ^{45}Ca from the medium, 4. aggregation, 5. release of lysozyme, and 6. generation of $O_2^{\bar{\cdot}}$. On these samples, we have also measured changes in membrane potential ($\Delta\Psi$, by uptake of $TPMP^+$), release of β-glucuronidase, degradation of phosphatidyl-inositol, and accumulation of cyclic AMP - all with a resolution of <5 sec. These have been ommitted for clarity.

phosphatidylinositol and formation of phosphatidic acid (55,56), net accumulation of Ca from the extracellular medium (57,58), rearrangements of microfilaments and microtubules (33,59), and the generation of cyclic AMP, but not GMP (47,59). After 10 seconds, the cells begin to aggregate (60). Generation of O_2^- and release of lysosomal enzymes follow after "lag periods" characteristic for each stimulus: 25-35 seconds (26,47,48). One such kinetic analysis is shown in Figure 1.

This general scheme does not hold for all secretagogues. Thus, PMA or concanavalin A do not provoke increments in cAMP (59) (Figure 2) and neither A23187 nor PMA induce breakdown of phosphatidylinositol (61). Although the actions of each agent (except, again, PMA) are augmented in the presence of

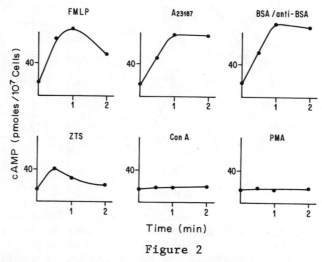

Figure 2

Effect of complete and incomplete secretagogues on cyclic AMP accumulation in PGE_2-treated neutrophils. Accumulation of cAMP (radioimmunoassay) in neutrophils exposed to PGE_1 and then to various secretagogues at 50% effective concentrations (ref. 59). Concanavalin A and phorbol myristate acetate ("incomplete secretagogues") differ from "complete secretagogues" such as F-met-leu-phe or zymosan-treated serum (C5a, C5a desarg) in neither causing accumulation of cAMP by themselves nor augmenting cAMP in response to PGE_1 (59).

extracellular Ca (58,62,63), influx of Ca is not necessary for enzyme release or O_2^- (63). Moreover, stimulus-specific pathways to secretion can be identified by means of impermeant reagents which blunt neutrophil responses to one, but not other, secretagogues. Anion channel blockers (SITS, DIDS) inhibit enzyme secretion induced by A23187, immune complexes, or C-opsonized zymosan (64,65) and inhibit O_2^- generation by C-opsonized zymosan, but have no effect on O_2^- generation in response to A23187 or immune complexes (64). The protein-in-activating reagent, p-diazobenzenesulfonic acid (DASA) blocked O_2^- generation induced by concanavalin A or FMLP, but not by A23187 or PMA.

Moreover, neutrophils remodel their membrane phospholipids when stimulated. Human neutrophils undergo stimulus-secretion coupling within the first minute after exposure to such chemotattractants as the peptide N-formyl-methinoyl-leucyl-phenylalanine (66,67). Within five sec after FMLP (10^{-7} M), neutrophils break down phosphatidylinositor (PI) and accumulate phosphatidic acid (PA) which contribute 5.31 ± 0.26 and 0.11 ± 0.07 percent of the total lipid phosphorous of resting cells (two-dimensional TLC). They also show rapid decrements (< 5 sec) in fluorescence of preincorporated chlortetracycline, used as an indirect probe of the mobilization of intracellular calcium, and take up ^{45}Ca from the extracellular medium. In response to FMLP, PA levels rose to 0.138 percent of total lipid phosphorous by 5 sec and were reciprocal to PI decrements. By 15 sec, PI breakdown (to 4.5 percent of total lipid phosphorous) far exceeded the amounts of PA generated, which continued to accumulate for 60 sec. Between 15 and 60 sec, PI levels returned to baseline values. No changes in PI or PA were found when neutrophils were activated by secretagogues phorbol myristate acetate (PMA, 1 µg/ml) or the ionophore A23187 (5 µM), although PMA induced rapid decrements in chlortetracycline fluorescence. Changes in PI and PA levels in

FMLP-stimulated cells were not influenced by chelation of all extracellular Ca (EGTA) nor by treatment with cytochalasin B, which enhanced ^{45}Ca uptake from the extracellular medium. However, in the absence of extracellular Ca (EGTA), PI breakdown induced by FMLP no longer exceeded formation of PA at 15 sec. Uptake of ^{45}Ca (0-60 sec) was induced by FMLP, but not PMA. Formation of phosphatidic acid and breakdown of phosphatidylinositol in response to FMLP significantly anteceded other aspects of neutrophil activation: aggregation, enzyme release, and O_2^- generation, but followed decrements in chlortetracycline fluorescence. The results indicate that initial PI breakdown and PA formation are not dependent upon uptake of extracellular calcium (EGTA; A23187) nor upon the prior mobilization of intracellular calaium (PMA). They support the general hypothesis that the "phosphatidylinositol response" is activated by specific ligand-receptor interactions and that products generated by this pathway may promote calcium fluxes, rather than vice-versa (66,67).

Neutrophils Generate, and Respond to, Oxidation Products of Arachidonic Acid

By 1971, it has become accepted, chiefly due to the important contributions of Vane's group (68) that prostaglandins caused inflammation and that non-steroidal anti-inflammatory agents (e.g., ASA and indomethacin) owed their therapeutic efficacy to inhibition of prostaglandin synthesis. But work in our own (69,70) and other (71,72) laboratories soon established that stable prostaglandins (PGE_2, PGB_2, PGI_2) inhibited lysosomal enzyme release and O_2^- generation by neutrophils (70), lymphocyte activation (73), elaboration of histamine by mast cells (71) and experimental inflammation of skin (74) and joints (74). Indeed, stable prostaglandins do not provoke the four cardinal signs of inflammation (75) and aspirin or indomethacin inhibit other steps in the metabolism of arachidonate (76,77). Attention was shifted to the newer derivatives: thromboxanes and lipoxygenase products.

In the next decade, based on studies of Zurier (78), Morley et al (79) and Malmsten, who had come from Samuelsson's laboratory to work with us (38), it became clear that neutrophils generated PGE_2 and thromboxane in response to phagocytosis - but that these products accounted for only a minority of arachidonate metabolites derived from stimulated cells (80,81,82). It was found that neutrophils contained active 5- and 15-lipoxygenases by means of which they formed a variety of mono-, di-, and trihydroxy eicosatetraenoic acids (83,84,85). The 5-lipoxygenase pathway was responsible for the generation of leukotrienes (LT's) (85,86), among which LTC_4 and LTD_4 (formed by basophils and macrophages) were found to constitute the slow-reactive substances (87,88,89,90).

It is now appreciated that in the neutrophil the sum of arachidonate metabolites formed via lipoxygenase(s) exceed that generated via the cyclooxygenase pathway (80,81,82). Incomplete evidence suggests that the bulk of precursor arachidonate in neutrophils arises from the action of a Ca-dependent phospholipase A_2 (82,91) and not a phosphatidylinositol specific phospholipase C, as described in platelets (92). Arachidonate is converted by an 11-cyclooxygenase to the endoperoxides PGG_2 and PGH_2 before their transformation (34,84) to PGE_2 and thromboxane A_2 (TxA_2) (determined as the metabolite TxB_2).

Alternatively, dependent on the stimulus, arachidonate is transformed to 5-HPETE, which is in turn metabolized to 5-HETE or to the 5-6 expoxide, leukotriene A_4 (86). In the neutrophil, which does not conjugte glutathione with LTA_4 to form LTC_4 and LTD_4 (the slow-reactive substances of mononuclear cells and

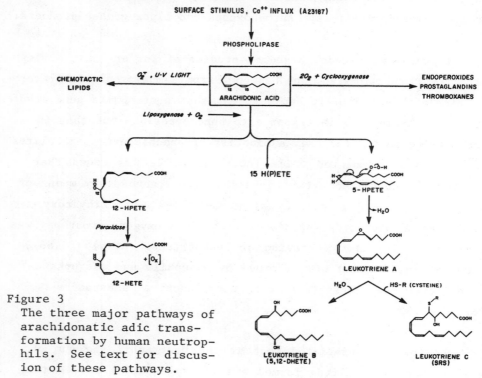

Figure 3
The three major pathways of arachidonatic adic transformation by human neutrophils. See text for discussion of these pathways.

basophils) another group of di- and trihydoxy eicosatetraenoic acids are formed. These include 5(S), 12(R)-LTB_4, generated enzymatically from LTA_4; this compound has a 6-<u>cis</u> in the triene portion of the molecule, whereas two all <u>trans</u> isomers (5,12-diHETEs) are formed by non-enzymatic means (84,85,86). Recently a 5(S), 12(S) di-HETE has been isolated from neutrophils stimulated by A23187 and arachidonate, together with trihydroxy derivates of LTB_4 and 5(S), 12(S) di-HETE (85). The third OH group of these compounds is at C-20, and can be further oxidized to yield the corresponding di-carboxylic acid (85). Whereas both LTB_4 and its 20-OH derivatives re formed from LTA_4, the 5(S), 12(S). di- and tri-HETEs arise after the double or triple oxygenation of arachidonic acid (86). Figure 3 indicates this pathway,

The biological functions of the metabolites of arachidonate in the neutrophil are not entirely clear: it is not known whether their predominant effects are intra- or extracellular (80,81). Acting via as yet poorly characterized surface receptors, which link to adenylate cyclase and raise the level within cells of cAMP (70), PGE_2 (PGI_2 from endothelium) inhibits neutrophil activation as judged by chemotaxis, aggregation, lysosomal enzyme release and O_2^- production (review in 75). Thromboxane B_2 (93) and some mono- and di-HETES are modest chemoattractants (65), whereas LTB_4 ranks in chemotactic potency with FMLP and C5a (65,94). Moreover, LTB_4 activates neutrophils with respect to aggregation (94), lysosomal enzyme release (95) and the local inducation of inflammation (94).

Whereas 5-HETE is reincorporated into neutrophil phospholipids and triglycerides (80,82), LTB_4 is released into the medium (80,82,96) after challenge of cells with A23187 (82), FMLP (97) and C-opsonized zymosan (95), but not PMA (82). There are contradictory reports (82,95) on the effects of zymosan.

In recent studies, carried out with E.J. Goetzl (98), we have found that LTB_4, but not its chemotactically inactive 6-_trans_ isomer, is a potent, permselective Ca ionophore in liposomes. Since stimulus-secretion coupling is mimicked by Ca ionophores of fungal origin (A23187, ionomycin) we sought evidence that products of the cell's own lipid metabolism might act as "endogenous" Ca ionophores. Of several phospholipids and lipoxygenase products tested, only phosphatidic acid (> 10 μM) and the secretagogue leukotriene B_4 (LTB_4 _cis_, or: 5(S), 12(R)-dihydroxy-6-_cis_,8,10-_trans_,14-_cis_-eicosatetraenoic acid > 0.15 μM) translocated Ca when added to performed liposomes. Other products of the 5- or 15-lipoxygenase (e.g., 6-_trans_-LTB_4, 5'- or 15-monohydroxy and hydroperoxyeicosatetrienoic acids) were not ionophores (Table 1). LTB_4 _cis_ was also two orders of magnitude more active in provoking

Table 1

TRANSLOCATION OF CALCIUM INTO LIPOSOMES BY LIPOXYGENASE PRODUCTS AND PHOSPHOLIPIDS

Agents Added to Liposomes	Ca Influx	Ca Influx After $SnCl_2$ Reduction	Mg Influx
LTB_4 cis	3.16 ± 0.26** (5)	0.00 ± 0.00 (5)	0.12 ± 0.15 (3)
LTB_4 trans	0.00 ± 0.00 (3)	—	0.25 ± 0.30 (3)
LTB_4 cis (OXIDIZED)	6.48 ± 0.27** (3)	0.00 ± 0.00 (3)	0.00 (1)
LTB_4 trans (OXIDIZED)	2.55 ± 0.22** (4)	0.02 ± 0.02 (5)	0.00 ± 0.00 (3)
Phosphatidic Acid	1.87 ± 0.16** (12)	1.18 ± 0.19** (5)	0.00 ± 0.00 (3)
tri-HETE	0.70 ± 0.19* (3)	—	0.00 ± 0.00 (3)
11-HETE	0.75 ± 0.73 (3)	—	0.00 ± 0.00 (3)
5-HPETE	0.15 ± 0.18 (3)	—	0.00 ± 0.00 (3)
5-HETE	0.00 ± 0.00 (3)	—	0.00 ± 0.00 (3)
15-HPETE	0.00 ± 0.00 (3)	—	—
Cardiolipin	0.00 ± 0.00 (3)	—	0.00 ± 0.00 (3)

Data are expressed as the Mean ± S.E. in mmoles M / mole Membrane Lipid / 5 min.
( ) = Number of Experiments
** $p < 0.001$
* $p < 0.02$

neutrophil migration than its 6-trans isomer or other lipoxygenase products. Phosphatidic acid, but not phosphatidylcholine, -inositol, -serine, or cardiolipin, promoted Ca translocation when preincorporated into liposomal bilayers at molar percentages above 0.5 of total membrane lipid. In liposomes with phosphatidic acid preincorporated, LTB_4 cis promoted increments in Ca influx equal to those it promoted in liposomes without phosphatidic acid: neither synergy nor competition were shown between phosphatidic acid and LTB_4 cis. Ionophoresis was permselective; neither influx nor efflux of Mg was induced by phosphatidic acid or LTB_4 cis. Loss of ionophoretic activity after $SnCl_2$ reduction of ionomycin, A23187, or LTB_4 cis indicated a role for oxygen functions in Ca translocation (Table 1). The order of ionophoretic potency was ionomycin > A23187 > LTB_4 cis > phosphatidic acid, with differences of an order of magnitude between each. The results suggest that phosphatidic acid (generated as part of the "phosphatidylinositol response") and LTB$_4$ cis (a product of lipoxygenation of arachidonate) may act separately or in concert to enhance the calcium fluxes of stimulus-secretion coupling. Indeed, LTB$_4$ shares many actions with the model ionophores A23187 and ionomycin, inducing stimulus-secretion coupling in neutrophils (95), provoking smooth muscle contraction (85) and promoting Ca entry into cells (99) - all at micromolar concentrations. We have therefore suggested that the capacity of LTB$_4$ to translocate Ca - to act, as it were, as an "endogenous ionophore" - may underly these extracellular actions (98).

Recent work from the laboratories of Oates (100) and of Samuelsson (101) has also demonstrated that products of the 15-lipoxygenase are formed by the neutrophil. In a series of reactions entirely analogous to those catalyzed by the 5-lipoxygenase, arachidonate is first transformed to 15-HPETE and to 15-HETE (itself a potent inhibitor of lipoxygenases), then via the 14,15 epoxide to 14,15- and 8,15 LTB$_4$'s, together with as yet unidentified isomers and 5,15-di-HETE (100). Finally, Hubbard and Oates (102), have found that lactoperoxidase-mediated iodination of arachidonate metabolites leads to the formation of iodinated macrolide derivatives in the thyroid. They have recently identified similar materials in neutrophils. Their effects on cells and tissues have not been studied.

Indirect evidence, utilizing a variety of inhibitors of lipoxygenases (5,8,11,14-eicosatetraynoic acid, or ETYA; 5,8,11-eicosatriynoic acid, or ETI; nordihydroguairetic acid, or NDGA; and of phospholipase A$_2$ (bromo-phenacyl bromide, or BPB) have implicated products of the lipoxygenase pathways in stimulus-secretion coupling in the basophil (103,104) and the neutrophil (76,105,106). Inhibition of chemotaxis, lysosomal enzyme release, O$_2^-$ production, and Ca accumulation by neutrophils exposed to these inhibitors "strongly suggested" participation of a lipoxygenase product in neutrophil activation (105). Our own view is that inhibitor studies constitute the lowest level of proof in biological studies: more work is clearly required.

As yet, no studies have been reported in which release of any of the oxidation products of arachidonate has been studied with respect to the earliest time intervals after stimulation: release of these products have been studied 1-3 minutes after neutrophil activation by one or another stimulus. Nor has an appropriate, stimulus-specific balance sheet been elaborated, in which the sum of the various products of the 11-cyclooxygenase, and the 5- and 15-lipoxygenase pathways have been compared. Finally, simple auto-oxidation (106) and active oxygen species made by activated neutrophils can transform unsaturated fatty acids, non-enzymatically, to chemoattractants (107) and Ca ionophores (108). It will be important to determine whether O_2^{τ}, or related species, formed by the stimulated neutrophil, can rodify arachidonate or its enzymatically-derived products.

REFERENCES

1. D.G. Palmer, Total leukocyte enumeration in pathologic synovial fluids. Am. J. Clin. Path. 49:812-818, 1968.
2. J.R. Winchester, H.J. Kunkel and V. Agnello, Occurrence of gamma globulin complexes in serum and joint fluid of rheumatoid arthritis patients: use of monoclonal rheumatoid factors as reagents for their demonstration. J. Exp. Med. 134:286s-293s, 1971.
3. N.J. Zvaifler, Breakdown products of C'3 in human synovial fluids. J. Clin. Invest. 48:1532-1539, 1969.
4. P.A. Ward and N.J. Zvaifler, Complement-derived leukotactic factors in inflammatory synovial fluids of humans. J. Clin. Invest. 50:606-612, 1971.

5. A. Cats, G.J.M. Lafeber and F. Klein, Immunoglobulin phagocytosis by granulocytes from sera and synovial fluids in various rheumatoid and non-rheumatoid diseases. Ann. Rheum. Dis. 40:55-59, 1975.

6. E.R. Hurd, J. LoSpalluto and M. Ziff, Formation of leukocyte inclusions in normal polymorphonuclear cells incubated with synovial fluid. Arth. and Rhuem. 13:724-733, 1970.

7. G. Weissmann, Lysosomes and joint disease. Arth. and Rheum. 9:834-840, 1966.

8. N.J. Zvaifler, Further speculation on the pathogenesis of joint inflammation in rheumatoid arthritis. Arth. and Rheum. 13:895-901, 1970.

9. A. Oronsky, L. Ignarro and R. Perper, Release of cartilage mucopolysaccharide-degrading neutral protease from human leukocytes. J. Exp. Med. 138:461-472, 1973.

10. N.M. Hadler, M.A. Johnson, J.K. Spitznagel and R.J. Quinet, Protease inhibitors in inflammatory synovial effusions. Ann. Rheum. Dis. 40:55-59, 1981.

11. H. Keiser, R.A. Greenwald, G. Feinstein and A. Janoff, Degradation of cartilage proteoglycans by human leukocyte granule neutral protease - a model of joint injury. II. Degradation of isolated bovine nasal cartilage proteoglycan. J. Clin. Invest. 57:625-637, 1976.

12. O. Ohno and T.D. Cooke, Electron microscopic morphology of immunoglobulin aggregates and their interactions in rheumatoid articular collagenous tissues. Arth. and Rheum. 21:516-527, 1978.

13. S. Shiozawa, H.E. Jasin and M. Ziff, Absence of immunoglobulins in rheumatoid cartilage-pannus junctions. Arth. and Rheum. 23:816-821.

14. P.M. Henson, The immunologic release of constituents from neutrophil leukocytes. I. The role of antibody and complement on nonphagocytosable surfaces or phagocytosable particles. J. Immunol. 107:1535-1546, 1971.

15. R.B. Zurier, S. Hoffstein and G. Weissmann, Mechanisms of lysosomal enzyme release from leucocytes. I. Effect of cyclic nucleotides and colchicine. J. Cell Biol. 58:27-41, 1973.

16. P.M. Henson, J.R. Hollister, R.A. Musson, R.O. Webster, P. Spears, J.E. Henson and K.M. McCarthy, Inflammation as a surface phenomenon: initiation of inflammatory processes by surface-bound immunologic compounds. In: Advances in Inflammation Research (Weissmann, G., Samuelsson, B. and Paoletti, R., eds.) 1:341-352, 1979.

17. H. Menninger, R. Putzier, W. Mohr, D. Wessinghage and K. Tillman, Granulocyte elastase at the site of cartilage erosion by rheumatoid synovial tissue. Z. Rheumatol. 39:145-156, 1980.

18. W. Mohr, H. Westerhellweg and D. Wessinhage, Polymorphonuclear granulocytes in rheumatic tissue destruction. III. An electron microscopic study of PMNs at the pannus-cartilage junction in rheumatoid arthritis. Ann. Rheum. Dis. 40:396-399, 1981.

19. W. Mohr, A. Wild and H.P. Wolf, Role of polymorphs in inflammatory cartilage destruction in adjuvant arthritis of rats. Ann. Rheum. Dis. 40:171-176, 1981.

20. S.M. Krane, Collagenase production by human synovial tissue. Ann. N.Y. Acad. Sci. 256:289-303, 1975.

21. J. Saklatvala and A.J. Barrett, Identification of proteinases in rheumatoid synovium: detection of leukocyte elastase cathepsin G and another serine proteinase. Biochim. Biophys. Acta 615:167-177, 1980.

22. C. May, B.B. Levine and G. Weissmann, Compounds inhibiting glucose utilization, antigenic release of histamine and phagocytic release of lysosomal enzyme from leucocytes of humans. Proc. Soc. Exp. Biol. & Med. 133:758-763, 1970.

23. G. Weissmann, R.B. Zurier, P.J. Spieler and I.M. Goldstein, Mechanisms of lysosomal enzyme release from leucocytes exposed to immune complexes and other particles. J. Exp. Med. 134:149s-165s, 1971.

24. E.L. Becker, Some interrelations of neutrophil chemotaxis, lysosomal enzyme secretion and phagocytosis, as revealed by synthetic peptides. Amer. J. Path. 85:385-394, 1976.

25. D.G. Wright, D.A. Bralove and J.I. Gallin, The differential mobilization of human neutrophil granules. Am. J. Pathol. 87:273-284, 1977.

26. G. Weissmann, Release of mediators of inflammation from stimulated neutrophils. N. Engl. J. Med. (Medical Seminars of the Beth Israel Hospital) 303:27-34, 1980.

27. J.E. Smolen and G. Weissmann, The secretion of lysosomal enzymes from human neutrophils: the first events in stimulus-secretion coupling. In: Lysosomes and Lysosomal Storage Diseases (Callahan, J.W. and Alexander-Lowden, J., eds.) Raven Press, New York, p. 31-62, 1981.

28. L.E. Hokin and M.R. Hokin, Effects of acetylcholine on the turnover of phosphoryl units in individual phospholipids of pancreas slices and brain cortex slices. Biochim. Biophys. Acta 18:102-110, 1955.

29. J.W. Putney, Stimulus-permeability coupling: role of calcium in the receptor regulation of membrane permeability. Pharm. Rev. 30:209-245, 1979.

30. L.M. Lichtenstein and R. DeBernardo, IgE mediated histamine release: in vitro separation into two phases. Int. Arch. Allergy Appl. Immunol. 41:56-71, 1971.

31. P.J. Edelson, Monocytes and macrophages: aspects of their cell biology. In: Cell Biology of Inflammation (Weissmann, G., ed.) Elsevier North Holland Press, Amsterdam and New York, pp. 470-495, 1980.

32. I.M. Goldstein, D. Roos, G. Weissmann and H. Kaplan, Complement and immunoglobulins stimulate superoxide production by human leukocytes independently of phagocytosis. J. Clin. Invest. 56:1155-1163, 1975.
33. I. Goldstein, S. Hoffstein, J. Gallin and G. Weissmann, Mechanisms of lysosomal enzyme release from human leukocytes: microtubule assembly and membrane fusion induced by a component of complement. Proc. Natl. Acad. Sci. USA 70:2916-2920, 1973.
34. I.M. Goldstein, S.T. Hoffstein and G. Weissmann, Mechanisms of lysosomal enzyme release from human polymorphonuclear leukocytes: effects of phorbol myristate acetate. J. Cell. Biol. 66:647-652, 1975.
35. I.M. Goldstein, J.K. Horn, H.B. Kaplan and G. Weissmann, Calcium-induced lysozyme secretion form human polymorphonuclear leukocytes. Biochem. Biophys. Res. Commun. 60:807-812, 1974.
36. S. Hoffstein, R. Soberman, I. Goldstein and G. Weissmann, Concanavalin A induces microtubule assembly and specific granule discharge in human polymorphonuclear leukocytes. J. Cell Biol. 68:781-786, 1976.
37. E.J. Goetzl and W.C. Pickett, The human PMN leukocyte chemotactic activity of complex hydroxy-eicosatetraenoic acids (HETEs) J. Immunol. 125:1789-1791, 1980.
38. I.M. Goldstein, C.L. Malmsten, H. Kindahl, H.B. Kaplan, O. Radmark, B. Samuelsson and G. Weissmann, Thromboxane generation by human peripheral blood polymorphonuclear leukocytes. J. Exp. Med. 148:787-792, 1978.
39. H.E. Claesson, U. Lundberg and C. Malmsten, Serum-coated zymosan stimulates the synthesis of leukotriene B_4 in human polymorphonuclear leukocytes. Inhibition by cyclic AMP. Biochem. Biophys Res. Comm. 99:1230-1237, 1981.
40. B.M. Babior, Oxygen-dependent microbial killing by phagocytes. N. Engl. J. Med. 298:(659-668, 721-725), 1978.

41. L. Fleck, Ueber Leukergie. Acta Haem 8:282-293, 1952.
42. J.R. O'Flaherty, H.J. Showell, P.A. Ward and E.L. Becker, A possible role of arachidonic acid in human neutrophil aggregation and degranulation. Am. J. Pathol. 96:799-809, 1979.
43. R.R. MacGregor, Granulocyte adherence. In: The Cell Biology of Inflammation (Weissmann, G., ed.) Elsevier North Holland Press, Amsterdam and New York, pp. 267-298, 1980.
44. D. Bainton, Sequential degranulation of the two types of polymorphonuclear leukocyte granules during phagocytosis of microorganisms. J. Cell. Biol. 58:249-264, 1973.
45. D.F. Bainton, The cells of inflammation: a general view. In: The Cell Biology of Inflammation (Weissmann, G., ed.) Elsevier North Holland Press, Amsterdam and New York, pp. 1-26, 1980.
46. R.B. Zurier, S. Hoffstein and G. Weissmann, Cytochalasin B: Effect on lysosomal enzyme release from human leucocytes. Proc. Natl. Acad. Sci. USA 70:844-848.
47. J.E. Smolen, H.M. Korchak and G. Weissmann, Stimulation increases levels of cyclic adenosine-3',5'-monophosphate in human polymorphonuclear leukocytes exposed to surface stimuli. J. Clin. Invest. 65:1077-1085, 1980.
48. J.E. Smolen, H.M. Korchak and G. Weissmann, Initial kinetics of lysosomal enzyme release and superoxide anion generation in human polymorphonuclear leukocytes. Inflammation 4:145-164, 1980.
49. H.J. Showell, P.H. Naccache, R.I. Sha'afi and E.L. Becker, The effects of extracellular K^+, Na^+ and Ca^{++} on lysosomal enzyme secretion from polymorphonuclear leukocytes. J. Immunol. 119:804-811, 1977.
50. H. Cohen, M. Chovaniec and W. Davies, Activation of the guinea pig granulocyte NAD(P)H-dependent superoxide generating enzyme: localization in a plasma membrane enriched particle and kinetics of activation. Blood 55:3, 355-363, 1980.

51. H.M. Korchak and G. Weissmann, Changes in membrane potential of human granulocytes antecede the metabolic responses to surface stimulation. Proc. Natl. Acad. Sci. USA 75:3818-3822, 1978.
52. H.M. Korchak and G. Weissmann, Stimulus-response coupling in the human neutrophil. Transmembrane potential and the role of extracellular Na^+. Biochim. Biophys. Acta 601:180-194, 1980.
53. S. Hoffstein, Ultrastructural demonstration of calcium loss from local regions of the plasma membrane of surface-stimulated human granulocytes. J. Immunol. 123:1395-1402, 1979.
54. P.H. Naccache, H.J. Showell, E.L. Becker and R.I. Sha'afi, Involvement of membrane calcium in the response of rabbit neutrophils to chemotactic factors as evidenced by the fluorescence of chlorotetracycline. J. Cell Biol. 83:179-186, 1979.
55. S. Cockcroft, J.P. Bennett and B.D. Gomperts, Stimulus-secretion coupling in rabbit neutrophils is not mediated by phosphatidylinositol breakdown. Nature 288:275-277, 1980.
56. G. Weissmann, C. Serhan, H. Korchak, J.E. Smolen, M.J. Brockman and A. Marcus, Neutrophils generate phosphatidic acid, and "endogenous calcium ionophores" before releasing mediators of inflammation. Trans. Assoc. Amer. Phys. (in press), 1981.
57. R.J. Petroski, P.H. Naccache, E.L. Becker and R.I. Sha'afi, Effect of the chemotactic factor formyl-methionyl-leucyl-phenylalanine and cytochalasin B on the cellular levels of calcium in rabbit neutrophils. F.E.B.S. Lett. 100:161-165, 1979.
58. P.J. Naccache, J.H. Showell, E.L. Becker and R.I. Sha'afi, Transport of sodium, potassium and calcium across rabbit polymorphonuclear leukocyte membranes. Effect of chemotactic factor. J. Cell Biol. 73:428-444, 1977.

59. J.E. Smolen and G. Weissmann, Stimuli which provoke secretion of azurophil enzymes from human neutrophils induce increments in adenosine cyclic 3'-5'-monophosphate. Biochim. Biophys. Acta 672:197-206, 1981.

60. H.B. Kaplan, H.S. Edelson, R. Friedman and G. Weissmann, The roles of degranulation and superoxide anion generation in neutrophil aggregation. J. Immunol. (submitted), 1981.

61. G. Weissmann, C. Serhan, H.M. Korchak, J.E. Smolen, J. Broekman and A. Marcus, Neutrophils generate phosphatidic acid, and "endogenous calcium ionophore" before releasing mediators of inflammation. Adv. Prost. Thrombox. Res. (in press), 1981.

62. I.M. Goldstein, S.T. Hoffstein and G. Weissmann, Influence of divalent cations upon complement-mediated enzyme release from human polymorphonuclear leukocytes. J. Immunol. 115:665-670, 1975.

63. J.E. Smolen and G. Weissmann, The roles of extracellular and intracellular calcium in lysosomal enzyme release and superoxide anion generation by human polymorphonuclear leukocytes. Biochim. Biophys. Acta (in press), 1981.

64. H.M. Korchak, B.A. Eisenstat, S.T. Hoffstein, P.B. Dunham and G. Weissmann, Anion channel blockers inhibit lysosomal enzyme secretion from human neutrophils without affecting generation of superoxide anion. Proc. Natl. Acad. Sci. USA 77:2721-2725, 1980.

65. A.I. Tauber and E.J. Goetzl, Inhibition of complement-mediated functions of human neutrophils by impermeant stilbene disulfonic acids. J. Immunol. 126:1786-1789, 1981.

66. G. Weissmann, C. Serhan, H.M. Korchak, J.E. Smolen, J. Broekman and A.S. Marcus, Neutrophils generate phosphatidic acid, and "endogenous calcium ionophore" before releasing mediators of inflammation. Clin. Res. 29:573A, 1981.

67. C. Serhan, H.M. Korchak, M.J. Broekman, J.E. Smolen, A. Marcus and G. Weissmann, Phosphatidylinositol breakdown and phosphatidic acid accumulation in stimulated human neutrophils: relationship to calcium mobilization and ^{45}Ca uptake. (submitted), 1981.
68. J.R. Vane, Inhibition of prostaglandin biosynthesis as a mechanism of action for aspirin-like drugs. Nature New Biol. 231:232, 1971.
69. R.B. Zurier and G. Weissmann, Effect of prostaglandins upon enzyme release from lysosomes and experimental arthritis. In: Prostaglandins in Cellular Biology (Ramwell, P.W. and Pharriss, B.B., eds.) Plenum Publishing Corp., New York, p. 151-172, 1972.
70. R.B. Zurier, G. Weissmann, S. Hoffstein, S. Kammerman and H.-H. Tai, Mechanisms of lysosomal enzyme release from human leukocytes II. Effects of cAMP and cGMP, autonomic agonists, and agents which affect microtubule function. J. Clin. Invest. 53:297-309, 1974.
71. H.R. Bourne, L.M. Lichtenstein, K.L. Melmon, C.S. Henney, Y. Weinstein and G.L. Shearer, Modulation of inflammation and immunity by cyclic AMP. Receptors for vasoactive hormones and mediators of inflammation regulate many leukocyte functions. Science 184:19-28, 1974.
72. H.R. Bourne, R.I. Lehrer, L.M. Lichtenstein, G. Weissmann and R.B. Zurier, Effects of cholera enterotoxin on adenosine 3',5'-monophosphate and neutrophil function: Comparison with other compounds which stimulate leukocyte adenyl cyclase. J. Clin. Invest. 52:698-708, 1973.
73. R. Zurier, J. Doty and A. Goldenberg, Cyclic AMP response to prostglandin E_1 in mononuclear cells from peripheral blood and synovial fluid of patients with rheumatoid arthritis. Prostaglandins 13:25-31, 1977.

74. R.B. Zurier, S. Hoffstein and G. Weissmann, Suppression of acute and chronic inflammation in adrenalectomized rats by pharmacologic amounts of prostaglandins. Arth. and Rheum. 16:606-618, 1973.
75. G. Weissmann, J.E. Smolen and H. Korchak, Prostaglandins and inflammation: Receptor/cyclase coupling as an explanation of why PGE's and PGI_2 inhibit functions of inflammatory cells. In: Advances in Prostaglandin and Thromboxane Research (Ramwell, P., Paoletti, R. and Samuelsson, B., eds.) Raven Press, New York, Vol. 8, p. 1637-1653, 1980.
76. J. Smolen and G. Weissmann, Effects of indomethacin 4,8,11,14-eicosatetraynoic acid, and p-bromophenacyl bromide on lysosomal enzyme release and superoxide anion generation by human polymorphonuclear leukocytes. Biochem. Pharmacol. 29:533-538, 1980.
77. M. Siegel, R. McConnell and P. Cuatrecasas, Aspirin-like drugs interfere with arachidonate metabolism by inhibition of the 12-hydroperoxy-5,8,10,14-eicosatetraenoic acid peroxidase activity of the lipoxygenase pathway. Proc. Natl. Acad. Sci. USA 76:3774-3778, 1979.
78. R.B. Zurier and D.M. Sayadoff, Release of prostaglandins from human polymorphonuclear leukocytes. Inflammation 1:93-101, 1975.
79. J. Morley, M.A. Bray and R.W. Jones, Prostaglandin and thromboxane production by human and guinea-pig macrophages and leukocytes. Prostaglandins 17:729-736, 1979.
80. W.F. Stenson and C.W. Parker, Metabolism of arachidonic acid in ionophore-stimulated neutrophils. J. Clin. Invest. 64:1457-1465, 1979.
81. E.J. Goetzl, Vitamin E modulates the lipoxygenation of arachidonic acid in leukocytes. Nature 288:183-185, 1980.
82. C.E. Walsh, M.B. Waite, M.J. Thomas and L.R. DeChatelet, Release and metabolism of arachidonic acid in human neutrophils. J. Biol. Chem. 256:7228-7234, 1981.

83. P. Borgeat and B. Samuelsson, Arachidonic acid metabolism in polymorphonuclear leukocytes: Effects of ionophore A23187. Proc. Natl. Acad. Sci. USA 76:2148-2152, 1979.
84. P. Borgeat and B. Samuelsson, Transformation of arachidonic acid by rabbit polymorphonuclear leukocytes. J. Biol. Chem. 254:2643-2646, 1979.
85. G. Hansson, J.W. Lindgren, S.E. Dahlen, P. Hedqvist and B. Samuelsson, Identification and biological activity of novel ω-oxidized metabolites of leukotriene B_4 from human leukocytes. F.E.B.S. Lett. 130:107-112.
86. O. Radmark, C. Malmsten, B. Samuelsson, R.A. Clark, G. Goto, A. Marfat and E.J. Corey, Leukotriene A: sterochemistry and enzymatic conversion to leukotriene B. Biochem. Biophys. Res. Comm. 92:954-961, 1980.
87. R. Murphy, S. Hammarstrom and B. Samuelsson, Leukotriene C: a slow-reacting substance from murine mastocytoma cells. Proc. Natl. Acad. Sci. USA 76:4274-4279, 1979.
88. S. Hammarstrom, R.C. Murphy, D.A. Samuelsson, A. Clark and C. Mioskowski, Identification of the amino acid part. Biochem. Biophys. Res. Commun. 91:1266, 1979.
89. L. Orning, S. Hammarstrom and B. Samuelsson, Leukotriene D: A slow reacting substance from rat basophilic leukemia cells. Proc. Natl. Acad. Sci. USA 77:4,2014-2017, 1980.
90. M.K. Bach, J.R. Brashler, C.D. Brooks, et al. Slow-reacting substance: comparison of some properties of human lung SRA-A and two distinct fractions from ionophore-induced rat mononuclear cells SRS. J. Immunol. 122:160-165, 1979.
91. R.P. Rubin, L.E. Sink and R.J. Freer, Activation of (arachidonyl) phosphatidylinositol turnover in rabbit neutrophils by the calcium ionophore A23187. Biochem. J. 194:497-505, 1981.
92. S. Rittenhouse-Simmons, Production of diglyceride from phosphatidylinositol in activated human platelets. J. Clin. Invest. 63:580-587, 1979.

93. E.A. Kitchen, J.R. Boot and W. Dawson, Chemotactic activity of thromboxane B_2, prostaglandins and their metabolites for polymorphonuclear leukocytes. Prostaglandins 16:239-244, 1978.

94. M.A. Bray, A.W. Ford-Hutchinson and M.J.H. Smith, Leukotriene B_4: An inflammatory mediator in vivo. Prostaglandins 22:213-222, 1981.

95. I. Hafstrom, J. Palmblad, C.L. Malmsten, O. Radmark and B. Samuelsson, Leukotriene B_4 - a stereospecific stimulator for release of lysosomal enzymes from neutrophils. F.E.B.S. Lett. 130:146-148, 1981.

96. H.E. Claesson, U. Lundberg and C.L. Malmsten, Serum-coated zymosan stimulates the synthesis of leukotriene B_4 in human polymorphonuclear leukocytes. Inhibition by cyclic AMP. Biochem. Biophys. Res. Comm. 99:1230-1235, 1981.

97. G.M. Bokoch and P.W. Reed, Stimulation of arachidonic acid metabolism in the polymorphonuclear leukocyte by an N-formylated peptide. J. Biol. Chem. 255:10223-10226, 1980.

98. C. Serhan, J. Fridovich, E. Goetzl, P.B. Dunham and G. Weissmann, Leukotriene B_4 and phosphatidic acid are calcium ionohpores: studies employing arsenazo III in liposomes (submitted).

99. P.H. Naccache, R.I. Sha'afi, P. Borgeat and E.J. Goetzl, Mono- and dihydroxyeicosatetraenoic acids after calcium homeostasis in rabit neutrophils. J. Clin. Invest. 67:1584-1587, 1981.

100. R. Maas, J. Oates and A. Brash, Novel products of the lipoxygenation of arachidonic acid. Adv. Prost. Thromb. Res. 9:(in press), 1981.

101. W. Jubiz, O. Radmark, J.A. Lindgren, C. Malmsten and B. Samuelsson, Novel leukotrienes: products formed by initial oxygenation of arachiodonic acid at C-15. Biochem. Biophys. Res. Comm. 99:976-986, 1981.

102. W. Hubbard and J.A Oates, personal communication, 1981.
103. G. Marone, S. Hammarstrom and L.M.Lichtenstein, An inhibitor of lipoxygenase inhibits histamine release from human basophils. Clin. Immunol. & Immunopathol. 17:117-122, 1981.
104. S.P. Peters, M.I. Siegel, A. Kagey-Sobotka and L.M. Lichtenstein, Lipoxygenase products modulate histamine release in human basophils. Nature 292:455-457, 1981.
105. H.J. Showell, P.H. Naccache, R.I. Sha'afi and E.L. Becker, Inhibition of rabbit neutrophils lysosomal enzyme secretion, nonstimulated and chemotactic factor stimulated locomotion by nordihydroguaiaretic acid. Life Sciences 27:421-426, 1980.
106. S.R. Turner, J.A. Campbell and W.S. Lynn, Polymorphonuclear leukocyte chemotaxis toward oxidized lipid componenets of cell membranes. J. Exp. Med. 141:437-446, 1975.
107. H.D. Perez, B.B. Weksler and I.M. Goldstein, Generation of a chemotactic lipid from arachidonic acid by exposure to a superoxide generating system. Inflammation 4:313-328, 1980.
108. C. Serhan, P. Anderson, E. Goodman, P. Dunham and G. Weissmann, Phosphatidate and oxidized fatty acids are calcium ionophores: Studies employing arsenzao III in liposomes. J. Biol. Chem. 256:2736-2741, 1981.

INDEX

ACR, 472
   and actin, 481,482
   cyclic nucleotides, 482,483
   definition of, 472
   gene regulation, 483, 484
   intracellular protease
      activator, 480
   neoplasm, 472,477
   plasma protease, 481
Adrenal glomerulosa cells, 449
Angiotensin, 435-450
   effect on K^+ transport, 436, 437
   effect on Na^+ transport, 436, 437,438
   and rpesence of ouabain, 441
   stimulation, 449, 450
   and vascular smooth muscle, 436, 447, 448
ACE, 79
   studies of activity, 79
A431 cell
   and EGF, 59-62
   growth, 54-62
   inhibitory factors, 54
   and serum absence, 55, 56
   and serum presence, 55, 56
   stimulatory factors, 54,55
Antibiotics, 4
Antibody
   antisera, 463
   F11 monoclonal, 460, 464, 466
   monoclonal, 457, 462
   and pH, 465
Antigens
   detection, 15
   expression in MuLV, 374, 375

Antigens (continued)
   expression of cell
      surface, 370
   histocompatibility, 366
   melanoma associated, 455, 456, 463-465
   species specific, 14,15
   surface, 365
Applied genetics, 2
Assay
   Immunofluorescence, 5
AII, 443, 448, 449
   and aortic smooth muscle
      cells, 446
   binding, 445
   and monensin, 443, 448
   NA^+-K^+ pump, 443, 447, 448
   Rb^+ uptake, 443, 444, 448
AIII, 447

Biopsies
   Skin, 8
Brain microvessels, 440, 447
Ca^{2+}, 449
CAD, 319
   cloning of gene, 325
   gene amplification, 320
   gene as marker, 331
   protoplast fusion 333, 336

Cancer
   and benign neoplasm, 472
   colorectal, 471-476
   extracolonic mani-
      festations, 473

Cancer (continued)
  Fata, 2
  and plasma protease activity, 474-477
  and polyposis, 472
Chromosome
  aberrations, 6, 101-106, 113
  analysis of variance, 106
  banding, 15, 20
  counts, 15
  of cyclamate exposed leucocytes, 103
CFU activity, 39, 40, 41
Cell
  epithelial, 2, 4
  HeLa, 18
  hematopoietic stem, 33, 34
  identification, 13
  in vitro, 2, 3
  lines, 5, 14, 17, 21, 25
  mammalian, 120
    cell culture system, 120
    cell growth media, 120, 124
    cell growth sensitivity, 125
  sensitivity, 3
Cell culture
  conditions, 3
  contamination, 3, 4, 26, 27
  feeding, 5
  hormones, 1, 2
  identification tests, 4
  laboratories, 3
  mammalian, 119
  monitoring, 14
  products, 1
  shipment, 8
  sterility, 3
  suspension, 139
  techniques, 2
  transfer, 5
Contamination
  cross, 4, 5, 14
  and measures to prevent, 3, 5
Cryopreservation, 508-516
  of pancreatic islets, 508
Cultures
  CFU-C, 41
  CFU-L, 41
  colony, 36

Cultures (continued)
  and electron microscopy, 38
  tissue, 243
Cyclamates, 91-94
  calcium, 99, 100, 107, 108, 115
  and chromosomes, 94
  and DNA synthesis, 92
  and in vitro effects, 91
  and leucocyte cultures, 92
  and radioactive thymidine, 111
  techniques of study, 91

Density gradient, 36
Diabetes mellitus
  insulin dependent, 489
  and microvascular disease, 489
  pancreatic islet transportation, 489-491
Diffusivity
  thermal, 128
DNA, 1, 92
  and microinjection, 315
  RHHV, 300

Endothelial cells, 67-86
  and culture density, 71, 72
  and growth factor, 179, 180
  and in vitro results, 67, 76, 77, 86, 180
  and in vivo results, 67, 68, 86, 180, 183
  isolation, 67
  large vessel bovine, 180, 195
  life span, 73, 76
  and migration assays, 183
  senescence, 77, 82
  and smooth muscle proliferation, 185
Epidermal growth factor, 49, 50
  and carcinoma cell line, 50

INDEX

Epidermal growth factor (continued)
  and fetal calf serum, 53
  and inhibitory effect, 49

Fibroblasts, 2, 6
  biochemical assays, 472
  cell cultures, 472
  human colon, 471, 474
Fibronectin, 160

Genetics
  markers, 13, 17, 18, 22, 26
  phenotypic, 14
Glycoproteins, 456, 457, 462
Grafts
  fabrication of blood vessels, 422
  living skin equivalent, 419
  procedure for fabrication, 420-422

HEK Cell Lots, 245, 246
  and assay of plasminogen activator, 245
  production of PA, 245
Hepa, 4
HLA, 16, 17, 22

Interferon, 176
  gamma, 221
Islets
  and cryopreservation, 508
    additives, 514
  transplants in man, 506
  transplantation of rat, 505
Isozyme, 24
  analysis, 22

K^+, 447, 448
Karyotype, 5

L14 lymphoblastoid cell, 460, 461
  growth, 461
Leukapheresis, 34, 35, 39
Liquid Nitrogen, 8
  and frozen ampules, 8
  storage, 4
  surveillance of level, 8

Lymphoblastoid interferon
  production, 272
  purificaiton of, 270
Lymphoblasts, 2, 6
Lymphocyte
  autoradiographic analysis, 97
  culture, 96
  morphology, 96
  preparation, 95
  proliferation, 96
Lymphotoxins, 205
  and antiserum inhibition, 212, 213
  and biochemical studies, 210
  and cell lysis, 205
  definition of, 206
  and growth inhibition, 205

Melanoma cells, 455
  antigenic expression, 455
  antisera, 457
  growth, 462
  serum free medium, 455
M14 cells, 455, 456, 459, 464
  cells and culture, 457, 460
  double-agar clonogenic assay, 460, 462
  growth, 455, 456, 467
  independent growth, 465
  indirect immunoprecipitation, 459, 463
  soft agar, 466
M. hyorhinis, 5
Melanoma
  metastic, 465
Microcarrier, 151, 164
  cell attachment, 152
  collagen-coated, 153, 154, 158
  desirable properties of, 171
  and interferon production, 169
  removal of medium components, 163

Microcarrier (continued)
  surface charged, 153
  surface for growth of, 151
Monad cell growth
  curve, 119, 121
  phases, 131, 136
Myelopoietic proliferation, 227
  thymic epithelial cultures, 230

Na+, 447, 448, 449
  permeability, 449
Neutrophils, 527-530
  immune complexes, 528
  lysosomal enzymes, 527-530
  mediator release, 527
  RA and neutrophils, 528, 529
  response to oxidation products of Arachidonic Acid, 534, 540
  stimulus secretion, 529-534

Pancreas
  enzymatic dispersion fragments, 493
  islet culture, 491, 492
    methodology, 496-498
  islet transplantation, 489-491
  sources of human, 492
Plasminogen activator, 243
  activity produced by Hek cells, 245
  expression of activity, 258
  production of body fractions of separated cells, 252

Radioimmunometric binding analysis, 459, 461, 462
Repositories, 6, 7, 8

Smooth muscle, 449
Staining
  hoechst, 24
Sterility, 3
Suspension
  bearings, 146
  interval timing, 148
  pure rotation of, 140
  secondary motion, 140, 141
  stirrer for cell culture, 139
  stirring interval, 139

Synmed-I composition, 458-461, 464, 465
  additional components, 458
  amino acids, 458
  inorganic salts, 458
  vitamins, 458
Synthetic, 468
  cell lines, 468
  tumor lines, 468

TCM inhibition, 233
T-cell clones, 347
  cytolytic, 348, 352-354
  methods of cloning, 350
  non-cytolytic, 348, 355-357
T-cell hybrids
  and chromosome analysis, 390
  functional analysis, 392, 394
  and helper activity, 384, 392
  repeated cloning, 395, 396
T-lymphocyte
  cultured lines and clones as immunogenetic tools, 405, 406
  isolation of human clones, 408
  source of gamma interferon, 219
  and specific memory proliferation, 410, 411
Transglutaminase inhibition, 449

Urokinase
  cloning of, 283
  initially observed, 282
  isolation of RNA, 284
  recombinant DNA synthesis, 286
  relationship between two species, 282